IET TRANSPORTATION SERIES 18

Collection and Delivery of Traffic and Travel Information

Other volumes in this series:

Volume 1	**Clean Mobility and Intelligent Transport Systems** M. Fiorini and J.-C. Lin (Editors)
Volume 2	**Energy Systems for Electric and Hybrid Vehicles** K.T. Chau (Editor)
Volume 5	**Sliding Mode Control of Vehicle Dynamics** A. Ferrara (Editor)
Volume 6	**Low Carbon Mobility for Future Cities: Principles and applications** H. Dia (Editor)
Volume 7	**Evaluation of Intelligent Road Transportation Systems: Methods and results** M. Lu (Editor)
Volume 8	**Road Pricing: Technologies, economics and acceptability** J. Walker (Editor)
Volume 9	**Autonomous Decentralized Systems and their Applications in Transport and Infrastructure** K. Mori (Editor)
Volume 11	**Navigation and Control of Autonomous Marine Vehicles** S. Sharma and B. Subudhi (Editors)
Volume 12	**EMC and Functional Safety of Automotive Electronics** K. Borgeest
Volume 16	**ICT for Electric Vehicle Integration with the Smart Grid** N. Kishor and J. Fraile-Ardanuy (Editors)
Volume 17	**Smart Sensing for Traffic Monitoring** N. Ozaki (Editor)
Volume 25	**Cooperative Intelligent Transport Systems: Towards high-level automated driving** M. Lu (Editor)
Volume 38	**The Electric Car** M.H. Westbrook
Volume 45	**Propulsion Systems for Hybrid Vehicles** J. Miller
Volume 79	**Vehicle-to-Grid: Linking electric vehicles to the smart grid** J. Lu and J. Hossain (Editors)
	Driver Adaptation to Information and Assistance Systems Alan Stevens, Corinne Brusque, Josef Krems (Editors)

Collection and Delivery of Traffic and Travel Information

Edited by
Paul Burton and Alan Stevens

The Institution of Engineering and Technology

Published by The Institution of Engineering and Technology, London, United Kingdom

The Institution of Engineering and Technology is registered as a Charity in England & Wales (no. 211014) and Scotland (no. SC038698).

First published 2020

The Institution of Engineering and Technology
Michael Faraday House
Six Hills Way, Stevenage
Herts, SG1 2AY, United Kingdom

www.theiet.org

British Library Cataloguing in Publication Data
A catalogue record for this product is available from the British Library

ISBN 978-1-78561-772-0 (hardback)
ISBN 978-1-78561-773-7 (PDF)

Typeset in India by MPS Limited

Contents

About the editors

Paul Burton is a freelance consultant and expert in intelligent transport systems, in particular, in traffic and traveller information. He has been the convenor of the standards groups ISO/TC204 Working Group 10 (Traveller Information) and CEN/TC278 WG4 (Traffic and Traveller Information) since 1996. Paul has been involved in several international-scale projects, as well as serving as a technical policy advisor for UK central government. Presently, Paul is working with the European Commission on mobility integration.

Alan Stevens is a chartered engineer and fellow of the IET, and a visiting professor at the University of Southampton, UK. He was previously research director of TRL UK with interests including connected and automated vehicles, human behaviour and evaluating the impacts of intelligent transport systems. He chaired ITS (UK), serves on IET's Transport Policy Panel and Automotive and Road Transport Systems Professional Network, and is advisor to the IET Transportation book program.

Biographies

About the editors

Paul Burton

Paul is a freelance consultant and expert in intelligent transport systems (ITS), in particular, the traffic and traveller information field. He is presently, and has been since 1996, the Convenor of the standards groups ISO/TC204 Working Group 10 (traveller information) and its sister CEN/TC278 WG4 (traffic and traveller information). Paul's experience in ITS goes back to 1991 when he was part of the team developing and implementing the SCOOT real-time urban traffic control systems across the world. Paul was one of the leading researchers developing the radio data systems – traffic message channel and developing it through a number of European test sites. Paul made led the large UK trial of RDS-TMC, which developed the business models that drive the implementation of RDS-TMC across the world. Paul was a technical policy advisor for the Central Government and its then Highways Agency bringing a policy understanding to all aspects of ITS, including satellite positioning and navigation systems. Presently, Paul is working with the European Commission on Mobility Integration, particularly in the urban setting.

Alan Stevens

Alan is a visiting professor at the University of Southampton. He has 30 years' experience working on the application of new technology to the transport environment and is internationally recognised for research on cooperative vehicle systems and for human–machine interaction concerning the behaviour and safety of drivers using in-vehicle technology. He contributes to international work through European projects, books and journals and is the previous Chair of ITS (UK). Alan's work has included input to policy development for the EC, UK Central Government and Highways England, including on intelligent transport systems, cooperative and automated vehicles, eCall, and standards. He is also involved in designing research studies carrying out specific technical investigations and supervising Ph.D.s. Topics include the design and assessment of driver information (both on-vehicle and off-vehicle) and the safety and responsibility of drivers when using connected and increasingly automated vehicles. Until recently, Alan was the chief scientist and research director at TRL (the UK Transport Research Laboratory) where he had responsibility for ensuring the technical quality of a wide range of projects in transportation.

About the authors

Paul Hutton

Paul has had what can be described as a 'varied career'. After gaining a degree in mathematics and a brief period in the city, Paul trained as a broadcast journalist on the prestigious course at Cardiff University before spending several years as a radio presenter, newsreader and TV and radio sports reporter. Paul then switched to concentrating on radio traffic news, overseeing the service to most stations in the United Kingdom and introducing new sources of information and working with developers to deliver an efficient traffic information gathering and dissemination tool. He then ran the European division of an American-owned global traffic information company and oversaw operations in Australia, Canada and the United States, also working on a fledgling app-based service as well as supplying the then-Highways Agency with its traffic radio station. Paul then left day-to-day traffic information and returned to journalism, editing SMART Highways for five years and now co-owns the transport-related news website Highways-News.com and, given his radio roots, introducing a popular ITS-related podcast which he presents from events around the world. Paul is also a keen supporter of the UK's Intelligent Transport Society, and acts as its part-time communications manager.

Jon Harrod Booth

Jon is an independent consultant with more than 25 years' experience within the field of intelligent transport systems (ITS) and highway-oriented systems specification and deployment. He has a long heritage in the specification and standardisation for traffic management data exchange and is the current convenor of CEN/TC278's Working Group 8, Road data, which is responsible for the standardisation of the European 'DATEX II' 16157 family of standards. He has been a member of that group for more than 20 years. Jonathan was part of the DATEX (I) task force in the mid-1990s, and one of the lead modellers for DATEX II. He also has broad spectrum of interests in location referencing and highway network modelling. Historically, he supported the Conference of European Directors of Roads (CEDR) Road Data Sub-Group addressing the issues relating to the exchange of traffic data, network data and network referencing approaches;

representing CEDR on two high-level European mapping initiatives – as the Public Sector representative to the e-safety group on Digital Maps (chaired by NAVTEQ & TeleAtlas) and as a reference group to the EuroRoads Project consortium (National Mapping Agencies). Continuing this work, he was recently a member of the CEN Project Team PT0701 developing this first version of the TN-ITS CEN Technical Specification (CEN/TS17268). Jonathan is a member of ISO/TC211 Geographic Information/Geometics, currently leading a work item that is seeking improved alignment between GIS standards and standards used for network model and map definition within ISO/TC204 ITS. He has consulted on network models and asset management for several large public sector organisations. For Highways England, Jonathan led the deployment of a new definitive network model used within their Integrated Asset Management system. He is a member of ISO TC204 WG3, the standards committee developing ITS location referencing and digital mapping standards.

Bev Marks

Bev trained at the BBC, England, qualifying as a broadcast and communications engineer; he spent 25+ years working on radio & tv studio and network systems, including the role of Project Manager RDS, BBC Radio with responsibility for the multi-disciplinary team developing RDS technology, installing transmission equipment nation-wide, central computers and network circuits.

Latterly in his BBC career Bev was also Project Manager Travel Information Systems, responsible for on-line computer based travel reporting in Police and Public Authority control rooms, linked to Broadcasting House.

Bev was involved internationally in RDS development representing the BBC on the European Broadcasting Union (EBU) RDS Technical Experts Group and Working Groups, including R/RDS-TMC and Universal Protocol. He is co-author of a number of RDS patents.

Following several years industrial experience with a professional broadcast equipment designer and manufacturer, he became a freelance broadcast engineer and spent 14 years, mainly working as a senior broadcast engineer, for the EBU working on many projects, including RDS, TMC and TTI Message Exchange Systems and DAB systems developments.

During this period he became the Secretary to the EBU/RDS Forum and the RDS specification, EN 50067, Working Group, Chairman of the RDS Guidelines Working Group and Organiser of the EBU RDS Registrations Office.

Bev worked as a broadcast expert for the EC DG VII DEFI project during 1995 and took on the broadcast systems expert role with the CEN TC 278 SWG 4.1 which undertook RDS-TMC standardisation and within the FORCE Project on RDS-TMC Implementation matters.

Meanwhile he worked as Senior Engineer on the EC DG XIII EBU-EPISODE project, to undertake broadcast sector co-ordination of the RDS-TMC projects of the 4th Framework RTD programme.

With the advent of the EBU initiative to establish the Transport Protocol Experts Group, Bev was its founding Secretary and quickly became its Chairman which he project managed through a period of change to become the TPEG Forum. Eventually he managed the joint decision to amalgamate with the TMC Forum into what is now the very well established TISA.

Bev was appointed the founding Executive Director of the Traveller Information Services Association (TISA). Bev established the structures and organisation of TISA that would serve efficiently and effectively the TISA membership, which comes from a very diverse industry. He is most pleased to see some 7 years after his retirement that TISA is still proving to be a key focus for TTI services. He was honoured on his retirement in 2012 in being awarded a Fellowship of the TISA.

Danny Woolard

Danny is an engineer by background specialising in survey and navigation and now runs his own consultancy business (Chiltech). He has been actively involved in the Traffic & ITS industry for over 25 years. His former business developed and deployed the first commercial broadcast traffic service based on RDS-TMC in the UK, finally selling this business to EU traffic service provider ITIS which, in 2011, was acquired by INRIX. After 8 years heading up ITIS's traffic business and with long family ties to Australia, Danny moved to Australia in 2007, where he continued his career as GM Operations with traffic and telematics service provider Intelematics. In 2012, Danny returned to Europe to work with INRIX.

Danny has played a significant role in the development of commercial Traffic Services in the European market, both commercially and in European industry groups, he chaired the TMC Forum for 4 years and oversaw its merger with the TPEG Forum to create what is now TISA (Traveller Information Services Association). He is still an active member of the TISA community as part of his consultancy role with GEWI Europe.

Danny now lives and works from Devon, UK. A devout 'Petrolhead' and owns a small collection of classic cars which he participates in Sprints and Hillclimbs.

John Mitchell

John is a senior researcher at the Transport Research Laboratory (TRL). He has over 18 years' research experience in the signing field, including fixed traffic signs, variable-message signs and temporary traffic management signs. He has been the technical lead for several research projects for the Department of Transport, the Highways Agency/Highways England and Transport for London. John has represented TRL in his specialist signing role at working group meetings and technical workshops forming part of the UK Government's Traffic Signs Policy Review. He was the technical lead for one of the Policy Review's major research projects on reducing sign clutter. John took a lead author role in a consultancy-based project for a Middle Eastern Government to update their *Traffic Signs Design Manual*. John was the lead researcher at the TRL for the off-road research trials conducted at TRL, in 2003, of the Motorway Signal Mark 4 (MS4). John's signing expertise has recently been used for a project commissioned by

Highways England to review their variable signs and signals (VSS) policy from a human factor's perspective.

Neil Hoose

Neil is a graduate civil engineer with an M.Sc. in Transport and a Ph.D. in the application of computer vision techniques to transport. He has a track record in leading transport-related projects in both industry and academia, including director-level appointments in consultancy and manufacturing. With over 30 years in the field, his experience ranges from academic research and teaching through project management to manufacture and supply of electronic systems to the traffic data and control industry. Since 1999, he has been an independent consultant in intelligent transport systems. His consultancy activities encompass the public sector and private sector clients. Alongside particular experience in traffic monitoring technologies and traffic control and data systems, he is involved in Smart Motorways, connected vehicle-infrastructure systems and travel information systems. In addition, Neil is a visiting professor at the Centre for Transport Studies, Imperial College London and is engaged in teaching at undergraduate and Master's level as well as supporting the centre's research into intelligent transport systems. In 2019, Neil was awarded a post-graduate certificate in cyber security management by the University of Warwick.

Nick Illsley

Nick spent 26 years as a career railwayman having joined the British Rail from School in 1976. During this period, the industry went through the privatisation process and Nick became part of the Board for the newly privatised Thames Trains company. Finally, he headed National Rail Enquiries before joining the Civil Service in August 2002 as the chief executive at Transport Direct. The Transport Direct portal was launched on 30 December 2004 and then serviced over 100 million user sessions. Highlights included working with the Olympics Family to deliver journey planning services for the London 2012 Olympic and Paralympic Games and embedding accessible travel as a GB-wide service. Nick was also responsible for driving the transport transparency strategy across the sector, which has been considered to be one of the success stories of Open Data in terms of both re-use and also new applications for end-users. When the Transport Direct portal closed on 30 September 2014, Nick decided to call time on life at the DfT and is now enjoying some more leisure time and is also helping organisations use data and information to drive efficiency and better customer experience. Outside work, his interests revolve around sport (an optimistic but always worried England and Leeds fan), family (a daughter of 35 and a son of 33 and three grandsons of 6, 4 and 2) and trying to keep in touch with as many friends and colleagues as possible.

Susan Grant-Muller

Susan is Professor of Technologies and Informatics at the Institute for Transport Studies, University of Leeds, and a Fellow of the Alan Turing Institute. Susan's research is at the multidisciplinary interface between digital technologies,

Big Data and transport under a low carbon energy future. She leads the 'Digital Futures' research theme at ITS, a research theme initiated by Prof. Grant-Muller in 2015 and building on rapid and recent developments in pervasive technology. Her specific interests are in incentivising behavioural change to reduce the carbon burden of the transport sector, the role of Big Data in sustainable transport paradigms, the evaluation of new technology-based intelligent transport schemes and the resilience of ICT-enhanced transport.

Frances Hodgson

Frances is a senior research fellow at the University of Leeds. She has 30 years' experience working on the social impacts of transport. She contributes to international work through European and nationally funded projects providing specialist input on the interaction between transport, social equity, behaviour, social organisation and transport policy and was co-investigator on H2020 EMPOWER, ESRC HABITs, Alan Turing Institute KARMA projects, all of which are concerned with the application and value of new data forms (e.g., mobile apps, social media). She was UK representative for the COST GenderSTE action on Transport and Gender, expert advisor to the Co-Motion project on older people's mobility. She has 15 years of experience in providing capacity building in the form of Masters training and short courses and Doctoral student supervision and is the programme director for the MSc Transport Planning. She has published for the Equal Opportunities Commission and in the many international journals.

Foreword

Travel is part of our modern world and ever since transport has been available there has been a need for information about service operations. The sustained growth in road travel has prompted a parallel wish for information about the performance of the road network, since congestion and other incidents can seriously affect journey times.

For many years the United Kingdom has been a leader in the field of traffic and travel information (TTI), taking early advantage of data collection and information technology to measure performance, to better manage networks and to keep travellers informed before they travel and during the trip. The UK Government was a pioneer in recognising the widespread benefits of up-to-date reliable information and led the way in passing legislation to encourage the involvement of private sector operators in the TTI field. The UK continues to be a major contributor to international standards and a sponsor of research into new or improved services.

I am delighted to see this book which is edited by two experts who have been deeply involved in the topic for many years and with whom I have had the great pleasure of working. It not only contains a wealth of technical information but also tells the story, for the first time, of how traffic and travel information developed in the UK and how it influenced European and worldwide developments. It looks back to the early days of journalistic broadcasting of traffic information, covers developments in digital and mobile technology and also comes up to date with connected vehicles and social media. Even with the expected increase in automation of transport operations, the need for traffic and travel information has not diminished and so this book is likely to remain relevant to engineers working on future integrated and seamless mobility services for many years.

Professor Eric Sampson CBE
Transport Operations Group
Newcastle University

Glossary of acronyms and technical terms

AA	A commercial UK provider of motoring services. Previously, the Automobile Association
ADAS	Advanced driver assistance systems
AGORA-C^{TM}	A dynamic location referencing method
AI	Artificial intelligence – software
ALERT C	A standard list of traffic-relevant messages
ALR	All-lane running – one aspect of Smart Motorways
AMI	Advanced motorway indicator – an early dynamic roadsign
ANPR	Automatic number plate reader – equipment to convert the images of number plates into digital character strings
API	An interface or communication protocol between different parts of a computer program intended to simplify the implementation and maintenance of software
AS-1	Additional service licence permitting the use of RDS groups on the Classic FM RDS datastream
ASTRID	Automatic SCOOT Traffic Information Database which stores and processes data from SCOOT sensors
AV	Automated vehicle
b2b	Business to business – a commercial transaction between businesses (rather than end-users) of good or services
BBC	British Broadcasting Corporation – A UK mass media broadcasting organisation
BCA	Benefit–cost analysis
BCR	Benefit–cost ratio
Bluetooth	A short-range wireless technology standard used for exchanging the data between fixed and mobile devices
Business model	A plan of operation including products, services, customers, financing and expected revenues
CAV	Connected automated vehicle (or Connected autonomous vehicle)
CCAM	Connected cooperative automated mobility
CCTV	Closed-circuit TeleVision – a system for sending pictures to one or more specific locations

CD	Compact disc – an optical disc data medium originally developed to store music
CEN	European Committee for Standardization
CENELEC	European Committee for Electrotechnical Standardization
C-ITS	Cooperative intelligent transport systems
CV	Connected vehicle
DAB	Digital audio broadcast
DATEX	A data exchange standard
DfT	UK Department for Transport (previously Department of Transport and Department for Environment, Transport and the Regions (DETR))
dGPS	Differential GPS – a technique for boosting the accuracy of GPS during selective availability operation
DRIVE	Part of the European Commission's 2nd Framework Research Programme (FP2) 1988–1991
DSRC	Dedicated short-range communication
EBU	European Broadcasting Union
EC	European Commission
eCall	A system for automatic notification of a vehicle crash to emergency services
EMS	Enhanced message sign
EN	European Norm – Full European standard
EON	Extended other networks, a method of signally the existence of a traffic report on a radio station to other stations in the same network
ERTICO	A public–private partnership organisation active in ITS and based in Brussels
ETA	Estimated time of arrival also Toyota's 'Electronic Traffic Avoidance'
ETSI	European Telecommunications Standards Institute
FM	Frequency modulation – a method for encoding signals used, for example, in radio broadcasting
FOT	Field operational test
FVD	Floating vehicle data
GCDP	Graphical congestion display panel
GDF	Geographic data files – a standard for defining road network data used in Intelligent transport systems
GDPR	General data protection regulation
GLOSA	Green light-optimised speed advice
GNSS	Global navigation satellite system

GPS	Global positioning system – a satellite-based radio-navigation system owned by the US government
GRIP	Graphical route information panel
HD	High definition – e.g. of a map or a display
Highways Agency	Former name of Highways England
Highways England	A UK government-owned organisation charged with operating, maintaining and improving most of England's motorways and major A roads (previously Highways Agency)
HIOCC	HIgh OCCupancy – a software algorithm which aims to detect the traffic congestion from road loop data
HMI	Human machine interaction (or interface), previously called man machine interface (MMI)
IEEE	Institute of Electrical and Electronics Engineers – a US-based standards organisation
INSPIRE	An abbreviated name for an infrastructure for spatial information in the European Community
IoT	Internet of Things – a network of physical objects capable of detecting and communicating information between each other using embedded sensors, software and connectivity
ISO	International Organization for Standardization
ITF	International Transport Forum (ITF) – an inter-governmental organisation within the OECD
ITS	Intelligent transport systems
ITU	International Telecommunication Union – an agency of the United Nations specialising in information and communication technologies
IVS	In-vehicle signage
LED	Light-emitting diode
LiDAR	A detection system that uses pulse-encoded laser light to measure the range, angle or velocity of objects (see also RADAR)
Location code	(supporting RDS) A 16-bit code that specifies a point (node), segment or area within a geographic region
LRM	Location referencing method
LRS	Location referencing system
MaaS	Mobility as a Service – the integration of various forms of transport into a single mobility service accessible on demand
Micromobility	The use of individual micro-transport such as cycles and scooters

MIDAS	Motorway incident detection and automatic signalling – a distributed network of sensors which measure traffic performance and help set signs to improve the traffic flow
MS	Member State
MS{n}	Matrix signal – a type of dynamic roadsign used on the motorway MS1, MS2, MS3, MS4, etc.
NaPTAN	National Public Transport Access Nodes – UK standard and database
NMCS2	National Motorway Communication System – used by the UK Highways Agency
NPTG	National Public Transport Gazetteer
NRES	National Rail Enquiry Service
NTCC	National Traffic Control Centre (later, National Traffic Operations Centre)
OBU	On-board unit
OECD	Organisation for Economic Co-operation and Development
OEM	Original equipment manufacturer – referring to manufacturer of a road vehicle
OpenLR™	A dynamic location referencing method
OSGR	Ordnance Survey grid reference
PDA	Personal digital assistant
PKI	Public key infrastructure – a system for secure transmission of data
PND	Portable navigation device
PTE	Passenger transport executive
PVD	Probe vehicle data
RAC	A commercial UK provider of motoring services. Previously, the Royal Automobile Club
RADAR	A detection system that uses radio waves to determine the range, angle or velocity of objects (See also LiDAR)
RCC	Regional Control Centre
RDS	Radio Data System
RTTI	Road traffic and travel information
RTTT	Road transport and traffic telematics (an older term for what was known as intelligent transport systems (ITS))
RWW	Road works warning
SA	Selective availability – a mode of operation which effectively downgraded the accuracy of GPS positioning
SCOOT	Split cycle offset optimisation technique – an area-wide system for coordinating and controlling traffic lights

SIM	Subscriber identity module – a microchip in a mobile device that connects it to a particular phone network
Smart motorway	A section of road with variable speed limits and hard shoulder running (previously called 'Managed Motorway')
SMMT	Society of Motor Manufacturers and Traders
SPaT	Signal phase and timing
SRN	Strategic Road Network – the UK's motorways and 'trunk' roads (major 'A'-class roads)
TCC	Traffic Control Centre
TDM	Traffic (or Travel) demand management
TfL	Transport for London – the UK capital's transport authority
Thick client	A computer system that runs an application which can locally access all the data and other resources needed to operate
Thin client	A computer system that runs an application which accesses most of its data and other resources from a remote server
TISA	Travel Information Services Association
TMC	Traffic message channel
TMC Forum	A non-profit organization of public and private organisations formed to discuss traffic information-related matters (subsumed into the TISA, in 2007).
TMS	Traffic Management System
TN-ITS	Transport networks for intelligent transport systems – a standard for the exchange of static road data
TOC	Train operating company
TPEG	Transport Protocol Expert Group – a data protocol suite for delivering traffic and travel-related information via multiple different transmission media
Traffic Officer	Employees of Highways England, who work alongside the police to help managing incidents on motorways and trunk roads
Transport Direct	A multi-modal transport planner for the UK
Travel Information Highway	Highways Agency initiative to exchange real-time traffic-relevant information to information-service providers
TRL	A UK research laboratory, previously Transport Research Laboratory and TRRL Transport and Road Research Laboratory
TS	Technical Specification – intermediate standard
TTI	Traffic and travel information
UML	A widely used software package for system design and specification

UTC	Urban Traffic Control
UTMC	Urban Traffic Management and Control
V2V	Vehicle-to-vehicle – indicating communication of data between vehicles
V2X	'Vehicle-to-Anything' – indicating communications between vehicles and other vehicles, infrastructure and mobile devices
Value chain	The series of linked activities that one or more organisations need to perform in order to deliver a product or service
VMS	Variable message sign
VSL	Variable-speed limit
WGS84	A standardised coordinate reference system for locating points on the earth's surface
XML	eXtensible markup language – software used to represent the data structures, particularly for Internet services

Introduction

This book is the result of a desire by a range of UK ITS professionals to record the progress in traffic and traveller information (TTI), particularly from its origins in the United Kingdom which then influenced and, of course, interacted with developments in Europe and indeed the rest of the world. The chapters have been contributed by leading exponents in the field of TTI and encompass collaborations during research and development through to implementation by industry; hence, the very different styles of the chapters and complementary overlap from different perspectives.

It should be appreciated that traffic information came much later than the need to provide travellers with information on their journeys, for instance, to markets or on pilgrimages, so TTI (or just traveller information) might be a better title as traffic information is just a contributary part, albeit a substantial one, of the whole travel picture.

Travellers have always needed information before and during their travels, whether it be which roads/paths are open, which are safe, facilities on route such as staging posts and hostelries, the availability and costs of ferries and bridges, and the time needed to complete their journeys. Originally this information was passed on by word of mouth and was of an informal nature; then, mileposts and road signs at intervals and decision points gave some further information to the traveller.

The collection of information became part of the TTI story from the turnpike operation which needed to assess the income from traffic over a toll bridge or toll road, to the stage-coach operator who wanted to assess the demand for their services.

Detecting and counting of vehicles has a long history in the developed countries since the rapid increase in road transport during the twentieth century; there has always been a need to count and classify! Of course, in the early days it was a totally manual operation with the toll keeper inspecting every vehicle and charging appropriately. Today, we have much more sophisticated methodologies with gantries and video detection systems to determine the type of vehicle and load in order to apply a charge. Emerging technology now allows the collection of a much broader range of data from vehicles and the technology is likely to extend even further with remote sensing from satellites. In between these extremes there is a panoply of technology used for a variety of purposes.

Traffic data is not only collected by instrumentation; visual reports and more 'journalist' information are collected from the ground or from the air as part of the media editorial contribution.

During the years up until the early 1990s, the provision of TTI, at least in the United Kingdom, was almost entirely the preserve of the public sector which ran train and regulated bus services and were responsible, totally, for the roads. Then, with privatisation of most travel services the private sector moved in to supply TTI as a service to travellers, and this has further increased following the advent of social media where data is crowdsourced and is increasingly being integrated with other services. In all these myriad offerings, business cases are key to providing sustainable services.

From the early 2000s, only a small proportion of the TTI available was able to be widely broadcast because time-slots on television, and even more so on radio, were limited. In addition, the broadcasting areas were so large that only travel information of national strategic importance was considered appropriate rather than local information that would be of real use, so only to a relatively small proportion of the broadcast audience.

The collection, collation and transmission of TTI has been transformed through digital coding with standardised applications such as DATEX 2, RDS-TMC and TPEG, and, increasingly, this coded information is now being integrated into intelligent agents and within in-vehicle services, such as satellite navigation.

Much of the research and development (R & D) of TTI systems in Europe has been supported by the European Union Framework R & D programmes and has been trialled in cross border projects. These developments have informed and been complementary with standardisation work in Europe and internationally. Whilst recognising this wider context, this book offers many examples of developments and deployments of TTI in the United Kingdom, which has both contributed to and benefitted from this wider international collaboration on TTI.

TTI services would be largely useless without providing geographical and other contextual information to allow users to make sense of the messages. As a key part of the TTI story, this book details the development and application of location referencing services.

TTI does not exist in isolation as an application but plays an integral part in the operation of many transport services. For example, in an inter-urban road context, TTI supports the development and rollout of 'smart' motorways where messages to drivers can have an immediate effect on their behaviour and help to ensure safe and efficient use of the road space. The backbone for immediate delivery of this traffic information to drivers is through the development and implementation of dynamic signage (variable message signs), which provides dynamic traffic information to road users on the UK's Strategic Road Network, but also the provision of traffic information on non-strategic roads, primarily in urban areas.

As we look to the future of road travel, a current focus to improve safety and increase capacity is for vehicles to actively cooperate with the road infrastructure and, eventually, with other vehicles. This 'Cooperative Intelligent Transport System' approach will use the whole spectrum of the TTI applications along with other intelligent transport technologies, particularly communications.

Much of the TTI described in this book focuses on road transport because that is where most initiatives begin as there is no inherent data collection and dissemination as in railway scheduling and signalling systems. However, the traveller requires information about all forms of transport provided as a service. For example, due to privatisation of the railways and buses there has never been a greater need for integration of information (both timetabled and dynamic) to allow seamless and efficient travel, both in everyday lives and for mega-events like the Olympics. These topics are therefore addressed in a specific chapter of this book.

Information being crowd-sourced and then made freely available is increasingly taking the place of more traditional collection and dissemination of data and information by public authorities. Of course, there is no such thing as free information and there is a chapter on the business models associated with social media provision of TTI. These are markedly different from the business models that had to be formed for TTI services like RDS-TMC.

But to the future.... We can only see where we are heading at present, knowing that the world and society is likely to change in ways we cannot yet imagine. To some extent we can foresee the range of technologies that we have available, but how they will be applied may surprise us. It seems likely that the nature of city and town centres will change from what we experience now so that TTI may look very different. At the time of writing this introduction we are in the midst of the COVID-19 pandemic and our society and interactions are changing our social and behavioural perspectives, probably permanently. However, the human race will always have the need to travel and TTI will always have a part to play.

The editors wish to acknowledge and thank the selfless contributions from each of the chapter authors and to Valda Stevens, who assisted greatly with all the proof-reading and consistency checking. Of course, the contributors to this book are only a small proportion of the many ITS professionals, in the public, private and international organisations who through their collaborative efforts have made ITS the industry it is today.

Chapter 1
Road traffic data collection

Paul Burton[1] *and Neil Hoose*[2]

1.1 Introduction

Detecting and counting of vehicles has a long history in the developed countries; there has always been a need to count and classify vehicles from the days of private roads and bridges with their manual toll charges for different classes of vehicles and goods. Of course, in the early days, it was a totally manual operation with the toll keeper inspecting every vehicle and charging appropriately. Today, we have a much more sophisticated methodology with gantries and video detection systems to determine the type of vehicle and load to apply a charge. Emerging technology allows the collection of a much broader range of data from vehicles and the technology is likely to extend even further with remote sensing from satellites. In between these extremes, there is a panoply of technology used for a variety of purposes.

This chapter considers different types of data detection and collection and suggests how they are used, rather than starting with an application requirement and determining the type of data required; it also concerns the collection of data for use in road transport and does not consider the detection and collection for rail and air.

1.2 Background and context

Data can be numerical, such as so many vehicles per hour, or it can be contextual and qualitative or a combination or repurposing of both. The data can be collected for a range of purposes including planning, revenue collection, Traffic management, driver hours, vehicle maintenance status and traveller information. Traffic data can be not only collected and used in real time but also stored and used to look at trends.

[1]CEN/TC278 WG4 and ISO/TC204 WG10 – Traffic and Traveller Information, Worcestershire, UK
[2]Bittern Consulting Ltd., Oxfordshire, UK

1.2.1 Collection of data for use in real time

Early examples of real-time collection come from the turnpike operators in their toll house detecting vehicles or people wanting to make a passage along the road or over the bridge; classifying the vehicle, for example, as cart with goods, cart with passengers, cart with one horse, cart with locals, cart with non-locals and charging them accordingly. Those attempting to avoid the toll would be passed over to the local authority, be that police or a local force.

Modern tolling systems can classify vehicles by, for example, number of axels or weight or weight per axel, type of vehicle, and account holding regular use vehicles. This allows a charge to be levied either to an account or determined to be paid at a later date. Often the registration plate of the vehicle is used to determine the type and ownership of the vehicle and to identify miscreants for penalty charging or further enforcement.

Motorway and urban traffic control systems generally use data in real time to operate their control and intervention systems. This data can come from direct observation, in-road surface detectors such as loops, CCTV, automatic number plate recognition (ANPR), reports and so on.

1.2.2 Collection of historical data for analysis

Authorities and other researchers often use traffic data collected for the purpose of off-line analysis or reuse data collected for other real-time purposes, for example, looking at real-time trends vs. historical data. Data that goes back many years may also be analysed to show the trends in the use of infrastructure.

1.3 Traffic reports

The earliest and most qualitative of traffic data is verbal reports by those out on the infrastructure, reporting back (to a base or control centre) what they observed whilst undertaking journeys or activities on the road. This section describes the range of reports from third parties rather than the authorities and road operators.

1.3.1 Motoring organisations' patrols

In the early days of motoring, travellers were often members of motoring organisations to get roadside assistance for vehicles that were not as reliable as they are today. The Automobile Association (AA) [1] was established in 1905 with the intention of warning fellow members of speed traps by a system of signals and salutes from a force of pedal cyclists! Within a few years, the AA were offering insurance policies and roadside breakdown assistance using motorcycle and four-wheel patrol vans.

It was a similar story with the Royal Automobile Club (RAC) [2] which started a few years earlier in 1897 in response to the restrictive 'Red Flag Act' of 1986 and campaigned for the 1903 Motor Car Act. In 1905, in a similar way to the AA, the

RAC provided assistance and guides on bicycles, then motorcycles and vans were gradually introduced.

Both the RAC and AA patrols out on the road were useful sources of motoring information and were the bedrock of the traffic situation services for the members of their organisations. Of course, there were only reports of traffic conditions where a patrol happened to be at the time; in essence very piecemeal.

Communications from the patrols in the early days, before personal two-way radio, was via the AA and RAC call boxes that used land lines. Even up to the late 1970s, the AA members' pack contained a key for the AA telephone boxes!

At the time of writing, the RAC do not directly provide traffic and travel information. The AA still do, via their *AA Roadwatch* services, but it is doubtful whether much information now comes from patrols.

1.3.2 Police and emergency centre reports

The police and other emergency services have always been a source of traffic information. Experience in the early days of digital traffic messaging such as the radio data system traffic message channel (RDS-TMC, discussed in later chapters of this book) showed that in the Germanic nations, a traffic incident didn't officially occur unless it was reported and logged by the police. Message sets for RDS-TMC differentiate quite clearly between Police Officers, Police Cadets and Others.

In the United Kingdom, the police and emergency services, however, are (quite rightly) more interested in dealing with the incident than in provide traffic information; this led in the 1990s to the provision of the National Traffic Control Centre (later National Traffic Operations Centre) and the Regional Control Centres which concentrated on network operation and traffic information.

Since the mid-2000s, the Highways Agency (now Highways England) Traffic Officers [3] have been patrolling the Motorway and Trunk Road network and, as well as providing assistance and ensuring the traffic flows freely during incidents, have been a valuable source of traffic data.

1.3.3 Radio audience ('jam busters')

With increasing numbers of drivers having access to a mobile phone in the course of their journeys, a new breed of traffic situation reporter was born.

In the 2000s, radio stations and motoring organisations recruited 'jambusters' – members of the public who phoned reports of abnormal traffic conditions (when it was safe to do so). To ensure the validity of the reports, many organisations registered their jambusters so that all information was attributed. Discussions over the air with jambusters quickly formed part of the editorial material of many radio stations in the same way that 'Sally Traffic' became an integral part of the BBC Radio 2 in the afternoons in the United Kingdom with almost 8 million listeners at its peak. Frequently, these jambuster reports were considered entertainment editorial, particularly by local radio stations who used it to define themselves as being involved in their local community. In the same category are reports from helicopter and plane-based surveillance – as discussed in Chapter 2.

1.4 Closed circuit television (CCTV)

1.4.1 Visual monitoring

For many years, CCTV surveillance of busy junctions and major features, such as long bridges and tunnels, has played a significant part in traffic management where there were control centres (covering cities, major bridges and tunnels). These early systems were monochrome and the communications to get sufficient bandwidth for an image were expensive. As late as the mid-1980s, many systems used 'slow scan' which meant that the image was only refreshed every second or so. This was sufficient for traffic control but hardly suitable for Traffic and Travel purposes.

The late 1990s and early 2000s saw the advent of network-based systems that allowed colour images in real time and the reduction of costs. This, and increased system reliability, led to the widespread deployment of cameras particularly on the motorway and trunk road network. By the early 2000s, the then Highways Agency made camera images available in real-time to radio and TV stations and eventually the general public. At the time of writing, these are available via the *Highways England Traffic Camera* website [4]. It is possible to decouple cameras from the system in the case of serious incidents where the cameras are used for monitoring and control and the images not suitable for the general public. At present the downloading of moving images is by permission only for agreed purposes, but still images are available after a simple registration.

Traffic control room staff can get an idea of the saturation levels and traffic speeds by referencing the passing of vehicles against fixed assets such as the lane markings.

1.4.2 Image-processing to form incident alerts

The installed traffic camera base is generally used for visual monitoring and typically a control room only has 20 camera images displayed at any one time. Images are only monitored by control room staff where there is a declared incident or to observe known congestion 'hotspots'; not the most efficient use of an expensive manpower resource when 'eyes' need to be everywhere.

It is now the practice in tunnels to employ the image analysis software to detect incidents. If the system detects slow moving or stationary traffic, it generates an alert for operators to look closely at the scene and determine a course of action. The software detects individual vehicles and analyses their progress across the image frame.

The managed motorway schemes [5] make similar, more complex and wider ranging use of automated image analysis to detect slow moving and stationary vehicles particularly on what was originally the hard shoulder.

As in interesting aside, the author has been told by a software provider is that image-processing is used on a number of stations to detect potential suicides. The system identifies and tracks individuals on the platform and if an individual lets 3 trains stop and move off, that they possibly might jump in front of the fourth; if so, an alert is raised and a member of the station staff dispatched to enquire about the wellbeing of the individual, ghoulish, but an illustration of the power of image-processing for incident detection.

No doubt that by the time of the publication of this book even more powerful software and systems will be developed to make the most of the CCTV resource, if one can still call it closed circuit in these days of complex wide area networks.

1.5 Road surface monitoring/counting

Traffic surveying has traditionally been undertaken manually using volunteers armed with a counter, clipboard and folding chair. This method is reasonably effective for counting and classifying vehicles but depends on the ability and tenacity of the surveyors. This section describes developments in technology which over time have been able to automate the process and allow better analysis both in real time and offline.

1.5.1 Census vs. real time

For many years, automatic counters and classifiers have been employed [6] that can be used for both offline and real-time applications. Firstly, detection and classification (the splitting of vehicle types into classes) can become census-type information for the purposes of offline analysis and planning. Generally, this data is used in time period slots such as every 5 min or even as large as every hour or day. Note that there are many systems on the road network recording and transmitting data every minute, or less, for the data to then be amalgamated into hourly packages for analysis – but in the future this finer time resolution data might be useful.

The same issue of data reduction is true for lane resolution. Most detectors can detect by lane, but the static data is often amalgamated to carriageway resolution.

The second application area is that of real-time control where data is needed at a much higher time resolution. For instance, the split-cycle offset optimisation technique (SCOOT) real-time adaptive urban traffic control system [7] takes data from its loop detectors 4 times per second. There are also systems that can transmit data (by vehicle type) on a vehicle by vehicle and lane by lane basis. Thus, traffic patterns at one point can be cross correlated with downstream patterns to get an idea of journey times. This sort of data is increasingly being provided by floating vehicle systems which will be discussed in a later chapter of this book, but these do not provide lane level resolution (at present).

1.5.2 Rubber tube traffic counters

The earliest traffic counters were based on pneumatic rubber tubes, sealed at one end and attached across the carriageway. The other end of the tube is attached to a box, generally attached by chain to a convenient lamppost; inside the box is a pneumatically activated switch that is attached to a simple numerical counter. This simple system counts axels in both directions. Of course, the rubber tubes are vulnerable to damage by vehicles and by vandalism.

Figure 1.1 Pneumatic tube system

More sophisticated systems were introduced with two tubes set at a known distance apart. This allows counting separately in both directions; the direction of the vehicle being determined by which tube is activated first. Again, the system counts axels, so derived vehicle count estimates will not be entirely accurate, depending on the mix of vehicles at the site and whether vehicles trigger the counter tubes simultaneously. A double-tube system can also calculate speeds – the time from triggering of the first tube to triggering of the second tube. The photograph of a rubber tube system in Figure 1.1 was taken in 2019 and was being used to record speeds in a country lane in Worcestershire due to residents' concerns over speeds.

Despite the shortcomings of tube detectors they are still very much used in the non-motorway trunk road network as they provide a cheap, low-tech, temporary solution where absolute accuracy is not required.

Before the widespread use of inductive loop detectors at traffic lights, pneumatic strips were used set into steel formers at the stop line to determine the presence of vehicles.

1.5.3 Inductive loops

Inductive loops work in the same way as mine detectors and essentially comprise three components; a loop of wire buried in the road surface which starts and ends at the same point, a cable from the ends of the loop to the roadside and a detector housed in a roadside cabinet. The induced current in the target object is in the opposite phase to that in the detector and suppresses the current in the detector by a small but measurable amount. Further details can be found in [8].

A slot is cut in the road surface, with a specific shape and dimensions depending on the particular application and what sort of vehicle is to be detected. Typical loop shapes include square, rectangular, chevron and diagonal.

A set number of turns of wire, defined by the application and detector technology, are wound in the slot in the road surface. The road surface is then made good with pitch which also makes a waterproof seal for the loop. The ends of the loop are returned to the side of the road via a long slot which, on multilane carriageways, will involve cutting the slot across lanes to join between the loop and the roadside cabinet.

Inside the cabinet there is a detector for each lane, although the detector circuit board cards can contain a number of channels, one for each lane. The detectors feed a current into the loop at a frequency so that the loop resonates (typically 20–80 kHz). A vehicle passing over the loop changes the field around the loop, so it resonates at a higher frequency; the closer the vehicle is to the road surface the greater the increase in frequency. This change of frequency is detected, indicating the presence of a vehicle.

A single loop installation can determine the occupancy, i.e. how much time a vehicle is over a loop, due to either its speed or length. The speed of a vehicle can be determined using two loops (from when it 'fires' the first loop to when it 'fires' the second loop). It is also possible to obtain the number of axels on a vehicle, but it is not always straightforward as the metallic bulk of a vehicle can mask the presence of axels. Classification by length is possible by combining the speed of a vehicle with the length of time the vehicle is present over the loop.

Inductive loop sets can be unidirectional or bidirectional in the same way as the rubber tube versions.

The power requirement for loops and detectors is relatively low and they are often powered by solar cells and batteries. The communication via mobile cellular radio is often the largest consumer of power and is often only powered when communication needs to take place to download data.

Installing inductive loops is a skilled and disruptive activity. Lanes, and in the case of multilane carriageways the whole carriageway, have to be closed to allow the sawing of slots into the surface and bringing the loop tails back to the side of the road. On single lane roads the tails of both carriageways are generally brought to a cabinet on one side of the carriageway only.

Being in the road surface, loops are generally reliable and the detectors are self-tuning so little day to day maintenance is required. However, the loops are vulnerable when there are roadworks and resurfacing works, which causes a problem when they have to be recut. Experience in hot climates, for example, in Hong Kong, is that loops are vulnerable when the road surface softens in the heat and is rucked up by buses braking at a junction. Additionally, they do not work well where there is a lot of metal in the surrounding infrastructure such as bridges, flyovers and reinforced concrete roads.

Inductive loops can be used for historical data collection for offline analysis and in real time for traffic control in systems such as SCOOT [7] or single sets of isolated demand dependant signals. Inductive loops are used in the MIDAS (Motorway Incident Detection and Automatic Signalling) motorway warning systems [9], but they are gradually being phased out in favour of side fire RADAR/LIDAR which are less disruptive to the road surface and where there is power readily available.

It is possible, using signal-processing techniques on the change in inductance from an individual loop, to obtain an estimate of speed without the need for a second loop as discussed in [10].

It is likely that inductive loops will be in general use for many years due to their low power consumption which means that they can be solar powered, making them feasible where there is no nearby mains power, and the certainty of the detection zone.

1.5.4 Piezoelectric and fibre-optic

Piezoelectric sensors can be placed across the carriageway that detect minor deflections in the road surface. When the sensor is stressed due to the road surface deflection it creates a very small current that can be detected. The amount of current is proportional to the deflection of the pavement and so that piezo-electric sensor can be used for axel detection and weight in motion applications [11]. Although not necessarily important for traffic and travel information services, piezoelectric sensors are part of the traffic engineers' armoury. Similarly, fibre-optic sensors have been used to detect the presence of vehicle axles where small distortions cause a change in scattering and reflection within the optical fibres.

1.5.5 Magnetometers

The main problem with inductive loops is the disruption to road traffic caused by their installation and maintenance. Magnetometer systems can alleviate this problem [12]. Magnetometers work by detecting the change in the earth's magnetism caused by metal objects passing by.

Magnetometers can be sourced that are around 150 mm diameter and 150 mm height that can be installed by removing a core from the road surface and placing the magnetometer in the resulting hole and then asphalt or pitch filled in to make the surface level. The magnetometers are battery powered and can last up to 8 years; they transmit their signal to the roadside cabinet by radio. The detection area of the magnetometer is relatively stable and definable so that they can be used instead of inductive loops.

Installation of magnetometer detectors is quicker and simpler than sawing slots in the road surface for loop detectors. A core drill mounted on the back of a moving lane closure vehicle can remove a core in a matter of a few minutes, the magnetometer placed in the hole and the surface made good. As there is no slot cutting needed to take a lead to the roadside only the lane being installed is affected.

Disadvantages of magnetometers are their expense and the lifetime of the battery. The cabinet also needs mains power to enable the receivers.

1.5.6 RADAR

Previous sections describe the benefits and disbenefits of road surface detection technology. A way of overcoming the installation and maintenance issues is to use technology remote from the road surface. Highways England are introducing Side Fire RADAR mounted on poles at the side of the carriageway that can detect vehicles in each lane with sufficient detail to fulfil the data needs for MIDAS [9] and the Controlled Motorway systems.

Radio detection and ranging (RADAR) technology was first applied to traffic sensing during the 1980s when advances in solid state technology allowed compact and affordable devices to be made, suitable for detecting the presence of traffic at signal junctions. These devices use 'continuous wave' (CW) radar to detect the shift in frequency between an emitted signal at a known frequency and the reflection of that signal from a moving object, a phenomenon known as Doppler shift. The technology can be calibrated so that the device can measure the speed of the object based on the magnitude of the shift. The direction of the shift can be used to discriminate between objects moving towards or away from the detector. Calibrated CW radars are used in some speed enforcement systems.

The limitation with CW radars is that an object will not be detected unless it is moving and so the distance to a target object cannot be determined. These problems can be overcome by the use of 'modulated' radar beams. Instead of using a continuous wave at a single fixed frequency, the emitted signal frequency can be changed continuously according to a predetermined configuration. This is called frequency-modulated continuous wave (FMCW) radar as shown in Figure 1.2. The technique compares the frequency of a returned signal with the frequency being transmitted at that time. Because the rate of change of the emitted frequency is known, the difference between it and the return signal is proportional to the time taken for the signal to reach the target and return. The speed of the signal is the speed of light and hence the range to the target can be determined from:

$$\text{Target range} = 1/2((F_2 - F_1)c)/R$$

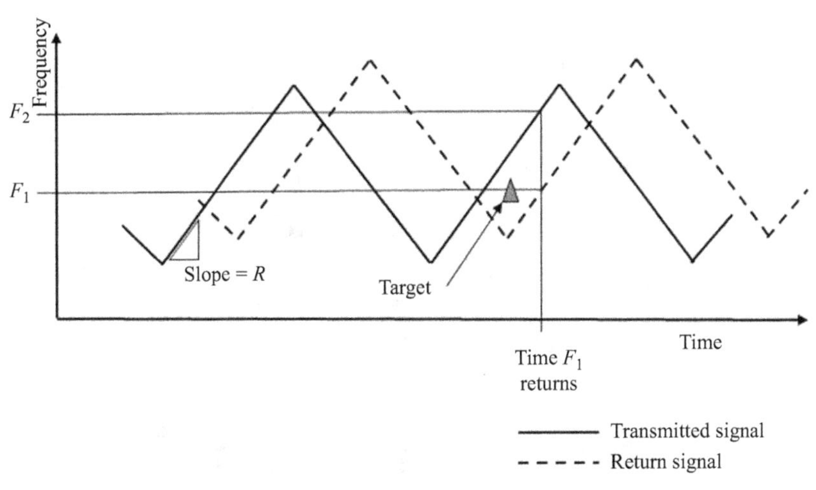

Figure 1.2 Frequency–time graph for FMCW radar

where

F_1 = Return signal frequency
F_2 = Transmit frequency at time return frequency F_1 is detected
R = Rate of change of transmitted frequency
c = Speed of light

Division by two is needed because time of flight is to target and back.

RADAR uses the microwave part of the electromagnetic spectrum and FMCW devices typically operate in the 18–26 GHz (also known as K-band) region. An individual device will operate in a narrow segment of this, for example the Wavetronix HD device operates in the range 24.0–24.25 GHz.

Doppler (CW) radars need to be mounted so that the beam is along the traffic stream as it is the movement of the vehicles that is important. If such a radar is mounted looking across the traffic it will not be able to reliably detect vehicles as the movement component along the radar beam will be very small. This is not the case or FMCW radars, as targets can be identified whether or not they are moving and so the device can face in any direction. Furthermore, because the range of the target is known, those targets at similar distances can be grouped together and a set of detection 'zones' can be set up. Thus, an FMCW radar mounted at the side of the road with its detection beam at right angles to the traffic can have a detection zone for each lane and this single device can monitor multiple lanes of traffic. As the radar beam is orthogonal to the traffic the Doppler effect cannot be used to monitor speed. If two beams are placed in parallel a known distance apart the speed can be determined by the arrival time of a target in each beam. This is analogous to the method used to measure speed by inductive loops.

A typical 'side fire' arrangement is shown in Figure 1.3. A device is mounted 4–6 m above the ground and a similar distance back from road edge looking

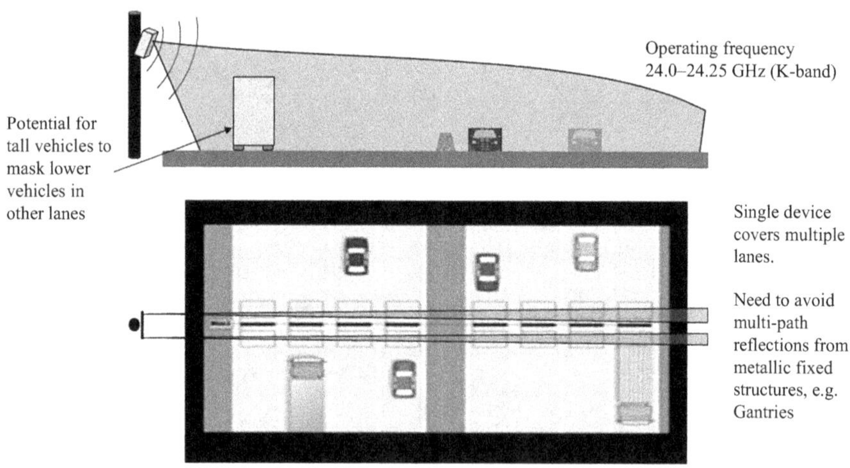

Figure 1.3 Typical arrangement for 'side-fire' FMCW radar

orthogonal to the direction of the traffic. The advances in radar electronics mean that the pair of radar beams can be enclosed in a single, compact housing easily mounted on a lighting column or its own pole. Such a device can monitor multiple lanes, enough to monitor both carriageways of a dual 4- or 5-lane highway from one side.

The data output from the device is the same type of traffic data produced by inductive loops. Therefore, this data can be fed directly into most incident and queue detection algorithms that are based on such data analysis.

Radars are very tolerant of poor weather and are only affected by extremely heavy rain or snow. They are vulnerable to electromagnetic interference from other devices transmitting in the same frequency range. The quality of the data can be affected by large vehicles nearer to the device obscuring smaller vehicles further away. This problem becomes more severe with increased traffic density and with increased flows of tall heavy goods vehicles or buses in the lanes nearest the device. Radars are also subject to multi-path reflections where the radar has reflected from multiple surfaces. These can create false targets in the detection zones.

Radars can be difficult to align and vibration or thermal effects in the mounting can shift the beam direction such that detection zones temporarily include fixed objects (e.g. safety barriers) that create false detections for as long as the beam is out of alignment.

1.5.7 LiDAR

LiDAR uses pulse-encoded laser light to measure the distance from the emitting device to any reflective surface. Unlike radar, the beam is very narrow in both vertical and horizontal directions. Devices usually comprise a number of emission–detection pairs mounted vertically. These can either be fixed or mounted on a rotating platform to provide a swept area of detection. The output from the sensing is referred to as a 'point cloud' which is a set of data points where each point is defined by a vector of horizontal angle, azimuth, range and reflected intensity. These devices have many applications including traffic measurement and are perhaps best known as part for their use in Google Streetview and Google-automated vehicles. They have been used for highly accurate counting and classification in toll booths and, in principle, could be deployed to provide target tracking and incident detection.

1.5.8 Validation of traffic data from loop sites

How does the traffic engineer know that a detector site is counting and classifying accurately? The reality is that it is impossible to know unless one counted manually at the site 24/7 to ensure that the counts were accurate at all times of day and climatic conditions. In general, loops either fail and give no count or produce a count that is unfeasibly high (e.g. well over 1,800 vehicles per hour per lane).

But it is useful to know that the site is producing sensible results, particularly after maintenance has taken place, where it is not uncommon that loop pairs can be crossed over, giving erroneous readings or readings for the wrong lane. It is also more common than one would imagine for there to be confusion in the naming of detector sites so that the outstation identifier is wrongly paired with the identifier at the instation. In some cases, it was discovered that outstations were on a completely different road to that recorded at the instation.

Traditionally, to check the accuracy of flow data, a video recording of the traffic has been made and a manual count made of the vehicles passing a reporting point. The counts are then compared with the flow data so that the accuracy of the data can be assessed. However, there are well over 6,000 reporting points on the Highways England network alone, so checking accuracy in this manner would take a very long time and, given that the majority of data are accurate, would be extremely inefficient. An additional problem associated with video assessment is the possibility of manual counting errors. These could result in an accurate site being assessed as inaccurate and vice versa. Where a site is initially assessed as inaccurate, the counts can be re-checked to eliminate the errors. However, if an inaccurate site is assessed as accurate, it is unlikely to be rechecked and therefore the inaccuracy will never be detected, unless it is selected as part of a future random sample.

In the late 2000s, CAVEMAN – Continuous Assessment of Validation Equations by Monitoring the Agency Network – was developed at the then NTCC (National Traffic Control Centre) at Quinton, UK.

CAVEMAN was based on the Long-term Integration Process (LIP), derived from Kirchhoff's law for electric current. In the case of traffic data, the currents are vehicle flows at upstream and downstream Reporting Points. The implication of this is that if the upstream and downstream flows are different, there must be either an unknown sink into which vehicles disappear or an unknown source from which vehicles can join the road. By integrating the flows over a long period of time and taking an average, the difference between the two should be extremely small. Where significant differences exist, one or more of the flows must be inaccurate. Where it is not possible to measure the entry and exit flows at unmonitored junctions, opposite flows can be compared (e.g. northbound and southbound are assumed to have equal flows). A small number of exceptions exist (notable amongst these is the Severn Bridge, where a toll was charged (at the time) to cross into Wales but not in the opposite direction). Once such a location is known and its effect quantified, it can be used in the same way as before but using the required factor to compare the opposite flows.

The network is divided into junction sets. There is some variation in the configuration of the junction sets, depending on the road type. The most common type of junction set is the Motorway grade-separated junction. An example is shown in Figure 1.4.

The flow at Link 1 is compared with the sum of those at Links 6 and 7, with the sum of the flows at Links 2 and 4 and with the flow at Link 12. These are known as validation equations. If all agree within a defined limit of accuracy, then all can be regarded as accurate. If they all disagree, the flow on Link 1 is inaccurate. If one is wrong and the others are correct, the flow on Link 1 is almost certainly accurate

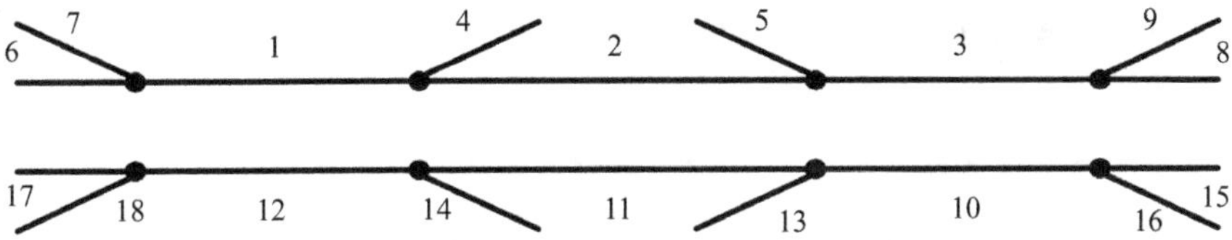

Figure 1.4 Typical junction set

and the inaccuracy will be in one of the other flows. All the comparisons can be verified against each other.

The concept of 'continuous assessment' was developed to allow an automatic measure of delivered service level on the English strategic road network. Previous to the introduction of CAVEMAN, the service level had been assessed on a monthly basis by using video survey results and using a statistical method. The single criterion was (and still is) that the product of the percentage of all reported data accurate to within defined levels of tolerance and the overall availability must be greater than 95%. For example, if 98% of all reported data were accurate to within the defined levels of tolerance, in order to meet the criterion, the overall availability of data must be 97%. The main disadvantages of the video survey method, apart from the cost, is that it depends on the ability of individuals to count and classify vehicles and it only considered a 2-h snapshot within the month. The consequence of the snapshot is that a data report that is in error for a short period of time is either reported as totally bad or totally good, depending on the timing of the snapshot.

CAVEMAN calculates the accuracy for each data report on a daily basis by applying each validation equation to the whole of the data from the previous seven days – this period has been determined as the optimum time span for validating traffic flows using the LIP validation model. This allows a total assessment to be made for every data report for the entire month. A great advantage of this is that every error that occurs will be reported, but only for the duration of the error. Daily, weekly and monthly reports of errors are generated, so that when errors occur they can quickly either be corrected or be marked as unavailable until repairs can be made. CAVEMAN has now been accepted by Highways England as the prescribed method for assessing service level for traffic flow data.

CAVEMAN validation equations define conditions where the net flow described by a validation calculation is zero.

The following is an example showing the derivation of a validation equation for a typical node in the project network.

Figure 1.5 shows a representation of three links joining at a single node.

If a, b and c represent the vehicle flow rate on the links, the test criteria are

$$a + b - c = 0$$

Figure 1.5 Links joining a single node

The measure of the variance from the test condition is

$$\mathrm{Err}(\%) = 2 * \frac{|(a_T + b_T - c_T)|}{a_T + b_T + c_T} * 100$$

This is used in the CAVEMAN model to calculate errors associated with unclassified (i.e. total) flows. Flow accuracy validation is also required for classified flows, where vehicles are divided into two length classifications: short vehicles less than 6.6 m in length and long vehicles greater than 6.6 m in length.

The traffic monitoring equipment categorises counted vehicles by length and supplies the count information in four categories. These four categories are used to identify vehicles as short (category 1 + category 2) and long (category 3 + category 4).

The Caveman model uses a similar error calculation to test for the accuracy of classified flows:

$$\mathrm{Err}(\%) = 2 * \frac{|(a_s + b_s - c_s)| + |(a_l + b_l - c_l)|}{(a_T + b_T + c_T)} * 100$$

where a_s, b_s and c_s are the short vehicle flow rates, a_l, b_l and c_l are the long vehicle flow rates and a_T, b_T and c_T are the total flow rates on each link.

The allowable deviation from the validation criteria before an exception error is raised is defined by the highway authority's link categorisation. The UK network classifies the different roads on the Motorway and Trunk Road network as 'A', 'B' or 'C', according to the importance of the routes. The required accuracy is shown in Table 1.1 for the three categories.

The CAVEMAN validation equations are designed to test the highway operator's acceptance criteria for each reporting point over a seven day period as this is the optimum time span for validation traffic flows. However, the single day errors are also used to provide a more granular measure which can be used to monitor step changes in flows. These errors give an early indication of faults such as equipment failure or changes to the network. The CAVEMAN criteria for raising exception errors use the acceptance limits are shown in Table 1.2.

Faults detected by CAVEMAN are:

- Poorly calibrated loop sensitivity results in one or more lanes either not counting all vehicles or counting vehicles in an adjacent lane
- Poor loop installation which may result in the same effects as poor calibration, but cannot be corrected without re-installing the loops in the carriageway
- Loop faults resulting in erratic counting, either when flows are low or when the faults are intermittent

Table 1.1 Accuracy requirements for flow data

Category	Allowed error for all traffic flows (%)	Allowed error for classified flows (%)
A	10	15
B	15	23
C	20	30

Table 1.2 Accuracy acceptance criteria for CAVEMAN error calculations

Category	Allowed error for all traffic flows		Allowed error for classified flows	
	7-day (%)	Single day (%)	7-day (%)	Single day (%)
A	5	10	15	20
B	10	20	20	25
C	15	30	25	40

- Equipment faults leading to inconsistent counting
- Incorrect loop configurations leading to traffic being miscounted
- Incorrect site configuration causing data for one site to be extracted from a different site
- Changes to the network itself – for example, when a two-lane road is reduced to a single lane or if a bypass is built.

1.5.9 Data from existing control systems

Real-time urban traffic management systems like SCOOT [7] work using an online link model. As vehicles pass onto a link they are detected, and the model moves them down the link until they get to the stop line. If the traffic lights are on red, the vehicle joins a 'vertical queue' at the stop line. If the vertical queue at the stop line is still discharging at a predefined saturation flow, the vehicle joins the back of that queue; if there is no queue at the stop line and the lights are still green the vehicle passed through. SCOOT uses all the data from the link to change the splits – the amount of time allocated to each movement at the junction, the offset of green times between adjacent junctions and the cycle time – the amount of time to run through all the stages in the region.

As part of its optimisation, the traffic optimiser calculates for each link the number of stops and vehicle delay; this data can be collated and ASTRID (automatic SCOOT traffic information database) performs this task [13]. The SCOOT loops also have an occupancy measure of congestion; if a loop which is at the start of a link is occupied for more than 4 s then the system flags up that the link is congested from the stop line to the start of the link.

The previous section explains that traffic and travel information can be sourced from the SCOOT urban traffic control system, but there are many other systems that have real-time models such as parking systems and public transport operations. As new modes of transport such as carpooling and car sharing become more established, then data useful for traffic and travel information may be available from them.

Incident detection systems are established both in the urban and interurban settings then data from them will be available.

Social media is covered in another chapter so will not be covered here except to note that there is a known correlation of cellular phone traffic and incidents. Environmental monitoring, particularly exceptional pollution, is an indicator of issues. Of course, weather (such as poor visibility) collected by environmental monitoring stations is a vital part of traffic and travel information.

Tolling systems that identify and record the presence and time of a vehicle passing through virtual tool areas provide a rich source of travel time data.

1.6 Number plate-based systems

ANPR systems [14] can be used as a source of traffic and travel information. Journey time measurements, i.e. the time it takes to get from one known point to another known point, can be derived from correlating ANPR data and the system has spinoffs such as origin and destination data, identification of foreign vehicles and in limited cases, the tracking of vehicles.

Number plates are recorded at one point in the network with a precise time (from GPS) and transmitted to a central instation. Number plates from downstream sites are also transmitted to the central instation with their precise times. The instation attempts to match number plates within predefined periods (up to twice the expected journey times). When a match is found the upstream time is subtracted from the downstream time and the resultant journey time is achieved. Infeasibility short journey time and overlong journey times are rejected; i.e. those more than one standard deviation from the expected are removed from the sample.

Most ANPR systems work by flashing an infrared light source at the front of vehicles (generally the illumination is close to coaxial with the camera lens). An infrared camera then records the scene through an infrared filter. As the scene is recorded in the infrared, all objects illuminated by visible light are rendered as black, only the number plates which are retroreflective and return the infrared light are rendered. This makes the scene relatively easy to analyse. An image analyser scans in a raster pattern from top to bottom and left to right. When the system recognises sharp changes in illumination from black to white over a number of raster lines it declares a plate and character recognition software decodes it. ANPR systems can operate in light and dark situations and are only defeated by low sun, particularly the 10 days before and after the winter solstice (where the low sun corresponds with the peak hour traffic) and where snow covers number plates. In the United Kingdom, most ANPR for traffic and travel information only monitors one lane (unlike police and security ANPR which is deployed in a curtain across many lanes). To protect privacy the outstation ‘hashes’ the number plate to make an irreversible hashtag. The same number plate always produces the same hashtag, but the hashtag can represent an unknown number of plates. In other words, the hashtag is not unique as the 24-bit hashtag is 16,777,215 variances whereas at the time of writing there are around 36 million vehicles on the road. This does not mean that for each tag there are two possible registrations; the hash tag software does not allocate the numbers uniformly. There are no images of the vehicle or number plate stored or transmitted by the system.

The ANPR system employed by the National Traffic Control System in 2011 had around 1200 cameras at over 400 locations and where most locations recorded two-way traffic.

A significant disadvantage of ANPR journey time systems is maintenance. ANPR sites are sometimes vandalised as they are attached to bridge parapets over the road, particularly in the school holidays. Failure of the cameras and processor boards can be frequent and the infrared light sources need replacing periodically. If an ANPR site becomes inoperable, then up to 8 journey time sections may be lost.

The data from ANPR can be used to create origin/destination matrices that can be used in traffic modelling studies.

Occasionally, for serious crimes, the system is used to identify the track and timings of target vehicles with a known registration (the registration is converted into its tag and matched with the database of tags for the day), across the network although the lack of a sighting (if it was in a non-covered lane) does not mean that the vehicle was not there or that it was that vehicle. However, it may be useful for leads and pinpointing what CCTV the police should look at.

As well as being able to recognise number plates and transmit them as tags, ANPR systems, from the format and semantics of the plate, can determine a country of origin which may be useful for statistical purposes in estimating the mix of vehicles from different countries.

1.7 Data collection using mobile devices

Rapid growth in the use of, and adoption of mobile devices, for example, cell phones, has created a fundamental change the collection of road speed and congestion data. This should be considered in the following categories.

- Mobile devices (phones), coupled with location determining technology such as GPS, so-called floating car data (‘FCD’)
- Bluetooth/Wi-Fi-connected devices for measuring travel time.
- Mobile phone network data for creating ‘movement’ information.

1.7.1 Floating car data

The concept of FCD goes back to the late 1990s with the advent of greater availability and lower costs of GPS and mobile phone data technology. A great deal of academic research was undertaken, spurning a plethora of concepts, demonstrations and indeed patents by many organisations globally.

The concept of FCD is fairly straightforward, that is measuring travel time between two points measured on the road network and communicating this data via mobile data, to a processing platform where speed is determined based on a number of aggregated travel time links. As they say though, the devil is in the detail.

In the early days, much of the debate that resulted from early implementations and questions from academia related to 'How many devices that were equipped with GPS and mobile modem, would be needed to create reliable, quality data?' Data that could reliably determine the road speed was regularly updated. Many figures and estimates were produced ranging from 5% to 25% of the vehicle population, the so called 'Car-Park'. In reality this was deemed unachievable at the time by many considering FCD's use as a means of creating usable data.

However, we should mention the pioneering work and developments undertaken by ITIS. What drove ITIS' concept was a business model based around seeding vehicles (Probes) with GPS/Mobile devices where the primary use was to offer 'other' telematics services, primarily fleet tracking. A number of partners were contractually signed up, either directly by ITIS' telematics business NavTrak, or by third party companies offering fleet tracking services. Remember these were in the days before Smartphones!

ITIS realised that the projected numbers being proposed to create quality data were flawed in a number of ways. By focussing on commercial vehicles, trucks, delivery couriers, coaches, buses and taxis, there was created a probe network that ensured high mileages at all times of the day. The issue of just using private cars is that their travel patterns do not ensure a good temporal distribution of data, with peaks only during commute times.

The other benefit of using truck and coach fleets is that the network they use is typically the primary network of interest and that where the major congestion occurs. Also, they do not deviate to avoid traffic.

From early tests, it was also discovered that there is a relationship between the traffic density and reducing road speed at during congestion times and incidents, with the latter acting as probe 'sinks', i.e. increased density of probes in a jam.

Other challenges related to a need for high-quality map matching and filtering techniques. Problems of parallel, lower class roads, off-ramps with queuing traffic, urban dual carriageways with denser side road networks and so called 'rogue vehicles', such as construction vehicles in road works, that could likely be probe vehicles had to be addressed and solutions developed.

In 2001, ITIS launched their FCD data collection as part of their early commercial offering, signing up and contracting with more companies, to acquire data. High profile fleets in the United Kingdom, such as Eddie Stobart trucking, National Express coach fleets, Addison Lee Couriers were also used as a good publicity method, which looking back, helped ITIS create a business model that is now adopted by many of the world's traffic service providers. On the story of ITIS' early development and focus on commercial fleets, the term 'FVD' became the norm as opposed to FCD.

Since these early pioneering days however, and following the advent of smartphones and mobile GPS devices, probe data is now the primary method of collecting road speed data for both commercial traffic services as well as public authorities who purchase processed data from commercial providers such as TomTom and INRIX which it forms part of their traffic product portfolio. This applies to both real-time data as well as historical data which is used for analysis purposes.

The best examples to mention in the use of FVD are of course Google Maps with data collected from millions of Android devices and Waze, a company owned by Google, whose maps and traffic are the ultimate example of crowd sources probe data.

1.7.2 Bluetooth/Wi-Fi travel time data

Another use of data collected from mobile devices is the use of anonymised data from a mobile's connectivity.

Bluetooth is a wireless technology standard (non-proprietary), that allows Bluetooth equipped devices to communicate directly with one another over relatively short distances using radio frequency communication. Bluetooth has become almost a standard feature on mobile phones, PNDs, and even the vehicles themselves. Bluetooth detection systems work by actively searching for in-range Bluetooth devices and capturing the unique media access control (MAC) address of each device. For a Bluetooth detection system to read the MAC address of a device, the device must be turned on and be in 'discoverable' mode (i.e., broadcasting its MAC address). Because each device has a unique and permanent MAC address, Bluetooth detection systems can determine vehicle travel times and speeds by calculating the time it takes for vehicles containing Bluetooth devices to travel between two or more Bluetooth sensors that are a known distance apart.

Although the Bluetooth standard is nearly 20 years old, Bluetooth detection for travel time data collection has only emerged as a viable option in recent years, largely due to the rapid growth in the adoption of Bluetooth-enabled devices. Despite its relatively recent emergence, Bluetooth detection is among the most commonly used technologies for calculating road travel times. A notable example of Travel time measurement using Bluetooth is a system and service developed by the Department of Planning Transport and Infrastructure, the South Australian State road agency. Their system called AddInsight is now used across all major state capital cities in Australia.

The use of Bluetooth as a technology however also has a number of downsides, specifically Bluetooth is not always enabled and 'discoverable', sometimes by user choice to conserve battery life. Also, the sample detection rate can be as low as 10%–15% of devices. Therefore, this type of wireless detection means of travel time measurement has now been extended to using Wi-Fi. Again, with more and more mobile devices becoming Wi-Fi enabled, including cars themselves means that this wireless method of measuring travel times.

1.7.3 Mobile network data

This chapter discusses the use of data contained within the infrastructure systems of a mobile phone network. It should be stated up front however that this is a topic which could be the subject of its own publication and incites many sub-topics such as data privacy and the 'Big Brother' issue.

There are a number of techniques used depending on the mobile network architecture deployed and the point in the network where the data is 'sniffed'. However, they are all similar in concept.

Mobile networks all require any connected device to be registered (Voice or data) on the network itself. This is a fundamental part of a systems operation. So, when a device is first powered on, a handshake exchange of information is communicated, including the mobile units ID. All of the time that the specific device is connected, whether active or not, the cell, the mast antenna sector and other parameters have to be known within the network. This forms the basis of cell handover (When a device is moving and changes the cell tower it communicates with). All of this intra-network data exchanges occurs all of the time the device is powered on. Of course, some of the more covert uses of this information can be used by the enforcement agencies (police) to detect movement across the network.

Using this network data has significant challenges, both technical as well as the aforementioned issues of data privacy. A number of companies in the traffic sector have however attempted to use this data for the purposes of generating traffic information to vary degrees of success. By using map matching and processing movement of a device through the network it has been shown that map matched devices can indeed be tracked. The real issue for using such systems for measuring traffic flow and speed is the lack of reliability of the processed data, particularly when the device is using the same cell tower sector antenna for an extended period. Longer distance travel time measurement however can be achieved as can origin destination of devices. However, again, this bought up many privacy issues, coupled with the fact that the availability and value of the data considered by the network did not make this a viable and cost-effective method.

The primary use today of this data is by the networks themselves who have created business divisions that create real time and historical data sets used to monitor the movement of people throughout the network. They have addressed the issues of privacy by aggregating and anonymising the data at a high level so no individual device can be tracked. Uses of this data include OD surveys for transport planners, retail and public venues who use the data to monitor people's movement habits and also public transport and event organisers who can use the data for security and safety applications.

Figure 1.6 shows the concept of cell handovers. This can differ depending on whether the device is actively on a call, or communicating, or whether it is in passive mode. Either way, the concept is to map these handovers onto the road network as shown in Figure 1.7.

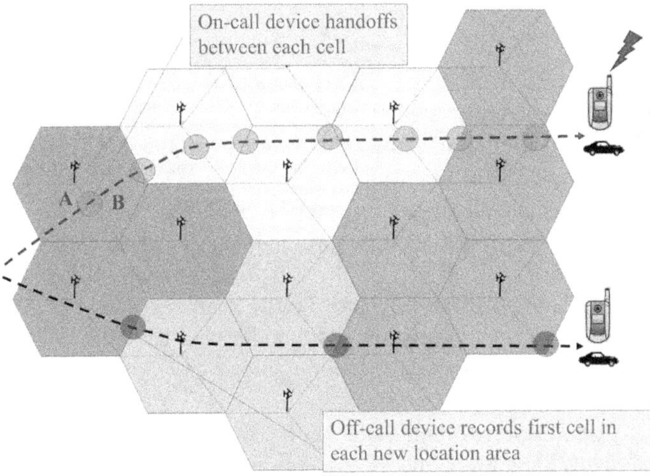

Figure 1.6 The concept of cell handovers

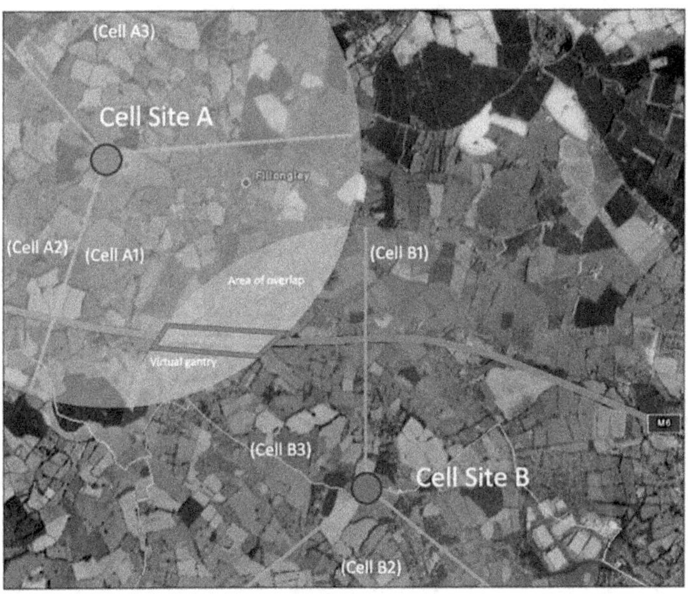

Figure 1.7 Cell handovers mapped onto the road network

1.8 Outlook

Many of the technologies described in this chapter are likely to be superseded by the social media and crowd-sourcing data technologies.

References

[1] The Automobile Association. The History of the AA. [Online] 2020. https://www.theaa.com/about-us/aa-history.

[2] The Royal Automobile Club. The History of the RAC. [Online] 2020. https://www.rac.co.uk/about-us.

[3] Highways England. Highways England Traffic Officers. [Online] 2020. https://www.gov.uk/government/organisations/highways-england/about-our-services#-traffic-officers.

[4] Traffic England. Traffic England Traffic Cameras. [Online] https://highwaysengland.co.uk/highways-england-and-traffic-cameras/.

[5] Wikipedia. Smart Motorways. [Online] https://en.wikipedia.org/wiki/Smart_motorway.

[6] Moore, R C, Davies, P and Salter, D R. *An automatic method to count and classify road vehicles*. London: Institution of Electrical Engineers, 1982.

[7] Hunt, P B, Robertson, D I, Bretherton, R D, Winton, R I. *SCOOT – a traffic responsive method of coordinating signals*. s.l.: Transport Research Laboratory, 1981.

[8] Dodsworth, J, Shepherd, S and Liu, R. *Real-time single detector vehicle classification*. 2014, Vol. Transportation Research Procedia. 2014, pp. 942–951.

[9] Wikipedia. https://en.wikipedia.org/wiki/Motorway_Incident_Detection_and_Automatic_Signalling. [Online].

[10] Pursula, M and Kosonen, I. *Microprocessor and PC-based vehicle classification equipments using induction loops*. London: IET, 1989. Second International Conference on Road Traffic Monitoring. London, UK, 7–9 Feb.

[11] Stevens, A, Spindlow, J R and Jones, G R *Fibre-optic axle sensors for vehicle data collection*. London: IEE conference publication no. 320, 1990. Third International Conference on Road Traffic Control.

[12] Clearview Traffic. Magnetometer Detectors. [Online] https://www.clearview-intelligence.com/products/m100-wireless-vehicle-detection-system.

[13] Hounsell, N and McLeod F. *ASTRID: Automatic SCOOT Traffic Information Database*. s.l.: TRL, 1990.

[14] Wikipedia. Automatic number plate recognition in the United Kingdom. [Online] https://en.wikipedia.org/wiki/Automatic_number_plate_recognition_in_the_United_Kingdom.

[15] Burton, P, Howe, L and Tate, P *Caveman – Continiuouse Assessment of Validation Equations by Monitoring the Agency Network*. New York: s.n., 2009. World Congress.

[16] Wikipedia. 1992.

Chapter 2

Road traffic news from the air

Paul Hutton[1]

For a period of about 20 years (long before the advent of satellite navigation with live traffic updates sourced from cellular floating data), traffic news from the air was seen as a trusted way for radio listeners to get the information about road incidents and traffic jams. However, in reality, traffic news from the air was more about showbiz and business opportunities than it was about information gathering and dissemination.

For this chapter, we asked Paul Hutton, journalist and former radio traffic news company Managing Director, to record his personal recollection of participating in the early days of collecting and broadcasting traffic information. This is the first time, to our knowledge, that this interesting aspect of early traffic and travel information has been documented and no other references seem to exit.

Editors

2.1 Early traffic information gathering

Traffic information gathering in the pre-Internet and pre-smart phone age, up to around the turn of the century, was not highly technical. Sources of information included the police, a number of traffic control centres, taxi firms and observers in a few static locations. Almost all the information was gathered by telephone; reporters, or their collators, would make calls to these sources and get information which they would then read out on air. The sources would not necessarily be accurate and, furthermore, the people contacted often felt they had better things to do so were not particularly interested in sharing information.

For example, the police contacts would often be in a control room with many incidents happening at one time. Calls into them for traffic information were not particularly high on their list of priorities, so these calls would be kept as short as possible. Usually the contact would check their command and control system to see if there were any major incidents and quickly terminate the call. This is not a

[1]PHInitiatives, Essex, UK

criticism of the police, simply a statement of reality that updating the media about a crash on a minor B-road in the middle of the day which may affect only a handful of drivers was not a major priority for them. Even less of a priority was regular congestion. If a motorway was usually jammed at that time of day because of sheer weight of traffic (as opposed to a crash) then that was not, to the police, an incident, so they would not report it to callers. Therefore, callers who relied on the police would not know the severity of that expected queue on any given day.

Taxi firms, which were persuaded to cooperate with the promise of a free name check on the radio, seemed a good way of gathering information, but the quality varied depending on the person in the taxi control room taking the calls. Some would radio around all their drivers and get every last jam; others were too busy and the incoming caller was fobbed off quickly with either 'nothing to report' or 'the usual'.

Traffic control centres were more helpful and they had access to traffic cameras that were used for monitoring major junctions; however, even they were not a perfect solution. Firstly, only a small proportion of towns and cities had a traffic control centre and those tended to monitor only a few key intersections (ones with traffic lights that they could affect by manually adjusting the phasing) rather than the whole network. Secondly, they could have their own reasons for not co-operating: when the author was responsible for information gathering for a UK-based traffic news company, the manager of one reasonably large city in England said that he did not wish to pass on information to be broadcast on the radio. When asked why, he replied that the only people who would normally know about a traffic jam are those affected by it, but if he told the radio traffic news station then everyone in the city would hear about it and so more people would know that he 'wasn't doing his job properly'. In another city, the control centre manager told me that his three-person team worked 8 am–4 pm every day so they could go home 'before the traffic built up' so there was no point calling during the afternoon/evening rush hour. (The author is not making this up!)

For some roads, the only sources were roadside cafes and petrol stations. These locations would be listed and called on a regular basis, asking the staff to look out of the window and report what the traffic was like outside, again in return for an on-air 'plug'. Listeners were also asked to call in with traffic reports and, as mobile phones became more prevalent, the 'jamline' was a useful source for tip-offs, although often callers did not have all the information, such as where they were or in which direction they were travelling. Sometimes they just made it up to 'wind up' the radio station. Hence, every call had to be checked with another source before it was put to air, reducing timeliness.

The information was originally processed through a series of Word documents for different areas – reporters would simply grab the relevant print-outs and use them for their broadcasts. Later, software was created to geo-locate the information and using a standard message list (actually an RDS TMC message code). Any extra information was added as free text. Whatever system was used, each piece of information needed updating manually and therefore could often be 30–60 min out of date.

2.2 US influence

In the late 1970s, a car dealer in Baltimore, USA, named David Saperstein commented to a staff member at a radio station on which he bought advertising that he had been surprised how little traffic information had been broadcast during a heavy snow fall. When he was told the station did not have any reliable source of information, Saperstein suggested his employees could spot traffic jams while out on the road. He could get it collated and provide it to the station in return for some free commercials. The station agreed, and he started providing the information. Other stations also wanted his reports so he began providing them too, again in return for airtime. Overwhelmed with airtime for his dealership, he began selling these advertising spots to other businesses which became hugely successful, leading to him selling his dealerships and going full time into the radio traffic news industry nationwide.

In 1982, the company, then based in Houston, Texas, began flying helicopters to provide information from the air for TV and radio. Local TV was a far bigger industry in the United States than in Britain and live shots of traffic (and breaking news) from the air is a staple part of morning and evening news programmes. In between TV reports, it made sense to maximise the value of the flying reporter by giving them radio reports to do as well.

2.3 Emergence of the 'Flying Eye' in the UK

For broadcasters, using aircraft was a useful way of gathering information, but – almost more importantly – it was a vehicle for sponsorship and a great platform to add a little extra showbiz to a radio programme. One of the best examples of a successful airborne traffic reporting operation was in London with Capital Radio's 'Flying Eye'. Back in the 1970s and 1980s, London only had two commercial radio stations, LBC and Capital, which both had an aircraft for traffic spotting. Bob Holness, who went on to be a famous TV presenter, had had a stint in the all-speech station LBC's 'London Lookout' while Capital had the 'Flying Eye' which even featured prominently in a car company's TV commercial.

Russ Kane took over the morning role in the 'Flying Eye', in 1984. An advertising copywriter and comedy scriptwriter, he had been convinced by his friend Peter Suchet to audition for the role, did so and got the job, as he told the author in an interview in 1991, because 'he was the only one who didn't throw up' when he was flown across the city in the twin-engine fixed wing plane. Expecting to do the job for a few weeks, he ended up reporting from the air for 20 years.

It was the arrival of Chris Tarrant to Capital Radio's flagship breakfast show in March 1987 that raised the profile of the 'Flying Eye' as Tarrant and Kane developed a quick-witted on-air partnership. This made the Flying Eye's slots much more than just about the traffic news. Tarrant had an extra sidekick ('Russ Kane in his funny little plane') to bounce off and the traffic slots could last for 2 to 3 min, much of it amusing banter and not to do with the traffic situation. Listening back to recordings of these reports, some of them could sound rather disjointed with the order of incidents bouncing from the M25 to North London, to central, back to north, down to south, back to the M25, Blackwall Tunnel and then somewhere in the east. But to the listener, it didn't matter – it was entertainment.

The whole 'showbiz' of having a key contributor to the show flying around in a plane was hugely popular. Members of the audience would occasionally see him from their cars while he was broadcasting, so Kane would generally start his report by saying where he was flying over at the time. This popularity led to highly lucrative sponsorship deals. Among the sponsors were Continental Airlines – building up their profile in the United Kingdom – and then British Airways who were reputed to be paying seven figures for the deal. They were replaced by Cellnet in 1993 paying an undisclosed, but significant, fee. The sponsors loved the high-profile 'lean-forward' content (i.e. involving highly engaged listeners seeking information), and with three reports an hour they got plenty of repetition.

2.4 The reality of traffic news from the air

So, here we are with a mental image of an airborne traffic report adding some extra 'sizzle' to a radio programme. It connected the listeners to the report because they may sometimes see the aircraft, and added to the trust that the reporter was providing accurate information because they were in the air to see it. However, the last of those points was actually, generally, inaccurate. Stopping to think for a moment, how could one plane gather information for the whole of London by looking at all those roads from 1,000 feet? The plane would fly around 120 mph; the M25 orbital motorway around London is 117 mi on its own – how could a reporter in the air have a timely view of all the roads? The answer was, of course, that he or she couldn't. Most of the information read out by a reporter was actually radioed up to them by an assistant on the ground. When the author worked at Capital Radio in the early 1990s, there was a specific job for a production assistant to sit in the corner of the newsroom reading the AA Roadwatch traffic information screen to the reporter in the plane.

This did not mean that the plane was useless for information. Having a wide view of an area and all the roads on it was great way of reporting. The plane came into its own when the reporter was told of a particular problem, and they could then fly there to see what was happening and give a live, bird's eye update on the situation. When there were no specific major incidents, the plane was able to fly to the places where there were always queues and the reporter could accurately describe whether traffic was better or worse than usual.

In the United Kingdom, in the 1980s, many stations were flying their own aircraft for their own areas, Metro Radio's in Newcastle, 'Starburst One', was one of the most popular and high-profile, while Essex Radio's 'Jambuster' Spotter Plane even flew before they could technically get the airborne reporter on air – they simply radioed down the information and it was read out by the studio DJ. This was expensive, though, and meant that they had to go to innovative lengths to save money. One station flew a helicopter that used to land on the fairway of a golf course midway through its hour in the air in order to save flying time which many would argue negated the value of airborne reports, given they were not gathering traffic intelligence between bulletins. Stations would make sure that their first report went out the moment they took off and the last report just as they were about to land, so the early reports could be top-heavy with information near their aircraft's base.

Spotting a crash from the plane was rare as it seldom happened. In fact, the author knows of only one specific incidence of this, and it happened in very strange circumstances in 2009. The author was at the time running a company which was flying spotter planes for traffic reports, one of which flew out of Oxford Airport. One day, the head of the radio division of broadcasting regulator Ofcom was a guest in the plane to understand how an airborne service benefitted a broadcaster. As the plane was a hundred feet or so up after taking off with the executive on board, the reporter that day, Neal Veglio, noticed a crash happen right before him on the road that ran alongside the airport. His first bulletin was timed to be almost the moment he was airborne (to get the showbiz of the airborne report in without spending any unnecessary money on fuel and pilot time) and he was able to immediately tell listeners to the local station in Oxford of a crash on a significant A road that was blocking it, and describe traffic building up by the second. The Ofcom executive in the back sat agog at the timeliness and accuracy of radio traffic news from a plane, and may, to this day, think this was a regular occurrence.

In due course, Neal Veglio did start to find information that others did not have, 'Most of the working information came from the ground, although with more experience, I learned a little bit of detective skill and was occasionally able to land the odd exclusive that made national news. One of our photos made the front page of all the local papers, and another was on the front page of the Guardian.'

2.5 Identifying the benefits of traffic news from the air

Neal Veglio, reflecting on his years reporting from the sky, says that the value of the aircraft to a radio broadcast was much more than just the sponsorship opportunity. 'The biggest difference for me, was the depth and colour it allowed you to add to the broadcast,' he explains. 'As radio personalities, our top-most requirement was to deliver effective storytelling, and that was a very useful skill for this particular job. It's much more engaging, for example, to describe how 'a small car seems to have gone off the road and there's now a recovery truck pulling it away from a row of trees' than 'there's been an accident on the Eastern bypass. Recovery is on the scene''.

Gauging the listeners' reaction to having traffic reports from the air was always difficult, if not impossible. One radio station which had an airborne service in the latter part of the 2000s, Eagle Radio in Surrey and Northeast Hampshire, did see its audience rise by more than 50 per cent in the two listening surveys after the Eagle 'Eye in the Sky' service was launched. Then group Programme Controller Phil

Angell told the author that nothing else much had changed on the station, or its competitors. 'It stands to reason,' he said, 'that if you have a traffic report from a spotter plane that listeners can actually see, then that will have a positive effect on the audience the station attracts.'

Jay Flood ran operations for the Australian Traffic Network which flew helicopters in its major markets of Sydney, Melbourne and Brisbane and a fixed-wing plane in Perth. 'In the late 1990s, it was an advantage to be aircraft based in cities around Australia,' he told the author. 'Airborne reporters had access to what I'd call unique traffic content. Often the air based reporters would exchange information with reporters on the ground and it felt like it was a disadvantage to be on the ground.' But, as in the United Kingdom, he said that the aircraft gave the value of showbiz and sizzle to broadcasts, rather than as an information gatherer, 'The only real information you ever gathered from the aircraft related to congestion points on arterial roads. In ten years of being in aircraft over major cities in Australia I personally only ever had three incidents that ground based reporters didn't have already.' Flood continues, 'The reason for the aircraft was two-fold – radio station image being number one. Having a station bother to put a reporter in the sky over their city showed they cared about the city and in the 1990s, it was even better. The other real reason was to create a barrier to entry for potential competitors.'

2.6 Amalgamation of flying services

The cost of flying and an explosion in the number of radio stations opened up a business opportunity to a third-party provider to operate services for multiple radio stations using the same aircraft, generating significant economies of scale. From the mid-1990s, both in the United Kingdom and Australia, versions of the US Metro Networks operation began and signed up radio stations in major cities. London's two radio stations, LBC and Capital, had a plane each but when new stations came into the London market, Kiss, Heart and Virgin, they were signed to the UK Metro Traffic Control's (MTC) traffic service. MTC flew a plane with three reporters, each doing a different London station. This obviously gave them their own distinct traffic service at a fraction of the cost – in fact no cost because their service was paid for by providing a small amount of advertising time which Metro then sold.

The three reporters would share information with those on the ground for other stations that did not qualify for an airborne service, and would, again, receive a lot of information from the ground that they would use in their reports. Their shifts were far from idle. Apart from the reports they did for their flagship stations in London, they would pick up reports for other provincial stations nearby, despite rarely flying over the areas to which they were reporting. This was before a scandal which hit broadcasting in the early 2000s where features were faked and, worse, competitions were run where people paid to enter despite not having any chance of winning. At that time, the idea of using the aircraft as a mobile studio was never seen as cheating the audience; it was merely adding some showbiz to the output. This, though, did not always work – reporters would sometimes make out they were in an area when they were not and report something so out of date or incorrect that it immediately lost the credibility of the report to anyone who happened to be stuck in a jam that wasn't reported, or cruising along a road where they had been told there was a jam.

One great example of mis-information happened to a good friend of the author who was holiday-covering an airborne shift which included a station on the English south coast more than 60 mi from London. The plane would rarely get to actually fly over the area it was covering, partly because of the time it would take to get there but more because it involved flying through Gatwick Airport's flight path and therefore under Air Traffic Control rules. The reporter therefore had to be deliberately vague about where he was and the information he was giving out. My friend, though, in his enthusiasm to do a good job, did not quite get that. He picked up his information pack before flying which included a list of roadworks in the area. He spotted that a major road yards from his parents' house in the area had lane closures so he 'went to town' on his reports, describing the road, the cones and the jam that he knew would be caused by halving its capacity. Because he knew the road so well, he described the cones being outside a fish and chip shop he knew well, glad he was able to give such excellent local colour to his report. He was very pleased with himself. There was just one problem – he had picked up next week's roadworks list. In the real world, there were no cones and no traffic jams. To this day, I'm not sure how you come back from that, apart from thinking like the city traffic official mentioned above: I suppose only those driving that route would have known the error, other listeners would have been blissfully unaware!

Mishaps such as those described above helped shape proper operational procedures to minimise the risk of broadcasting incorrect information. Reporters were supported by editorial teams on the ground who were getting information from a greater range of sources, including, in London, access to what was then Scotland Yard's traffic camera network (before it was handed over to Transport for London). Not only did that mean the reporters were able to convey accurate traffic information but also that they would then know where to fly to find the jams to circle over and see and be seen. The reporters also got better at conveying information from the air to the point that some stations began asking for more elaborate reports. One relatively small station on the outskirts of London which covered two medium-sized towns asked for different airborne reporters for each. One would do the traffic for one town and then hand over to a colleague sitting next to him to do the other town. They never said on air that they were over either town, but just allowed the listener to infer it from the way they reported.

2.7 Further challenges to the flying traffic services

As well as the significant costs, even in shared operations, there were other operational issues as radio shows including traffic bulletins tended to be 6 a.m.–10 a.m. and 3 p.m.–7 p.m., longer than any plane would fly. Therefore, there was always a need for a ground reporter to do reports when the aircraft wasn't flying, as well as passing on the information to the airborne reporter when he/she was in the air. Additionally, the planes could only fly when the weather was good enough. On many days the airborne reporter would get a call from the pilot telling him or her that they would not be flying that day (which became known to the hard-working people in the main control centre as 'duvet cancellations'). Some weeks a plane might not fly at all which, while saving fuel, had a negative effect on the audience and sponsors.

There was another issue with the planes that was unexpected, and that related to listener perception. As described above, the three newer stations in London were keen to have airborne traffic reports in order to compete with the established Capital Radio when it came to traffic reports on their music stations. However, during research, the other stations discovered that as Capital was so famous for its 'Flying Eye', some people who were listening to them actually thought they were listening to Capital because they perceived it to be 'the' station with an airborne traffic report. When listenership is calculated by people remembering which station they were listening to at any time, the idea that they may be incorrectly ticking a competitor was a major issue. One station therefore dropped the use of the aircraft and, instead, concentrated on promoting how it was using all the cameras in London to gather its traffic information.

Other stations began questioning the use of the aircraft as well. After 9/11 in 2001, the cost of insurances and the paperwork increased and around that time fuel also became more expensive. Stations therefore began to question the value of the planes. Eventually most of the aircraft, and even the 'Flying Eye', were grounded as the quality of live traffic information from the ground began to improve.

However, that was not the end of the use of aircraft for traffic reporting in the United Kingdom. In 2007, the author was asked to run a UK version of the traffic network of which Jay Flood was Operations Manager in Australia. One of the key things that the Global Traffic Network UK (as it was named) wanted was to bring back the showbiz of traffic information by flying planes again. The author could write an entire book about his experiences running that business but, to cut a long story short, GTN UK began flying two single-engine Cessna aircraft, one based in Oxford and the other in Plymouth. The Oxford plane served a station in Oxford, another in Reading, and then one in Guildford with the route meaning that all three areas got a reasonable amount of time with the plane above them. This required that the main reporter, Neal Veglio, played co-presenter on three different radio stations with three styles and different subjects of conversation each day, meaning he not only had to keep across three stations' traffic but their personalities. 'Thankfully this wasn't my first gig,' he says, 'And talking to time, and in different styles was a skill that has come quite naturally to me. My previous job actually had more stations on a shift, so my time at GTN was actually relatively relaxed!' And he agrees that in a varied career, this was a highlight, 'There are not many people who have talked live on the air, telling compelling (occasionally newsworthy and controversial) stories, ridiculous dad jokes, and voicing morning show characters, from 2,000 ft in the air, and I'm proud to say I'm in an exclusive and small membered club!'

Once again, though, money talked. The idea of flying the aircraft was to create a demand again for commercial stations to sign up to the GTN UK service against a competitor, but stations were no longer excited by the concept of airborne reports. One station in London listened to a pitch for an airborne service and at the end asked, 'if we take a service from you that doesn't have an airborne element, will you pay us the difference?' Money talked and after less than four years the planes were grounded again.

2.8 Anecdotes from flying operations

There was one amusing story to tell from the Cornwall operation which flew out of Plymouth Airport. Non-UK readers might not know that the journey from Oxford, where the main management team was based, to Cornwall is around four hours so when things went wrong it became a logistical hassle to get there to fix it. It was necessary on three occasions to go to Plymouth because, inexplicably, during live broadcasts from the plane there the reporter simply could not be heard so the DJ in the radio station crossed to the report only to hear engine noise. Trips down there, extra flights and tests, did not find the problem. It turned out that the reporter, keen to get a full-time radio job, happened to be scared of flying; hence sometimes, after possibly a bump or two in the thermals a thousand feet up, was so terrified he was incapable of speech. Not something he wanted to admit when he landed, hence his blaming of the equipment. Finally, he took a ground-based job and a new reporter took his place in the air.

However sometimes there were reasons to be scared, once the author remembers being in the London office when the reporters in the plane alerted everyone to a problem. The exact nature of that problem is forgotten but it may well have been to do with the plane's undercarriage. The situation was serious enough for the plane to have to circle to lose fuel and for the emergency services to be on standby at the airport, so the Head of News at the operation asked one of the reporters to commentate as he came in to land in case there was a crash which would be newsworthy. It is to the reporter's credit he did so, although, as it happened, they managed to get down safely.

Many of the pilots supplied for the job were potential commercial aircraft pilots working to get their flying hours up. One forgot to check the fuel before taking off and only realised there was a problem when the engine 'coughed' quite a significant distance from the airfield. Again, disaster was averted as the plane landed on fumes, but the reporters all refused to fly with him again.

The author remembers a particularly filthy day in London when, surprisingly, Capital's 'Flying Eye' was in the air when the reporter let out a shriek. Apparently, flying much lower than normal because of the weather right at the base of the clouds, they had flown just a few feet past the newly built Canary Wharf, 800 feet in the sky.

'The moment we lost alternator power and had to glide land will be in my memory for the rest of my life,' says Neal Veglio. 'It's no exaggeration to say I genuinely thought I was going to die. I returned the next morning with a hangover, I won't lie.'

In Australia, Jay Flood had similar issues, 'One morning taking off from [Brisbane's] Archerfield Airport we were heading out on our usual morning run. This was prior to the tower opening and the pilots needed to communicate with each other. We had built up take off speed, the pilot was about to lift off and right in-front of us another plane touches down – like a head on we'd see many times reporting traffic. We had a very good pilot who managed to steer up left off the runway and came to a skidding halt. The other plane failed to make any radio calls and was coming in from the country – it was a really close call.'

2.9 Conclusions

Technology moves on and the use of floating vehicle data to gather information means live traffic reporting is now generally correct, unlike the situation in the 1980s when roadside diners were the best source you could have, and listeners would no doubt now question why a radio station needs an airborne service. 'These days jumping in an aircraft is a massive disadvantage,' says Jay Flood. 'The information is going the other way – with data streams, CCTV and so many other forms of traffic information it feels a disadvantage to be in the plane. Ground based reporters can tell you where the traffic is back to and how it's changing by the minute. The aircraft is now a tool of the past. More specifically, the use of drones is much more cost-effective and delivers better pictures and videos.'

So, into history has gone the 'Flying Eye', 'Thunderbird One', the 'Eye in the Sky', 'Starburst One', 'London Lookout', 'Flying Jack', 'Flying Fox' and innumerable other names for airborne traffic services, but it is worth noting that despite satellite navigation devices and other solutions with live traffic information, radio traffic news continues to be relevant and relied on by drivers across the world, and a key part of most local stations' information output.

Chapter 3
Location referencing

Jon Harrod Booth[1]

3.1 Introduction

Location referencing is a subject which is familiar to many of us. Some people see this as a very technical and specialist domain, which it certainly can be. It is also a diverse domain. In reality, all of us interact with it on a daily basis during navigation, often unconsciously; after all you found the front door this morning, found your way to work, school or the shops, and your way home. But asking you to travel to an unfamiliar location, or by an unknown route, or a new mode of transport introduces a new set of skills, tools and use of services to help you find your way.

This chapter seeks to provide a simple introduction to topics relating to location referencing, especially in its applications within intelligent transport systems (ITS). It assumes little prior knowledge from the reader and provides an introductory, overview text on what is a detailed subject and upon which many books and guides can be, and have been, written.

This chapter starts with a brief introduction concerning location referencing concepts and prerequisites. It then moves on to discussing and introducing some of the basic concepts and terms that are commonly in use. These provide a basis to discuss various approaches that have been created. The following section covers the evolution of the various location referencing methods (LRMs) that are commonly applied in ITS. Finally, there is a short discussion of the future influences that will affect location referencing and location-based services.

3.2 Historic context

A historian's view of maps and map-making reaches back to the first evidence of cave-painting representations from France and Spain dating from some 14,000–19,000 years ago, with scratched marking of local terrain and features. The first forms of map we are aware of include the Babylonian Map of the World (or Imago Mundi) from modern Iraq, which is a clay tablet, dated to roughly the sixth-century BC.

[1]Harrod Booth Consulting Ltd., UK

This continues with a long heritage of ancient map-makers in Egypt, ancient Greece, the Middle East and the much later religio-centric maps such as the Mappa Mundi from around AD 1300.

On a more mundane level, looking back into days of yore, we have the impression that, in general, people did not tend to travel beyond their immediate locality very often. Local knowledge, helpful directions from other travellers and local citizens, and the erection of 'finger-post' (permitted under UK legislation in 1697), and the erections of milestones or markers (117 Roman ones still exist in the United Kingdom) all aided the traveller with navigation.

We see that general-purpose land maps for navigation appear relatively late. In the United Kingdom, although maps for military purposes were starting to be developed from 1745 (initially of the Scottish Highlands as a response to Scottish Rebellion) maps from Ordnance Survey being used for civil purposes did not appear until the early 1800s. These maps were not widely accessible and were expensive.

We now step forward in time, and as an individual, you don't have to be so old to remember days when essentials tools for non-habitual travel would be:

- a paper road atlas – which was typically published annually
- paper-based maps (1 inch to 1 mile was a common scale)
- A–Z Gazetteers for specific cities and towns.

These were not just tools for the private individual. In the 1990s, there were certainly examples of emergency service response management systems using apparently strange coding to uniquely identify the location of an incident, such as 69-A6, which initially baffled the curious investigator. This turned out to be a key and critical location reference, which was a specific gazetteer publication and dated version page number and grid reference to a page in a paper book. Such archaic systems would be bewildering to the 'Millennial' reader.

Before going further, it is helpful to address some key terms and concepts to ensure that the way we discuss location referencing is addressed consistently.

3.3 Location referencing terms and concepts

Before discussing location referencing, it is helpful to make a distinction between location references and maps. Location referencing is a process by which a location (a real-world place) is referred to by a reference of some description; the nature of the reference depends upon various factors including the referencing method and system that is used. Perhaps, the most commonly known reference will be a grid reference such as an Ordnance Survey Grid Reference (OSGR) in the United Kingdom or a latitude and longitude.

A map is a collection of information or data with a spatial content arranged within some convention, typically designed for a specific set of uses. Some maps support transportation, such as road maps and atlases; others support other specific purposes, or may be general purpose. Any specific map or map dataset can be considered an object. Maps often are distorted either for cartographic purposes, such as road roundabouts being made much bigger than they should be on a 1:50,000 scale map, because they otherwise would be invisible to the reader; or distorted to aid reader interpretation – a good example of this is the standard map of the London Underground network, which is logical and schematic rather than a true spatial representation.

The most prominent exposure to maps for many people today is in the form of a map rendered on the screen of a smartphone, digital device or computer. Such maps consist of a digital spatial data set that provides some spatial context, against which things are referenced by some form of location referencing.

As the creation of spatial data sets can be expensive and time-consuming, there are often costs associated with the purchase, use, or distribution of map content.

3.4 Basic concepts of location referencing

Nearly all transport and ITS applications need some form of description of the location of features (such as physical objects, networks, vehicle stops, points of interest, restrictions, events) in a spatial context, both in absolute relationship to the surface of the Earth, and/or in relation to other features.

The term 'location' has been formally defined in various ways, e.g. in several standards from ISO and CEN. A very basic definition is provided by ISO 19111 [1] as an *identifiable geographic place*. A 'place' is defined in ISO 19155 [2] as an *identifiable part of any space*. Such a part may be a single point, a segment, an area, a volume or any other part that the space in question may be divided into.

The terms 'location' and 'position' can also be confusing. One distinction considers that a 'place' is referred to as a *position* when that place is identified using coordinates, while a 'place' is referred to as a *location* when that place is identified using geographic identifiers. ISO 17572-1 [3] defines in a more general way a location as a *simple or compound geographic object to be referenced by a location reference.*

Various forms of definition of 'location reference' exist, such as a 'description of position in the real-world', a 'description of an identifiable geographic place' or a 'label which is assigned to a location'.

A location reference is described using a LRM, and within a location referencing system (LRS) based on the LRM. A 'LRM' is a *methodology of assigning location references to locations*, and a 'LRS' is a *complete system by which location references are generated, according to a location referencing method.* (ISO 17572-1 [3])

For example, an LRM could be a linear referencing approach that defines a location by a measure along a defined linear object, from a defined starting location. In traffic engineering terms this is often referred to as a 'chainage'. The LRS would define the scale and units of measures used, how the measure is measured, coordinate systems used. This can be considered as the specific configuration of use of the LRM. It is helpful if the definition of the LRS also provides some viewpoint on the accuracy of location references in creates.

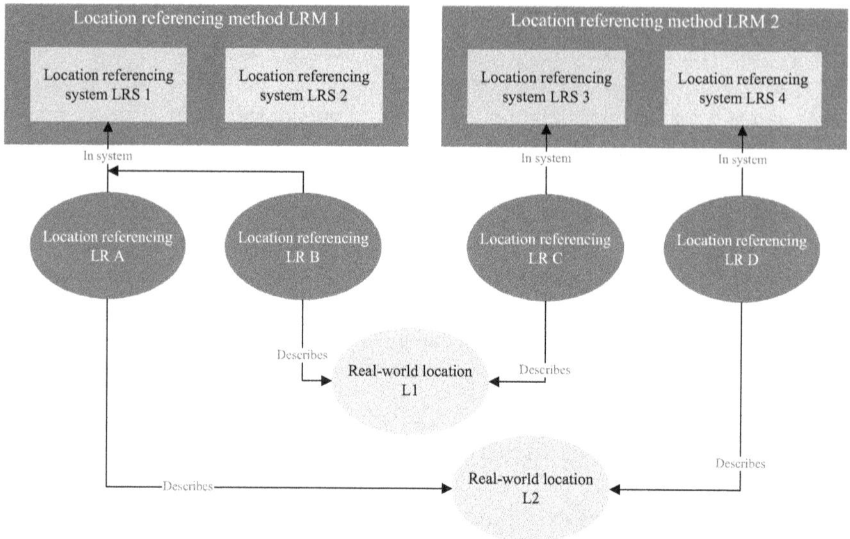

Figure 3.1 Location referencing concepts – adapted from CEN/TR17297-1 [4]

From these definitions, it follows that location referencing is about describing a location within an LRS, according to an LRM. Applying location referencing to a real-world object results in a description of that object's location.

In Figure 3.1, two LRMs are given, i.e. LRM 1 and LRM 2, each with two LRSs, e.g. LRS 1 and LRS 2 conforming to LRM 1; and LRS 3 and LRS 4 conforming to LRM 2. Two real-world locations are described, i.e. L 1 and L 2. As shown in Figure 3.1, the same location can be described with location references in different LRSs. Both, LR A and LR B with different LRSs and LRMs, describe location L 1. Of course, different locations can be described with location references from the same LRS; Both real-world locations L 1 and L 2 are described with location references (LR A and LR B) from LRS 1. This leads to an important observation that there is no universally applicable commonly used LRM or system.

Throughout the remainder of this chapter, several different approaches are described, each of these is a different LRM. Within each method different LRSs may be applied, dependent on the specifications or standards that define and control them.

The location reference can be a reference to a tangible physical object such as a vehicle, a pedestrian, a road side unit, a road sign, or it can be a more intangible concept such as a restriction or regulation such as speed limits or access restrictions; it can be an event like a traffic incident or a road closure. Location referencing of these features results in descriptions where these features are in relation to each other and in relation to the real world.

Several different methods and sub-methods exist for location referencing. Three main classes of LRMs exist:

- Location referencing by coordinates.
- Pre-coded location referencing; that is applications that message using a location reference to a pre-determined location (with a pre-determined location reference or ID). For example, the UK's NaPTAN (National Public Transport Access Node) data set provides unique identifiers for each listed public transport access point (e.g. bus stop, train station). Messaging concerning status or event information can indicate location by reference to the unique NaPTAN location references.
- Dynamic location referencing; LRMs that do not pre-code specific locations, but provide attribution within status or event messages that enables the information receiver to seek to derive the intended location by algorithms that support map-matching. Each of these methods typically provide coordinates to achieve coarse location matching and then additional attribution to try to improve the location fix on the basis of things such as the shape of road sections, road names, direction of travel. These methods do not assume that the spatial database of the message creator is the same as that of the message receiver, and therefore apply some concepts of fuzzy matching. Indeed, different map data sets do differ and therefore simply passing a set of coordinates is not robust for a good contextual locational fix. The goodness of the receiver's derived location against the sender intended location depends on a large number of factors including the localised similarity of the spatial data sets, the nature of the location sought to be described, the specifics of the LRM and system, the required accuracy and resolution of the receiver's systems processes.

3.5 Location by coordinates

Description of location references by use of coordinates is relatively well known. Since the advent of location-based services on smartphones and tablets, it is not uncommon to see the GPS-derived longitude and latitude. However, it is also worth saying that most citizens cannot easily interpret this information, and although this is an efficient means to pass details of a specific location between our devices, it is much less common for direct human to human conversations to prefer use of coordinates to describe a location. Of course, device-presented GPS longitude and latitude are just one form of coordinate; many others exist.

Coordinates have no meaning unless the system to which they are related to has been defined. Using coordinates implies the use of a defined coordinate system (how the x, y, and possibly z measure change in space) and a coordinate reference system, which is how the x, y, z measures are described and encoded. ISO 19111 [1] defines these terms more strictly. A coordinate system will be tied to a datum which defines a set of *parameters that define the position of the origin, the scale, and the orientation of a coordinate system.*

The Earth is almost a sphere, but with a very irregular surface. Most maps throughout history have sought to represent a portion of the surface of the Earth in a 2-dimensional flat presentation. Hence, various projection systems have been made for different purposes to support the translation of features on the curved Earth into coordinate systems for mapping. The nature of the projection depends on many factors, and many projections exist. Each projection introduces distortions and the use of the specific projections depends on whether the scale of the distortions is acceptable for the intended purpose.

The 'International Association of Oil and Gas Producers' (IOGP) maintains a registry of coordinate reference systems known as the 'EPSG Geodetic Parameter Registry' [5]. Coordinate reference systems are often identified by their identifier from the registry – the 'EPSG code'. By way of example, the projection EPSG code 3857, refers to the WGS84/Pseudo-Mercator projection which is commonly used by many smartphone/web location-based services such as used for Google Maps, Open Street Map and Bing maps.

The GPS system supplied by the US military is just one of several available Global Navigation Satellite System (GNSS) and the use of GPS-coordinates, although prevalent, is not universal. Precise positioning used to be the domain of geodesy, surveying and infrastructure construction, with most transport applications requiring location referencing accuracy of perhaps tens of metres. As the demand for the precision of location referencing increases, for example, with the advent of automated vehicles, the choices between different coordinate referencing systems and projections become more apparent. The GPS coordinate system is based on a coordinate reference system called WGS84. Within this system, baselined in 1984, tectonic continental drift is a factor. In general terms the North American continent is drifting westwards relative to Europe by about 1.5 cm per year. In other parts of the world this effect is more extreme, with reference locations shifting by much larger amounts. The European (Eurasian) plate is drifting by about 2.5 cm per year, against the WGS84 frame of reference. Hence, for some applications, it is not sufficient to provide information in terms of coordinates and their coordinate referencing system/projection, but also when measurements were made [6].

3.6 Pre-coded location referencing – RDS-TMC ALERT C

Audio radio broadcasts of traffic information have been around since the 1930s, but more commonly from the 1950/1960s. For specialist surveying and geodetic purposes, digital encoding of maps has been available since the 1970s. In 1985, the opportunity arose to use the 'radio data system' (RDS) to silently transmit digital traffic information as part of FM radio broadcasts (data is actually encoded on the FM sub-carrier). This was the birth of digital traffic information provision and of RDS-TMC (radio data system – traffic message channel). A key part of the specification to support RDS-TMC was the ability to identify the location of a traffic incident or event.

A significant challenge for the developers of RDS-TMC was the limited data rate available for broadcast. Therefore extreme care was taken to ensure the component elements of the TMC message were as compact as practically possible. Trials of various approaches for location referencing were undertaken and this ultimately resulted in a system which is called ALERT-C which is still in wide use today. ALERT-C uses a pre-coded, pre-distributed list of defined locations within a table, using a standardised format and structure. The subsequent broadcast messages then refer, by reference to a location or a combination of locations, by way of a compact location ID. The approach is set out in the ISO Standard ISO 14819-3 [7].

Each TMC message references to a specific location coding table, and then to a specific pre-defined location within that table. Each table supports 65,563 location code options. (Of these, 63,487 can be used for normal location referencing purposes, whereas the remainder being allocated for other purposes.) The identification of a specific location coding table is specified in ISO14819-3 [8] and uses a combination of extended country code (8 bits), country code (4 bits), and location table number (6 bits). The location code itself is defined within a specific table in 16 bits, and is unique worldwide.

The allocation of specific location tables is managed by an industry association, the Travel Information Services Association (TISA) [8], but, in general, these are allocated on a country basis with larger countries having a greater number of tables available. So, for example, in the United Kingdom we have the capability of having 16 tables.

If these tables are organised to define each location once, in the United Kingdom, just over one million locations could be defined. However, in the United Kingdom, the allocation and organisation of location tables is done differently with specific tables being allocated to specific information providers. With this approach, an information supplier is limited to 63,487 locations which creates a practical limitation to the granularity of traffic information that can be delivered by this means.

Locations are defined as one of three main types:

- area locations
- linear locations
- point locations.

These locations are arranged into a hierarchical structure, with some data creator discretion as to how the hierarchy is formed. In general, point locations belong to linear locations and linear locations belong to area locations. With the hierarchy in a table, locations may also belong to larger locations; for example, a region may belong to a country, and a street segment or link may belong to a road. It is typical to see a structure which provides a length of road within a region; and within a road the definition of a series of road lengths (or links) that segment the road into major traffic sections between major interchanges. So, for example, in the United Kingdom, a road link could be the M1 motorway junction 6A to 10 northbound.

Location referencing in ALERT-C is done with a combination of predefined locations, direction information and extents. Different location coding methods are presented in Annex C of ISO 14819-3 [7]. These coding methods are:

- Pre-defined primary location + extent
- Pre-defined primary and secondary locations
- Distance markers (primary location + extent)
- Primary and secondary locations using pre-defined location, extent and distances.

Taking the example of the first coding method above, the location affected is defined by providing one primary location, which is usually upstream of a traffic event or incident, and then defining the extent of the event/incident by use of an integer number extent figure, which 'steps' along the route using the next and previous location reference values that are presented in the location coding table. This method provides a defined location which envelopes the actual (probably smaller) location and extent of the event or incident, but supports driver information and driver routing choice. The later location coding methods defined in Annex C provide means to define the event/incident location more accurately, using distance measures from the primary location and the secondary location (whether that secondary location is derived from the stepwise extent or explicitly referenced itself). This, in principle, supports definition of locations within a subsection of road to a reasonable degree of accuracy.

However, a clear limitation of the ALERT-C location referencing mechanism and its table based approach is that these tables are predefined and typically pre distributed in advance, often being loaded to a receiver within a vehicle with no obvious mechanism to update that data without direct support from the vehicle manufacturer and the receiver equipment supplier.

Due to the limitations in the location code available within a specific location table, ALERT-C location tend to be defined at a generalised level. For example, with the notional location code for a complex motorway junction being a single point approximately at the centre of the geometry of the junction, this point may well not represent any specific location within the road carriageway passing through the junction. In many cases, ALERT-C referencing is not able to support the full spectrum of turning motions and identification of transitioning carriageway sections found in complex junctions. Thus, the information contained within an ALERT-C message is often delivered for an abstracted, strategic driver information purpose.

3.7 Dynamic location referencing

3.7.1 Underpinning concepts

Different commercial and public sector map dataset suppliers create, maintain and support different, competing data sets that support, amongst other applications, location-based services and road-based navigation. These map data sets are not identical; nor are the means of updating them and distributing them shared and synchronised. Dynamic location referencing has been designed as a means to support location-oriented services without the need for the synchronised distribution of pre-coded location tables, and recognising that there may be differences between the information sender's map data set and the map data set of the receiver.

Dynamic LRMs seek to supply sufficient attribution to enable the receiver's system to correctly decode the intended location, matching the exchanged attribution to the receiver's map data set in an acceptable proportion of cases. A combination of a simplified network geometry and a set of specific road characteristics is exchanged, with each dynamic LRM specifying different requirements.

Such differences in map data sets can arise from data sets from different suppliers, different versions of the same data set, or dated partial updating of dynamically managed data sets. Dynamic location referencing has also been referred to as 'on-the-fly referencing' or 'map agnostic location referencing'.

Unlike the pre-coded RDS-TMC ALERT-C location referencing based on pre-coding locations and location tables, the dynamic location code is created when needed from a map database in the sending system, transmitted in a message, decoded with the map database in the receiving system, and then discarded.

Two main dynamic LRMs have been developed and deployed:

- AGORA-C™ which is standardised in ISO 17572-3 [9]
- OpenLR™ which is deployed as an open specification [10], and also standardised in ISO 21219-22 [11].

These two approaches are briefly outlined in Sections 3.7.1.1 and 3.7.1.2.

3.7.1.1 AGORA-C™

The AGORA-C™ method relies on common attributes that are mostly available in current digital road map datasets. A dynamic location reference according to AGORA-C™ is constructed as a set of information elements, which consists of a set of points and related characteristics.

The points represent a simplification of the road geometry down to a specific threshold, and each point is one of:

- Location point, i.e. 'A core point that represents the start, an intermediate, or the end point of the real-world location to be referenced'
- Intersection point, i.e. 'A core point representing an intersection, located at places where the road section signature at the location changes'
- Routing point, i.e. 'A core point used to reconstruct the location by route calculation'
- Extension point, i.e. 'A point belonging to the location reference extension. All points in the set of points in the location reference extension are by definition extension points'.

The coordinate values for the points are given in the International Terrestrial Reference System latitude and longitude. It should be noted that this earth-centred reference will produce different coordinate references to other reference frameworks, including WGS84 (World Geodetic System 1984) which underpins many GNSS coordinates systems such as that used by GPS (Global Positioning System) and the multitude of GPS location-based applications which we are all used to in our internet browsers, on our smartphones or guiding us through our vehicle's Satnav system.

The additional characteristics in AGORA-C™ that are provided to identify the location are:

- topological geometric characteristics such as bearing at point, connection angles and location
- road descriptor, e.g. the full national road number of the road
- form of way, i.e. classification of the physical properties of the road as defined in ISO 14825 [12], e.g. Slip Road (parallel road, slip road of a grade separated crossing, slip road of a crossing at grade), Freeway, enclosed traffic area (parking place, parking building, unstructured traffic square, pedestrian square)
- functional road class, i.e. classes 0 (main roads) to 9 (the least important roads of a given network); it should be noted that although the general ranking approach of functional road class is clear, how one should code different roads into the hierarchy of classes is not specified and therefore is open to interpretation by different data sources and map suppliers
- driving direction, i.e. the allowed legal driving direction.

ISO 17572-3 [9] provides the rules for encoding each of the forms of attribute, and the rules for selecting options within the approach to be used.

As a way to facilitate open access to AGORA-C™, the creators of the AGORA-C™ concept have come together to pool their essential patents under a single joint patent licence agreement handled by Via Licensing Corporation [13]. There are licensing fees associated with encoding locations according to AGORA-C™ and the licensing fees are listed in [14]. The Licence holders are Panasonic, Robert Bosch/Blaupunkt, Siemens, TeleAtlas and Toyota (although some of these organisations have subsequently been subject to mergers, name changes, etc.)

The AGORA project and other testing efforts sought to improve code size and 'hit rate' (correct location matches). Testing demonstrated a success rate of about 98% and an average code size not larger than 34 bytes, which is well below the generally agreed number of 50 bytes for maximum acceptable average code size. It is interesting to compare this to the size of ALERT-C location codes given above.

3.7.1.2 OpenLR™

A second 'on-the-fly' LRM developed by TomTom is OpenLR™. The OpenLR™ method is based on the same principle as AGORA-C™, with a combination of a simplified geometry and road characteristics. The simplified geometry is given with coordinates in WGS84 longitude and latitude. Ideally, the geometry will contain only the points that are needed to describe a shortest path calculation via the selected points. The additional characteristics that are provided to identify the location are:

- topological characteristics such as bearing and distance to next point
- form of way as a subset of form of way from ISO 14825 [12]
- functional road class, i.e. classes 0 (main roads) to 7 (the least important roads of a given network).

OpenLR^{TM} has also been included in ISO 21219-22 [11], as a possible approach for describing locations in the TPEG location container (TPEG-LRC) (see ISO 21219-7 [15]), which is described briefly below. OpenLR^{TM} also appears as a LRM in a number of other standards.

There will always be a risk of mismatch of a decoded dynamic location referencing message into the receiver's data set. This can arise because of differences of the road characteristics (such as road classifications, road numbering or naming, topological deviations, geometrical deviations within different map data sets, and of tuning of the encoding and decoding software). However, the ROSATTE project and others have performed test suites for both AGORA-C^{TM} and Open LR, and reported that satisfactory results could be obtained.

As OpenLR^{TM} is described in an open source royalty-free specification, there are no licencing fees related to the use of the method. In 2014, TomTom transferred the rights and licensing of OpenLR^{TM} to the OpenLR association (see www.openlr.org). OpenLR^{TM} is the most widely adopted on-the-fly LRS.

3.7.1.3 Intersection location

It is worth mentioning a third dynamic LRM which is the ILOC (Intersection Location) approach that is actually a predecessor of the AGORA-C^{TM} method. The ILOC referencing method was developed by the ERTICO-led EVIDENCE project. The ILOC method defines road sections by identification of the limits of the section through its bounding intersections. Location references for intersections were identified by coordinates (WGS84), up to three road section descriptors (indicating roads joining at the intersection) and some other attributes such as the road class. However, experimentation showed that this method did not perform acceptably in complex road environments.

3.7.2 Further methods

There are other dynamic LRMs that are under development or initial deployment and this will continue to evolve. One example is TPEG-ULR, developed by TISA [8], called universal location referencing, which is a new, open and royalty-free method for dynamic location referencing, for use in the TPEG location container (TPEG-LRC). Technical considerations and commercial issues related to other dynamic LRMs, such as AGORA-C^{TM} and OpenLR^{TM}, motivated the development. There are plans for this method to be standardised in Part 11 of the TPEG 2 ISO 21219 series, but at the time of writing this process has not started.

3.8 Other LRMs

As well as location referenced by coordinates, pre-coded and dynamic LRMs it is worth mentioning another couple of approaches. These should not necessarily be seen as competitive with the methods mentioned above, as it is possible to mix the feature and capabilities of these different approaches in some cases.

3.8.1 Linear referencing

Linear referencing is a method for pre-coded location referencing that uses a combination of references to predefined identifiers representing linear features and coordinates on the referred features (e.g. in an existing map data set). The concepts for linear referencing are defined in the widely-adopted ISO 19148 [16], where the basic term 'linear referencing' is defined as a *specification of a location relative to a linear element as a measurement along (and optionally offset from) that element.*

Two of the basic concepts in linear referencing are:

- a 'linear feature', defined as a 'one-dimensional (linear) object that serves as the axis along which linear referencing is performed'
- a 'linearly located event', defined as an 'occurrence along a feature of an attribute value or another feature'

A linear location is specified with a reference to a linear feature and a coordinate along the known feature. Point locations ('at locations') are specified with one single coordinate or measure, while segments ('from/to locations') are specified with one coordinate/measure for the start point and either a coordinate for the end point or measure from the start point coordinate.

Locations are referenced according to a 'linear referencing method', within a 'linear referencing system', analogous to other location referencing approaches. Linear referencing methods can be interpolative (values ranging from 0 to 1 or percentages along the linear features), absolute (following a measured unit) or relative (e.g. distance from a reference milepost, kilometre post, hectometre post). The linear referencing system can, for example, be a road network model in an operator's database, using an interpolative linear referencing method.

Figure 3.2 illustrates the concepts of linear referencing, with one linear feature (Link ID 864392) and linearly located events; a warning sign and a speed limit. The extent of the linear feature is from location 0 to location 1. The warning sign is located at position 0.28 along the linear feature, while the particular speed limit goes from location 0.44 to location 0.78.

Linear referencing can also contain more complex location referencing, including horizontal and vertical offset measures. Linear referencing is widely used for location referencing in networks of roads, railways and utilities, for example, in public road databases. The strength of the method is that the network can remain stable while characteristics related to it may change. As different features such as speed limits, road widths, traffic volumes, accidents, culverts, and signs are individually located with linear references, no change need to be applied to the road network if a located feature changes. The weakness is that a user of the linear reference system will need to have access to the network to be able to translate the linear locations into real-world positions.

The INSPIRE Generic Network Model [17] defines a model for linear referencing that is being used in the INSPIRE Transport Networks Model [18]. With this model, the network elements in the data set creates a linear referencing system that uses absolute measures in length along the transport links as the linear referencing method. Linear

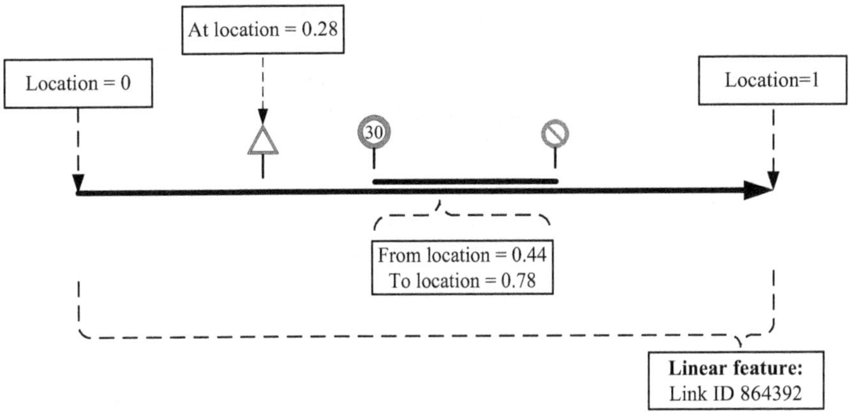

Figure 3.2 Illustration of linear referencing (taken from CEN/TR17297-1 [4], created by this author)

referencing is also used for location referencing in several CEN standards, in particular in DATEX II (EN 16157-2 [19]), and in TN-ITS (CEN/TS17268 [20]).

The linear referencing concept, described in ISO 19148 [16], is widely adopted in many commercial software systems for managing various aspects of road networks.

3.8.2 What 3 words

An interesting, novel and somewhat eye-catching approach is the What 3 Words service [21]. The service has subdivided the entire surface of the world into 3 m squares and associated a three-word combination as a stable identifier to each square; for example *Limes.canny.coins*. This approach has the advantage of being somewhat easier to communicate and recall at a human verbal reasoning level than a series of digits representing, say, a longitude/latitude. It is sufficiently novel that it is not yet widely embedded in many services and is a proprietary commercial solution, but usage can be expected to increase. Unfortunately, *where.my.home* or *here.I.am* are not allocated!

3.9 Usage of LRMs in some standards for ITS

The previous sections cover several important LRMs that are used in the ITS sector, but others exist; it is not possible to be comprehensive in a single chapter such as this. This section shows instances of where these methods are referenced and incorporated in widely used application-domain European standards.

3.9.1 DATEX II

DATEX II, as standardised in the CEN 16157 series, is intended to be used for the exchange of traffic information between traffic information centres, traffic control centres and service providers. The DATEX II specifications are supported and promoted by the DATEX II stakeholder community [22]. DATEX II provides mechanisms to support location references using several different LRMs within their provision of information. EN 16157-2 [19] details location referencing used in the DATEX II series, which supports location references by:

- Point location:
  - Coordinates
  - ALERT-C – Methods 2 and 4, see EN ISO 14819-3 [7]
  - TPEG point location, see CEN ISO/TS 18234-6 [23];
  - Point within linear reference, which defines a specific profile of EN ISO 19148 [16].
- Linear location:
  - ALERT-C – Methods 2, 4 and linear by code, see EN ISO 14819-3 [7]
  - TPEG linear location, see CEN ISO/TS 18234-6 [23]
  - Linear along linear reference, which defines a specific profile of EN ISO 19148 [16].
- Area location:
  - ALERT-C area, see ISO 14817-3 [7]
  - TPEG area location, see CEN ISO/TS 18234-6 [23].

EN 16157-2 [19] also includes OpenLR™ and GML LineString.

It should be noted that although DATEX II provides a framework for describing locations using all of the methods above, it does not mean that these solutions are interoperable. If a data supplier uses one LRM, the receiving party has to be able to interpret that specific method.

3.9.2 TPEG

TPEG provides a specification primarily designed for the broadcast of traffic and travel information. More details can be found in Chapter 4 on digital coding, collation and exchange of traffic information. Standardised versions of TPEG have evolved over time and several families of standards exist. The CEN ISO/TS 18234 series, largely oriented towards binary encoding, and CEN ISO/TS 24530 series, largely oriented towards XML encoding, defined a suite of specifications which is now known as TPEG generation 1. These have largely been replaced and superseded by ISO/TS 21219 series generation 2 specifications.

TPEG technology has two key demands to satisfy: mobility of access and language independence. With mobile client devices, it is clear that any location information given to the end-user is both human understandable and machine-readable. TPEG originally satisfied this need with the TPEG-LOC methodology by delivering both types of content with all messages. Subsequent developments created the concept of a 'Location reference container', such that TPEG messages could describe the location with a choice out of potentially multiple LRMs (both pre-coded and dynamic LRMs), to suit both 'thick' and 'thin' clients. ('Thick' clients run computer applications which can locally access all the data and other resources needed to operate; 'thin' clients have applications which accesses most of their data and other resources from a remote server).

TPEG 2 supports the following:

- Location referencing container which supports whichever LRS is used for a TPEG2 service, see ISO/TS 21219-7 [15];
- GLR, which is a type of location referencing by coordinates, see ISO/TS 21219-21 [24];
- TPEG2 OpenLR™, see ISO/TS 21219-22 [11];
- TPEG2 Multi-modal Routing, see ISO/TS 21219-23 [25];
- TPEG-LOC, which is a straightforward on-the-fly coordinate and description-based method, see CEN ISO/TS 18234-6 [23], which uses modified ILOC referencing concepts. This methodology is also included in ISO 17572-3 [9]. It is intended that this technology is incorporated into TPEG2-ULR, *Universal Location Referencing* which is still being developed and not ready for standardisation. TPEG-LOC is used in many DATEX II services.

3.9.3 TN-ITS

Transport networks for intelligent transport systems (TN-ITS) is a standard for the exchange of static road data, with a focus on changes, directly from the source of the changes, specifically from the public road authorities and/or public/private road operators, to data users including mapping organisations.

CEN/TS 17268 [20] is the CEN Technical Specification for TN-ITS, which supports location referencing by:

- AGORA-C™
- OpenLR™
- encoded geometry
- use of references to pre-coded locations such as supported by INSPIRE, and linear referencing defined as a profile of EN ISO 19148 [16].

3.9.4 INSPIRE

INSPIRE is the abbreviated name for an infrastructure for spatial information in the European Community which is being established under Directive 2007/2/EU [26]. Under this directive, public authorities provide road network geospatial data sets and many other forms of geospatial data in an INSPIRE-compliant way. The form of modelling of the road network in INSPIRE is specified within the INSPIRE implementing rules and technical guidelines for transport networks [18].

The transport network specifications in INSPIRE [18] (which are not legally binding) are based on the INSPIRE 'Generic Network Model [17], with linear referencing as a core element. The network is based on a set of network elements, such as road links and nodes. Network properties, such as speed limits, are connected to the network through linear referencing.

The linear referencing model in [17] was designed before EN ISO 19148 [16], and is limited to absolute linear referencing based on the geometry of the network elements. However, it is stated in the specification that the model will be updated in accordance with EN ISO 19148 [16].

3.9.5 Geographic data files (GDFs)

GDF is a long-standing but evolving ISO standard for the definition of road network data for ITS. The latest published version is EN ISO 14825 [12], which is identified as v5.0. Revised versions (v5.1) are being prepared for publication at the time of writing.

GDF itself defines the base map data set describing the road network and its characteristics. Many forms of location referencing are utilised, but this is substantively underpinned by the common concepts related to locating by coordinates.

As part of the revision and updating of the GDF standard, a new representation of elements of the road such as lanes, carriageways and potential evasion areas (which provide options for routing in exceptional and emergency situations) are being introduced into the GDF model. Many of these new inclusions are using a novel approach which is referred to as the 'Belt Concept'. This is a step-away from most existing ITS navigation type road models and road map data sets, which tend to use linear reference models that broadly represent a notional road centreline. The belt concept introduces elements of the road that have width, edges and other structure rather than just being a centre line.

3.10 Location accuracy

When users receive location information, the nature, objective and delivery means of the service they are using implies approximate levels of accuracy. For example, audio radio traffic broadcasts tend to describe location in relatively coarse terms, often defining locations on a well-known route either at or between well-known landmarks or intersections. This provides appropriate information for the service, but is low resolution in terms of accuracy. By contrast, serving digital data into connected or automated vehicles will seek to provide location information that is accurate to a few metres, or possibly better.

Few data services are delivered with a complete view of data quality throughout the information chain, from initial capture to ultimate delivery and use. If the quality characteristics are unknown or poorly described, the quality of the final delivered data, and perhaps the quality of the service, is unknown. Service providers are recommended to make greater use of standardised means of assessing and characterising data quality, and to publish service data characteristics, to provide users with informed choice. For example, in the sphere of geographic information, EN ISO 19157 [27] provides a framework for describing quality characteristics.

3.10.1 Location referencing accuracy example

Many ITS applications have historically evolved within their own environment, often developed by distinct teams and suppliers, with a strong focus on operational constraints, requirements and objectives set for the related service. Most of these applications are rarely deployed to meet specific regulations or legislation, beyond general duties of care in respect of the safe, efficient and environmentally effective management of a road network. Such legislation rarely places specific requirements on the quality with which data is characterised or geo-located.

For example, in the provision of traffic information, road link location referencing mirrors the typical form of audio traffic broadcasts (e.g. 'there is an accident on Ford Road, between the junction with Hilman Street and Anglia Road'; or fuzzy well-known locations such as 'the White Post roundabout'). This form of information provision and its related location referencing has value supporting route guidance, whether performed by machine or human, by potentially indicating a driver decision point (junction), at which route choice is required to avoid the stated issue. There is not a strong correlation between the stated link end of the road link – or fuzzy location – and a well-defined spatially identifiable point or area in a known reference frame (i.e. the described location would be difficult to precisely locate at a detailed level on a map). However, it is sufficient for many applications.

The location accuracy of these forms of location reference is difficult to quantify. It can be of the order of several tens of metres. This approach underpins the link-oriented location referencing of ALERT-C.

ALERT-C referencing also supports a more detailed ability to locate at a sub-link level, by use of offset measures; indicated start and end locations within the bounds of one or more links. This is specified in EN ISO 14819-3 [7]. Use of this approach tends to rely upon inputs from various information sources such as:

- traffic management operatives identifying incident locations
- end of queuing traffic using CCTV traffic cameras
- automated incident and queue protection systems (which use a variety of detection mechanism such as image recognition)
- various forms of traditional vehicle detection systems (including inductive loops and radar detectors)

The identified location may be subject to human interpretation, inferred from a point detection, or other means.

The location accuracy of these traffic events may be of the order of a few metres to several tens of metres within the source systems where the traffic event is first recorded. In many places it is common practice to tie the traffic event locations back to the operational physical location system of marker posts (e.g. typically at 100 m nominal spacing), and identify the event location as being at the marker post location closest downstream from the event location in the direction from which the traffic approaches, or between a pair of marker post locations that encompasses the traffic event location. The location of these marker posts is then translated into an offset measure within the link-based referencing system. The location accuracy of this form of referencing is typically in the order of several tens of metres, up to a maximum of perhaps 100 m, discounting gross human or machine error.

Figure 3.3 illustrates forms of location referencing that may be used to locate an event on the road network. The examples given in Figure 3.3 show how the

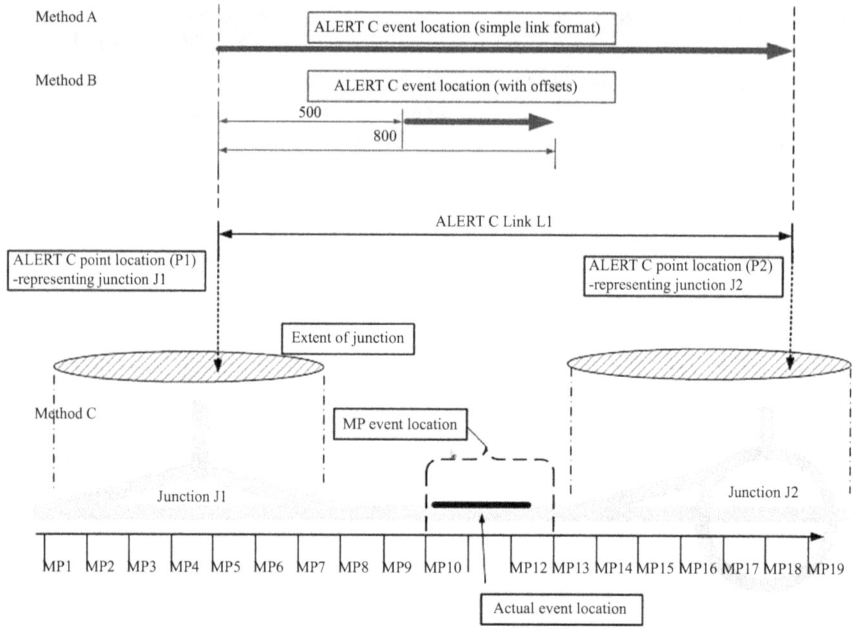

Figure 3.3 Location referencing accuracy illustration (taken from CEN/TR17297-1 [4], created by this author)

reported location of the event will differ depending on the LRM used, but also where the event is in respect of 'known' reference points in the specific LRS and operational practice used to encode the event location within each system.

Method A in Figure 3.3 illustrates use of simple link-based ALERT-C references. The derived location of the network event is given as on the link between Junction J1 and Junction J2. This link (L1) is defined in the ALERT-C location table as being between two representative ALERT-C point locations (P1 and P2). The location of Points P1 and P2 are location points within a potentially complex geometry and ill-defined extent of the boundaries of Junction J1 and Junction J2 respectively. They are just representative, potentially having no particularly meaningful real-world position in the respective junction.

Method B in Figure 3.3 shows a more accurate method of indicating the event location. This shows it being defined by means of reference to an ALERT-C link (L1) and offset measures for the start and the end of the event with reference to the start of Link L1. Often operationally, the start and end location of the network event will be passed from the road operator's network management system to be encoded in ALERT-C; the road operator's network management system will often use reference post or marker posts as a LRS for operational purposes. These are shown as MPn in Figure 3.3.

Method C in Figure 3.3 shows marker posts being used to define the extent of the event. Typically operators will record the marker post upstream of the actual event's start location and the marker post downstream of the actual event's end location to define the event's location in the marker post system.

Although the use of ALERT-C link-based referencing remains important, many traffic centred services are moving to a more granular level of detail. Several traffic services and applications are providing quality of service/congestion applications giving users information against short lengths of road typically using colour-coded intensity thematic mapping, or related data feeds. Depending on the commercial choices of each of these services the road lengths represented are much shorter than the typical length of a link within an ALERT-C location referencing table. Some such services report data on road segments of 500 m length, perhaps down to a granularity of 25 m; this is considerably shorter than the typical inter-urban ALERT-C road link length and is often presented in the form of:

- spatial coordinates that are derived directly from the underpinning network models used by the commercial service
- rendered as a polygonal image for display on in-vehicle systems and smart devices that can be referenced to links within an ALERT-C location referencing table.

Tying such data to an alternate reference road network model introduces translation/transformation errors.

As applications and services improve there is an increasing move towards reporting data at a lane level, e.g.:

- reporting incidents
- speed control
- flow characteristics on a lane-by-lane basis.

Hence, there is the challenge of correctly locating vehicles to the correct carriageway and potentially the correct lane; a location accuracy of the order of two or three metres, is not sufficient.

With the advent of automated vehicles, a new era of location referencing and location accuracy is emerging. The specific requirements of information concerning the positioning of vehicles within the context of the roads they are using, and asset and safety features on and around those roads is not yet fully clear. However, the expectation is that the demands for quality-source reference information for connected and automated vehicles can be expected to require location accuracies of better than 25 cm. This evolution over time is challenging the way location information is created, managed, exchanged, related and used.

3.10.2 Timeliness and currency

Many people consider road networks to be static, but in reality they are almost continuously changing. Therefore, when creating data associated with a reference network or map of the road network there is a need to take account of this on-going change. Users report problems when translating data between systems, even using the same reference network data set, when the maintenance of those data sets is not done in a coherent manner. This can result in the failure to translate data and inappropriate associations.

The use of techniques such as TN-ITS permits road change information to be distributed from the road authorities to commercial map vendors on a weekly or daily cycle. However, significant challenges exist in many authorities concerning the updating, control and availability of the at-source captured data. Again, a greater focus on use of formal means to characterise quality is recommended.

3.11 Outlook on current developments and future needs

As can be seen through the earlier sections of this chapter, different disciplines, needs, objectives and suppliers have created different approaches to supporting location-based services within different geographies and for different modes of transport. This creates a diverse landscape. One of the challenges for the future is that as we move towards more integrated mobility solutions that perhaps blend the use of private vehicles with public transport and alternative novel modes of transport (such as Uber or Lyft-style services, e-scooters, car club vehicles, demand responsive transport, lift sharing) the need to interoperate between LRS becomes stronger. There is no single common system – nor is there likely to be in the future.

Systematic means for converting location references in one system to location references in a second system are not commonly adopted. CEN/TR17297-2 [28] proposes one approach, but this is relatively new and rather untested. It is not purely a question of converting one location reference to another but also understanding what effect this has upon the quality of that location reference (i.e. how much confidence can be placed in the accuracy of a derived location reference, and whether this is sufficient for the intended use).

The need to blend different forms of location reference exists in many circumstances. For example, to support some smart parking applications there is a need to be able to bring together information sourced from the vehicle about its location with information concerning where parking bays are, which is typically sourced from a local authority or supplier database. This requires conversion between different systems.

Many commercial approaches are being deployed or are already in widespread use. Google Maps and Waze, for example, are providing high-quality smartphone-based user services, often as a more granular level of resolution than can be provided by services provided by road operators. These services are located within a mapping system that is used and often rendered on the provided device (e.g. smartphone, tablet). Such location-based services often helpfully link to the device's location either by the device's own GPS receiver, maps of Wi-Fi hotspots, or network services.

Approaches for creating and sharing navigation databases for the road are also evolving. Alongside the continuing development of formally standardised approaches such as GDF, industry groups and consortia are developing what they hope will be commonly accepted, widely deployed, new approaches. One, being developed by the Open Auto Drive Forum (OADF) [29], with many major automotive interests represented, is their Navigation Data Standard (NDS) specification [30] for the sharing of next generation navigation databases. This is an evolving effort and, at the time of writing, discussion between OADF and standardisation bodies are on-going concerning NDS as a future, formally recognised standard.

Additionally, as vehicle sensing technologies improve and become more widely deployed, the capability to crowd source change information concerning the road network, road features and furniture strengthen. This will bring an important development in the arsenal of methods and tools that are used to maintain road 'map' datasets.

There is a strong recognition that with the emergence of automated vehicles and wider deployment of connected vehicle services, some traditional aspects of managing the road network require modernisation. It is perhaps universal across all jurisdictions, states and countries that information concerning traffic regulations and local restrictions are not freely available in a robust digital format. Interesting examples include the work that is currently being undertaken in development, testing and standardisation of systems to support automated valet parking [31] that are vehicle-infrastructure systems that support largely automated vehicles (SAE Level 4 – [31]) to park without a driver in suitably equipped parking facilities. Some variants of proposed automated valet parking systems rely heavily on fine-grained detail HD maps.

In the United Kingdom, we have Traffic Regulation Orders (or Traffic Management Orders in London) which largely are paper-oriented, even if held in digital form. Current work by the Department for Transport is arguably at the forefront of many administrations looking to fully digitise traffic regulations in a machine readable, machine comprehensible form. This is becoming a necessary requisite input to the deployment of many future digital services as well as automated vehicles. The needs and precision of location information related to data resources such as digital traffic regulations are not fully defined, but the expectation is that location resolution of a few centimetres will be required. In the near future this will challenge current widely deployed and used map dataset approaches.

References

[1] CEN/ISO. EN ISO 19111:2007 – Geographic information – Spatial referencing by coordinates.

[2] CEN/ISO. EN ISO 19155:2012 – Geographic information – Place Identifier (PI) architecture.

[3] ISO. ISO 17572-1:2015 – Intelligent transport systems (ITS) – Location referencing for geographic databases – Part1: General requirements and conceptual model.

[4] CEN. CEN/TR 17297-1:2018 – Intelligent transport systems – Location referencing harmonisation for urban ITS – State of the art and guidelines.

[5] International Association of Oil and Gas Producers. EPSG geodetic parameter registry. [Online] https://www.epsg-registry.org.
[6] National Geographic. "Australia Is Drifting So Fast GPS Can't Keep Up". [Online] September 2016. https://www.nationalgeographic.com/news/2016/09/australia-moves-gps-coordinates-adjusted-continental-drift/.
[7] ISO. ISO 14819-3:2013 Intelligent transport systems – Traffic and travel information messages via traffic message coding – Part 3: Location referencing for Radio Data System – Traffic Message Channel (RDS-TMC) using ALERT-C.
[8] Traveller Information Services Association. [Online] http://www.Tisa.org.
[9] ISO. ISO 17572-3, Intelligent transport systems (ITS) – Location referencing for geographic databases – Part 3: Dynamic location references (dynamic profile).
[10] TomTom International B.V. OpenLR Whitepaper Version: 1.5 revision 2. 2012.
[11] ISO. ISO/TS21219-22, Intelligent transport systems – Traffic and travel information (TTI) via transport protocol experts group, generation 2 (TPEG2) – Part 22: OpenLR location referencing (TPEG2-OLR).
[12] CEN/ISO. EN ISO 14825:2011, Intelligent transport systems – Geographic Data Files (GDF) – GDF5.0.
[13] Via Licensing Corporation. [Online] http://www.vialicensing.com.
[14] Via Licensing. Via Licensing – AGORA-C licensing fees. [Online] http://www.vialicensing.com/licensing/agorac-fees.aspx.
[15] ISO. ISO/TS21219-7, Intelligent transport systems – Traffic and travel information (TTI) via transport protocol experts group, generation 2 (TPEG2) – Part 7: Location referencing container (TPEG2-LRC).
[16] CEN/ISO. EN ISO 19148:2012, Geographic information – Linear referencing.
[17] European Commission. INSPIRE D2.10.1 2013, INSPIRE Data Specifications – Base Models – Generic Network Model.
[18] European Commission. INSPIRE D2.8.1.7 version 3.2 2014-04-17, Data Specification on Transport Networks – Technical Guidelines.
[19] CEN. EN 16157-2:2018, Intelligent transport systems – DATEX II data exchange specifications for traffic management and information – Part 2: Location referencing.
[20] CEN. CEN/TS 17268:2018 Intelligent transport systems – ITS spatial data – Data exchange on changes in road attributes.
[21] What3Words. [Online] https://what3words.com.
[22] CEDR. DATEX II. [Online] http://www.datex2.eu/.
[23] CEN/ISO. CEN ISO/TS 18234-6:2006, Traffic and Travel Information (TTI) – TTI via Transport Protocol Expert Group (TPEG) data-streams – Part 6: Location Referencing application (TPEG-Loc).
[24] ISO. ISO/TS21219-21, Intelligent transport systems – Traffic and travel information via transport protocol experts group, generation 2 (TPEG2) – Part 21: Geographic location referencing (TPEG-GLR).

[25] ISO. ISO/TS21219-23, Intelligent transport systems – Traffic and travel information (TTI) via transport protocol experts group, generation 2 (TPEG2) – Part 23: Roads and multimodal routes (TPEG2-RMR).

[26] European Commission. INSPIRE Directive 2007/2/EU. [Online] https://eur-lex.europa.eu/legal-content/EN/ALL/?uri=CELEX:32007L0002.

[27] CEN/ISO. EN ISO 19157:2015, Geographic information – Data quality.

[28] CEN. CEN/TS 17297-2:2019 Intelligent transport systems. Location Referencing Harmonisation for Urban-ITS. Transformation methods. [Online]

[29] Open Auto Drive Forum. NDS Association. [Online] https://nds-association.org/.

[30] Open Auto Drive Forum. Open Auto Drive Forum. [Online] http://www.openautodrive.org/.

[31] ISO. ISO/AWI23374 – Under Development – Intelligent transport systems – Automated valet parking systems (AVPS) – System framework, communication interface, and vehicle operation.

Chapter 4

Digital coding, collation and exchange of traffic information

Paul Burton[1], Bev Marks[2] and Jon Harrod-Booth[3]

4.1 Introduction

The first chapter of this book covered the collection of road traffic and travel data. This chapter is about technology for collating that data to deliver traffic and travel information (TTI) to end-users. A subsequent chapter will discuss the development of European standards (EN) to encourage the take up of travel and traffic information and to enable Europe-wide solutions.

4.2 Background and context

With greater mobility across Europe and borders between countries crossed with ease, there has been a drive for solutions that enable seamless delivery of TTI in the preferred language of the traveller.

However, the delivery and reception of language independent information is not the whole story. Following the collection of traffic data and travel data, the information needs to be collated, that is combined from a number of sources, filtered by importance, location and relevance; it then needs to be coded and delivered to the dissemination media.

In the early days of TTI, the main method of communication was the written word, such as in newspapers (traffic reports and road closures) or by spoken voice over the radio and rarely on the television. This information was often not specific to a particular location and, in many ways, was seen as 'editorial filler' rather than informative content.

As traffic increased rapidly in the 1960s and 1970s, it became apparent that more traffic information was needed and UK radio stations (the BBC at the time) started to dedicate staff to the collection and collation of traffic news. It should be remembered that this was before the days of mobile phones. VHF/FM radio was in its infancy and so the only broadcastings was on AM national radio. By 1967, BBC local radio was being offered with the start of Radio Leicester. Local radio became

[1]CEN/TC278 WG4 and ISO/TC204 WG10 – Traffic and Traveller Information, Worcestershire, UK
[2]TISA (Traveller Information Services Association), Brussels, Belgium
[3]Harrod Booth Consulting Ltd., Somerset, UK

an ideal channel for locally focused traffic information although traffic bulletins still continued on national radio. Traffic bulletins were delivered by voice and at set times but still required the traveller to be tuned into the station delivering the travel broadcast at the time.

A more direct and focused approach to TTI was needed.

4.2.1 Switching radio listeners to traffic and travel broadcasts

4.2.1.1 ARI

In the mid-1970s, German broadcasters (led by their research institute IRT) and a receiver manufacturer (Blaupunkt) developed the ARI (Autofahrer-Rundfunk-Information – driver radio information system). ARI signalled the presence of traffic information in VHF/FM broadcasts used by the German public network of VHF/FM radio stations using a 57 kHz subcarrier added to a station's VHF/FM signal. ARI was the forerunner of the radio data system (RDS) travel flags (see later) and was replaced by RDS in the late 1980s (implemented by the BBC and Swedish Radio).

4.2.1.2 RDS traffic programme and traffic announcement flags

RDS is a communications protocol standard for embedding small amounts of digital information in conventional VHF/FM radio broadcasts. RDS standardises several types of information transmitted, including time, station identification and programme information. Further information about RDS is provided in a later section of this chapter.

The RDS data stream includes two flags carried on the 57 kHz RDS data stream which essentially do the same as ARI. They are the:

- Traffic programme (TP) flag that signals which stations are likely to carry traffic information announcements;
- Traffic announcement (TA) flag which signals that the tuned station is currently broadcasting a traffic announcement; this allows an in-vehicle receiver to either un-mute the receiver or switch from a CD player or other media to play the announcement.

These flags were only capable of alerting the traveller if the in-car receiver was tuned into the station transmitting the TA and TP flags. However, later on in the development of RDS, the 'extended other networks' scheme was developed that signals which other stations are linked (from the same broadcaster) to the present station, and hence enables access to TAs on the linked stations. For example, a receiver tuned to BBC Radio 4 could automatically tune to BBC Hereford and Worcester to pick up a spoken traffic broadcast and then return to BBC Radio 4 afterwards.

Most importantly, these systems only alert travellers to travel and traffic bulletins read into a microphone by a reporter. These generally take place at a maximum of 3 per hour, and then only during the peak hours. A system where the traffic bulletins were independent of the sound broadcasting on radio was needed; hence, the development of coded traffic information which would be independent of the microphone.

4.2.2 *Route planners*

From the early days of personal computer, route planners from software companies such as SIA enabled a road traveller to plot the optimum route from one location to another. The early planners only gave routes from one centre to another, albeit providing 'way points'. First and last mile route planning did not exist for a number of years. It was possible to optimise the route by time or by distance but, again, the planners did not take account of any traffic conditions en-route, either by time of day from historical data or real time data. It took a number of years, into the twenty-first century, before real-time data was available to such systems.

4.2.3 *Early navigation systems*

In the very late 1980s, in-car navigation was starting to be developed; this was before the availability of civilian satellite-based global positioning systems which didn't become available until the mid-1990s, and were initially very expensive with limited accuracy.

In the late 1980s, Siemens developed the ALISCOUT navigation system, later to be renamed EUROSCOUT. This system could supply turn by turn instructions (and lane positioning) by using dead reckoning from inertial and distance travelled data. The system relied on a series of pole-mounted infra-red transmitter/receivers which informed the in-car unit where it was (positionally) and how to get from that point to subsequent transmitters along its route. The transmitters could be updated with real-time traffic information and also receive travel time data from the vehicle, thus building a picture, in real-time, of the traffic conditions.

In 1988, a demonstration Autoguide system based on EUROSCOUT technology was implemented in central London and in the corridor out to Heathrow Airport. In July 1989, the announcement was made that, subject to successful negotiations on terms and to the implementation of enabling legislation, a licence should be awarded to GEC to operate Autoguide in the London area. The announcement envisaged that following the successful implementation of a major pilot system, commercial Autoguide operations could begin by the end of 1993, and that a licence might be considered for a second pilot area outside London. However, due to commercial and legal issues the London pilot never went ahead, and the emergence of satellite position systems superseded infrared beacons and inertial navigation systems.

4.2.4 *The Road Traffic (Driver Licencing and Information) Act 1989 and its order of 1990*

The legislation referred to in the previous section is the UK Government's Road Traffic (Driver Licencing and Information) Act 1989 and its order of 1990 [1].

The Road Traffic (Driver Licensing and Information Systems) Act 1989 and its accompanying Driver Information Systems (Exemption) Order 1990, which modified the scope of the Act, served two main purposes. The first was to enable operators of driver information systems to install system apparatus in the highway where the Secretary of State decided that such powers are necessary. The second was to ensure that new driver information systems did not prejudice road safety or good traffic management.

A licence to operate a driver information system could therefore be granted to provide (a) powers to install apparatus, or (b) to impose conditions on the operation of the system, or (c) for both of those things. It could confer a right to operate a particular system in any area of Great Britain of the operator's choosing or restrict the operation of the system to a particular area.

In general, the operator of a driver information system would need a licence if it:

(a) required the installation in, upon, under, over, along or across a public road of system apparatus capable of transmitting data to, or collecting information from, motor vehicles or both; or

(b)

 (i) gave positive and dynamic route guidance (defined as detailed junction by junction turning instructions and other advice as to routes to be followed which took into account both the status of the road network and also current traffic conditions collected by a driver information system) providing specific directions to individual drivers which could be updated by information received from outside of the vehicle;
 (ii) used equipment in motor vehicles which had been designed to give the specific directions referred to above; and
 (iii) would be available for use by the public (a system which was operated, for example, by a police force in order to give guidance only to police vehicles would not therefore require a licence).

The Act did not apply to conventional radio traffic broadcasts and these (including RDS) would not be licensed under the Act; they transmit exactly the same information to thousands of motorists regardless of their location, nor to autonomous navigation systems such as in-car electronic maps, which do not receive updated information from outside of the vehicle.

Many thought that this legislation was heavy handed and served to stifle the introduction of navigation systems, and there were arguments in the early 2000s as to whether a navigation system that modified its route choice due to external information fell under the act. In the end, it was agreed that they did not, in part because of the complexity of identifying which organisation or organisations would require a licence as the complete system was comprised of a number of operators, collectors, collators, service providers, broadcasters and in-vehicle equipment providers.

However, obtaining a licence certainly benefitted companies like TrafficMaster which was enabled to install equipment on the public highway. It is likely that, without this government legislation and support from the Department for Transport, regional bodies responsible for roads would have proved an awkward barrier to equipment being attached to their infrastructure.

4.3 Collation of traffic data and information

4.3.1 The BBC motoring and travel units

Originally, traffic information at the BBC was the preserve of the News Department which produced traffic bulletins as part of the news output. At weekends, the BBC produced a regular 10-min slot on Radio 4 that looked at the travel prospects for the week.

In the mid-1980s, responsibility for traffic broadcasting passed to the Radio Outside Broadcast Production Department which developed a system for traffic information based on the revolutionary Newsroom script production system BASYS. BASYS took text input from multiple terminals and helped a professional Traffic and Travel reporter to quickly produce a script for the newsreader as an output. BASYS was used for traffic bulletins fed from a number of BBC provided terminals placed in police control rooms across the country, and notable control rooms like the Dartford River Crossing. With the onset of local radio, BASYS produced scripts that were relevant to the geographic area covered by the radio station. BASYS input was based on the Broadcast for Motorists Lists mentioned in Chapter 5 of this book. BASYS also produced output that could be used in the BBC CEEFAX teletext services [2].

Other organisations such as AA Roadwatch and the RAC also collated traffic information, but generally in a more manual way where available data was collated by skilled journalists and broadcasters.

Presently there are a number of collation systems provided by commercial organisations such as GEWI.

4.4 Exchange of traffic information

In the early 1990s, the need to progress traffic information from data collectors to data aggregators and service providers was becoming more pressing, particularly in the European context of cross-border cooperation. True to the European ideal of access to all services from all countries it was important that traffic situations passed on from centre to centre language independently in the same way as RDS-TMC. The information to be exchanged was limited to traffic and road situations only. There was not the intention to exchange public transport information which is covered by TRANSMODEL.

4.4.1 DATEX

In the early 1990s, a task force was set up to combine several experimental approaches for a centre to centre traffic and road related data exchange service. This ultimately resulted in the data exchange standard DATEX. The task force drew on contributing projects and on previous work which had agreed a list of traffic messages, called ‘ALERT C’, in order to offer a traffic message channel (TMC) feature within RDS. This is further described in a later section of this chapter. It was agreed that a modified set of RDS-TMC ALERT C codes would be used as they were designed to be language independent. The majority of RDS-TMC ALERT C messages were used and a Trigram code assigned to each one, such as Accident = ACC. Location referencing for the data was based on the RDS-TMC ALERT C location coding procedure which in many countries was being adopted as a national set, owned and maintained by the public authorities. Other location referencing approaches were also permitted.

It was further agreed that that the Trigram codes and location data would be exchanged based on an international Electronic Data Interchange (EDI) standard developed under the United Nations called EDIFACT (United Nations/Electronic Data Interchange for Administration, Commerce and Transport) [3], which was widely used in other applications.

Most of the development of DATEX was undertaken by the CETE Méditerranée (France), MIZAR (Italy), Heusch Boesefelt (Germany) and Castle Rock Consultants (UK). The EDIFACT proved challenging. The first instance of the use of DATEX was in the European PLEIADES project [4] which demonstrated that Traffic Information Centres (TICs) in England and France could have real-time knowledge of travel conditions on the main routes for cross channel trips. Several other projects were deployed elsewhere; the European CENTRICO project [5] took this concept forward. Although a number of implementations of DATEX were trialled between different neighbouring countries they were very limited.

4.4.2 Open Travel data Access Protocol – OTAP and Travel Information Highways – TIH

DATEX, based on EDIFACT, was developed at a time when the Internet was not widely available and most communication from point to point was by private telecoms circuits. When the Internet arrived, data exchange could be modernised and this led to development of the Open Travel data Access Protocol (OTAP) which was a joint initiative between the two European projects CENTRICO [5] and STREETWISE [6]. OTAP became popular and led eventually to the development of DATEX II web services (see below). The network model dataset used in the United Kingdom comprised the motorways and trunk roads and was based on the National Traffic Control Centre network where RDS-TMC codes were assigned to points and links.

Travel Information Highways (TIH) was an initiative from the United Kingdom. The Highways Agency which exploited the Internet to disseminate and exchange real-time information on traffic and road conditions to third-party information service providers, who in turn could provide information services to the public. However, once DATEX II was established there was little justification for further investment in TIH.

4.4.3 DATEX II

Hypertext Markup Language (HTML) is the standard markup language (i.e. a human-readable computer language) that annotates, or ‘tags’, elements within a document to suggest how they could be displayed in an Internet web browser. XML stands for eXtensible Markup Language and has some similarity to HTML but can also be used to represent data structures such as those used in web services.

With the internet becoming more pervasive and XML based systems being developed, a project was launched in the early 1990s, sponsored by the European Commission, to develop an XML based version of DATEX. This was undertaken using conceptual models of operations and involved developing XML schemas i.e. detailed descriptions of data types and content including order of elements, grammatical rules and range constraints. The initial work was continued by the CENTRICO [5] project and its subsidiary projects, such as STREETWISE [6], and continued by the EasyWay projects (2007–2012) [7]. The outcome became known as DATEX II [8].

The DATEX II modelling approach is based on model-driven architecture principles with modelling undertaken using the Unified Modelling Language [9]. Since the second half of the 2010s, Model Driven Architecture (MDA) and Unified Modelling Language (UML) have become a widely used, well established and a stable environment for system specification. It provides a standard way to visualise the design of a system and proved an ideal environment to capture the DATEX II domain model. Harmonising ITS concepts at a European level takes a long time and it would be unsuitable to capture the results from this effort in a short lived, technology dependent way. The UML offers exactly the required stability, with the concrete mapping to exchange artefacts.

The current implementation platform for messaging is the W3C standard for XML schema definition [10]. The mapping is well defined in the specifications and has been implemented in a tool that users can download together with the model itself, the whole specification and further supporting material.

DATEX II can be used for all data exchange applications with dynamic information on transport systems and in particular the road system. The main uses are:

- Rerouting, network management and traffic management planning both on motorway and urban networks;
- Lane or queue control systems and related applications like ramp metering (controlling vehicle access onto motorways to smooth integration of new traffic), dynamic speed limits and overtaking control;
- Linking traffic management and traffic information systems;
- Applications for data exchange between multi-modal management and information systems;
- Applications for the exchange of traffic data;
- Provision of services in the framework of road management with a strong link to network safety or performance like Truck Parking;
- Applications where information exchange between individual vehicles and traffic management infrastructure takes place.

For all these domains, DATEX II ensures that interoperability and cooperation can take place between diverse actors to allow the unhindered exchange of data or information. However, DATEX II is also designed for use within single operator systems. Users are able to extend the DATEX II model according to application specific needs, but they are also able to select those elements for schema creation that are actually used in their services. Thus, slim services can still work with slim schemas without having to carry the full burden of the huge DATEX II model.

Users that have created extensions that they feel could be of wider use can submit them in a dedicated section on the DATEX II website. Here they can be downloaded by other users and then discussed for future integration into the main standard. The DATEX II organisation has defined an approval process that deals with these user community provided inputs as well as with all other user feedback (forum discussions, issue reports) via the website.

The DATEX II organisation not only supports DATEX II users and maintains the standard, it also monitors and supports the deployment of DATEX II at the European level. A deployment guideline has been produced that aims at steering the use of DATEX II, and of monitoring its uptake, in the scope of the EasyWay initiative.

At the time of writing there are seven parts to DATEX II standardised in Europe to technical specification (TS) level with moves towards taking them to full EN:

- Part 1: Context and framework;
- Part 2: Location referencing;
- Part 3: Situation publication;
- Part 4: Variable-message sign (VMS) publications;
- Part 5: Measured and elaborated data publications;
- Part 6: Parking publication;
- Part 7: Common data elements.

References to these standards can be found on the DATEX II website.

DATEX II, like its predecessor DATEX, has its roots in the RDS-TMC ALERT C message lists and structure. The mapping between RDS-TMC ALERT C phrase codes and DATEX (1) phrase codes was very strong, enabling conversion of data transmitted (centre to centre) using DATEX II for forward dissemination by RDS-TMC. In DATEX II this relation is less direct and straightforward. Various location referencing systems can be used (defined in the profile for a service). Originally it was the RDS-TMC ALERT C 16 bit predefined location referencing system. At the time of writing, Open LR [11] is becoming the preferred method, but any agreed location referencing system can be used.

At the time of writing this work, the DATEX II community goes from strength to strength with 60 nodes operational across Europe, and is playing its part in the development of Cooperative Connected Automated Mobility (CCAM) applications. The DATEX community can be found at www.datex2.eu [12].

Further parts of the DATEX II standard series are at various stages of development. Additionally, different forms of data encoding are now under development and early deployment. These include JSON which is popular with mobile app developers and binary encoded ASN.1.

The DATEX (I) standards have been withdrawn.

4.5 Delivery of traffic and travel information to end-users using RDS-TMC

At the time of writing, Radio Data Systems – Traffic Message Channel (RDS-TMC) using ALERT C is one of the standard methods of transmitting traffic information to road travellers using end-user terminals. RDS-TMC ALERT C was also used as the basis for both DATEX and DATEX II (see previous sections).

4.5.1 Background

RDS standardises several types of information transmitted, including time, station identification and programme information.

To offer a TMC feature within RDS, RDS-TMC was developed in the late 1980s and early 1990 in the RDS-ALERT project [13], part of the EU FP2 DRIVE programme. At the end of the project a decision was made to combine the two competing protocols (ALERT A and ALERT B) to form ALERT C which was taken forward for standardisation in 1992. ALERT C was used in various large field trials but did not come to 'commercial' fruition until 2000 when Toyota agreed to add RDS-TMC to their TNS200 receiver in various grades of the Toyota Avensis. The service was provided and transmitted by ITIS (now INRIX), the location referencing by the then Oscar Faber, with the equipment and system integration by ITIS. The receivers were developed by Aisin. The service drew on the concepts developed in the large UK RDS-TMC trial (described in the topic 'European Developments in Traffic and Travel Information' (Section 5.3.5.1)).

4.5.2 RDS-TMC

RDS-TMC is a one-to-many unidirectional broadcast systems. Its objective is to broadcast TTI messages as data on VHF/FM transmissions using RDS. This allows delivery of accurate, timely and relevant information without the need to interrupt the radio programme – just the opposite of the common practice with spoken traffic messages. Thus, TTI can be broadcast inaudibly in the background of existing VHF/FM radio programmes to be later presented to the user.

RDS-TMC relies upon a simple encoding process which requires the use of predetermined databases for event descriptions (and location information) to be present at both the Service Providers and all their customers or subscribers. Thus, a simple number can be transmitted via RDS-TMC, and by using a look-up method, it can be interpreted into an understandable message in the RDS-TMC receiver. Whilst this has some disadvantages, e.g. the receiver must have an up-to-date copy of the database, it has one big advantage: the customers can use a receiver which presents the information in their own language. This is a big advantage in Europe, with so many languages in everyday use in very close geographical proximity. The coding also permits users to receive only those messages that are relevant to their needs. Thus, it may be possible for tourists to travel, for example, from London to Rome, through Belgium, Germany, Switzerland, France and Italy, whilst always getting the traffic information via RDS-TMC in their own language; the location codes being on a smart card or CD-ROM that was purchased in London.

The TTI messages that are distributed will, of course, be conveyed to and stored in receiver memory which may subsequently permit traffic information to be presented via speech synthesis in the language required or inserted onto a user understandable graphic display. Messages can be filtered on the basis of criteria derived from the needs of the end-user (e.g. location, direction, route). This filtering aims at enabling the user to receive only relevant information, selected from all the messages that are available on an RDS-TMC service, at any given time.

4.5.3 Open data architecture in RDS

The TMC service is but one form of data that RDS is able to transmit using its Open Data Applications (ODA) feature. This was developed to allow data applications, not previously envisaged to be conveyed in an RDS data stream, with minimal difficulty. The ODA feature, specified in the RDS standard EN 50067:1998, permits a number of pre-determined 'RDS Groups' to be used and these have to be indicated in any RDS transmission using the ODA-application identification (AID) feature, to allow decoders to monitor for them and then decode the relevant data. A wide choice of groups are available for use by ODA data services and these may be selected by transmission operators at their convenience. The service is identified in the associated RDS type 3A groups, in the same RDS data stream.

RDS is divided into 16 Group Types divided into A and B, each of which are 104 bits wide of which 64 bits carry data (the rest is used for error correction, etc.). RDS-TMC standards use a predefined ODA which had been agreed for simplicity of decoder design, in advance of the service introductions. Thus, the ALERT C application uses type 8A groups and is so signalled in the type 3A group.

4.5.4 The RDS-TMC ALERT C protocol

The RDS-TMC ALERT C protocol is standardised in ISO EN 14819-1:2020.

A single group message is made up of 37 bits from the type 8A group. It is possible to have messages made up of more than one group – a multigroup. The presence of a multigroup message is signalled in the first group.

A single group 37-bit message is made up of the following parts:

Location	16 bits
Event	11 bits
Extent	3 bits
Direction	1 bit
Duration	3 bits
Diversion advice	1 bit
F (0 = Single group message)	1 bit
T (0 = User message)	1 bit
TOTAL	37 bits
If F = 1, then a multi group message is indicated.	

Multi-group messages are sequences of between two and five type 8A groups which constitute a detailed TMC message. The presence of a multigroup message is signalled in the first 8A group. A continuity group is provided in subsequent 8A groups which is incremented for each 8A group of the multigroup messages.

Multigroup messages can contain:

- The length of the route affected;
- Speed limit advice;
- Additional more detailed quantifies;
- Supplementary information codes;
- Explicit start time;
- Explicit stop times;
- Additional events;
- Detailed diversion instructions;
- Destination;
- Precise location references;
- Cross linkage to a source of the problem on a different route.

4.5.5 Location referencing

Most RDS-TMC messages provide information about a location (e.g. a stretch of road, an intersection, a region) and they refer to it by using a location reference. This is an identifier that can be interpreted without ambiguity by the receiving system. In RDS-TMC, locations are pre-defined and pre-coded, and the codes are stored in location code tables. The maximum number of codes in one table is determined by the field length for location codes in RDS-TMC, i.e. 16 bit, which corresponds to 65,536 possible codes.

One disadvantage of these tables is that they need to be created and maintained. Also, the receiving system must use exactly the same location code table as the one used for encoding of the message – otherwise it cannot be interpreted.

The rules for location reference coding are specified in ISO EN 14819-3:2020.

RDS-TMC uses a hierarchical structure of pre-defined locations. A system of pointers provides upwards references to higher level locations containing the specified location. For example, Kent would have an upward area reference to South-East England which will be upwards referenced to the United Kingdom, then the British Isles, then Europe.

Junction 25 on the M1 motorway in the United Kingdom would have a section of route referenced to a motorway segment, e.g. Leicester – Sheffield. This segment will then be referenced upwards to the whole road, i.e. the motorway M1.

4.5.6 Primary and secondary locations

The primary location reference is the nearest downstream location in the direction of travel to the event. The secondary location is indicated in terms of extent, i.e. the number of steps back along the road, through other pre-defined locations.

In many cases, events affecting road traffic cover a number of locations, such as where accidents result in long tailbacks. The ALERT-C protocol defines such occurrences by addressing the location of the incident as the primary location, then identifying the end of the tailback by using the direction and extent fields. These fields consist of 4 bits in total, one direction bit and three extent bits. The direction bit indicates the queue growth and not the direction of traffic flow. The extent bit identifies the number of locations along the road that are affected by the problem with a maximum of 8 (primary location and seven related locations). An extent of 1 would identify the secondary location (the end of the event's extent) as being the next location along the same road from the primary location. An extent of 3 would force the receiver to search the database for the third location along the same road from the primary location.

4.5.7 The event tables

The ALERT-C protocol event list is standardised in ISO EN 14819-2:2020. These event phrases total over 1,400 from a possible 2,048 events, and these are divided into 31 update classes for coding and message management convenience. The Event list in this standard is written in so-called CEN-English to allow it to be translated by knowledgeable traffic engineering interpreters into user-friendly phrases in the languages of Europe.

4.5.8 Duration

Event duration can be one of two types. 'Dynamic' – lasting only a short time or 'Longer lasting'; each type of duration can have a forecast element. Each and every event description has a flag determining whether it is dynamic or longer lasting. Duration for each of the two types can be one of 8 values. A value of 0 indicates that 'no explicit duration' is given.

Dynamic durations start at '15 min' through to 'the rest of the day'.

Longer term durations start at 'for the next few hours' through to 'for a long period'.

4.5.9 Diversion advice

The diversion flag determines whether a receiver should recommend a diversion or not.

4.5.10 Message management

Message management in RDS-TMC is complex because it is a uni-directional transmission. The service provider cannot know what messages each receiver has it its memory so cannot manage it. The service provider can send out a 'cancel' message for a particular message or location (or all messages at all locations), but the service provider will not know whether the receiver is turned on or in its service area. This problem is overcome by the receiver deleting messages when their currency is over; this is determined by the duration value. Messages can be updated by sending another message at the same location with the same direction and from the same message class.

4.5.11 Implementations

At the time of writing, it is estimated that there are RDS-TMC services in more than 50 countries over 4 continents. Services can either be publicly funded or be offered by private operators. In the case of private operators, the funding follows a 'Business to Business' model where the system provider takes a small fee for each customer unit from either equipment suppliers or car manufacturers.

The original intention of RDS-TMC was to replace voice traffic broadcasting by using specially adapted standard (DIN)-sized car radios to continuously give traffic information by synthesised voice on the road or route selected by the driver. It was not envisaged in the early 1990s that in-car entertainment systems would include many other features and that full colour map-based navigation systems would be in widespread use. Now, almost all TMC messages are displayed on navigation systems and in many cases the recommended route is altered due to ALERT C messages.

With the advance of social media and crowd sourced data/information through the use of smartphones, RDS-TMC is likely to become less attractive to drivers although still favoured by public authorities and vehicle manufacturers due to its language independence and stable standards; however the crowd sourced applications need to gain more reliability before they take over from RDS-TMC.

4.6 TPEG

4.6.1 Background

In 1997, the European Broadcasting Union (EBU) invited experts from the national broadcasters, transmission operators, consumer electronics manufacturers and transport authorities to come together in an EBU project named B/TPEG [14] to develop a new protocol for TTI that would be suited for a multimedia broadcasting environment using high speed broadcast bearers. It was planned that applications, service and transmission transport features would be developed which would enable travel-related messages to be coded, decoded, filtered and understood both by humans (visually and/or audibly) and by agent systems.

At that time, RDS-TMC was beginning to be implemented but the EBU saw a number of shortcomings. Firstly, the low data rate of 104 bits/sec meant that only two messages could be broadcast every second. Secondly, the pre-coded predefined location referencing had limitations: there was a limit to the number of locations (only junctions could be coded) and the location table in the receiver had to be synchronised with the location table at the service provider. Thirdly, the fixed and rigid event table was not suitable for analysis by an intelligent agent in the vehicle. The EBU therefore wanted a faster and more flexible system with an 'on-the fly' method of location coding.

4.6.2 TPEG binary

Beginning in 1997, the B/TPEG group met for two days, once per month, to develop a new protocol. The first protocols coded in binary for efficiency.

TPEG was seen as providing a pipeline for data delivery such that users need not be aware of the mechanisms and communication technologies (e.g. DAB, Internet) involved; the Transport Layer in the OSI 7-layer conceptual model of communications [15].

The TPEG Specifications comprised a number of 'Parts', defining the mechanisms that permit Service Providers to operate services which can use one or more delivery technologies (e.g. DAB, Internet) from one message generation process. They allow a range of receiver types to be used simultaneously, ranging from sophisticated agent receivers serving navigation systems, through to simple receivers (e.g. a personal digital assistant plug-in receiver/decoder card) only able to decode 'top level' information.

Further information can be found in ISO/CEN TS 18234 parts 1, 2 and 3, which comprise introduction, numbering and versions, syntax, semantics and framing and service and network information, respectively.

4.6.2.1 TPEG applications

The first priority for the Project Group was to develop an end-user-oriented application for road traffic messages, together with the core protocol, network and services layers. The road traffic messages (RTM) application was based on the DATEX I Data Dictionary to ensure that much of the prior knowledge regarding RDS-TMC and such services were reflected in this application.

The TPEG specifications were designed to allow an existing service provider (e.g. of RDS-TMC), to migrate towards multimedia delivery. By employing the

TPEG Specifications, they could deliver several services through a single message generation process, and be able to offer services for significantly different end-user situations. One of the key objectives is to free the end-user from the need to have a location database, on a smart card or CD-ROM, before using a service.

RTM was the first TPEG application to be developed and was standardised in CEN and ISO as TS 18234-4 for the binary version. Messages using RTM were designed in a hierarchical fashion so that they provided a detailed view that could be reduced down for simple receivers and still have the same meaning. For instance, a message could be *Road Traffic Collision* between *a double decker green service bus* and a *small yellow diesel hatchback car*. A complex receiver could analyse and present that, but a simpler receiver might reduce it to *Road Traffic Collision between a bus and a car* and a very simple receiver could just display *Road Traffic Collision.*

Early experience with TPEG-RTM showed that for initial TPEG services, RTM was too complex for the significant technology that would be needed to implement it; hence a TPEG 'lite' application (TPEG-TEC) based on the RDS-TMC event list was developed.

Further TPEG applications were developed; however, not all were published in CEN and ISO due to issues with the numbers of countries prepared to support the developments. Some of those more relevant to TTI are briefly described below.

TS 18234 Part 5: Public transport information
This standard specifies an application which describes public transport information which is intended to cover all modes of public (i.e. collective) transport as well as inter-urban and intra-urban travel. The application is designed to allow efficient and language independent delivery of public transport information directly from service provider to end-users.

TS 18234 Part 7: Parking information application
This standard specifies an application to deliver parking information to a variety of receivers using a number of different channels, foremost digital broadcasting and internet technologies. Parking information may be presented to the user in many different ways including textually, voiced and graphically using standard formats.

TS 18234 Part 8: Congestion and travel time information application (ISO only)
This standard specifies an application to provide information in relation to events and some status information on the road network and on associated infrastructure affecting a road journey. For example, limited information about abnormal operation of links in the network may be included, e.g. ferries, lifting bridges.

TS 18234 Part 9: Traffic event compact
This standard specifies the application traffic event compact (TEC). It has been specifically designed to support information about traffic events, e.g. road works, traffic jams. A specific form of traffic events are local hazard warnings which, as safety-related messages, are sent with high priority to assist a driver in encountering dangerous situations (e.g. black-ice, accident behind curves, obstacles on road).

TS18234 Part 10: Conditional access information
This standard contains the definition of the TPEG conditional access information (CAI) application. It enables dedicated conditional access data, such as management

messages (e.g. control words and entitlement control messages) to be delivered to recipient client devices. This TPEG application is designed for a service provider to establish set-up, prolongation or revocation of services to a specific client device, using a limited capacity unidirectional broadcast channel and without recourse to service–client handshaking. This TPEG application defines the logical channel for the transmission of the additional CA information (CAI) and how the CAI is linked and synchronised to the scrambled content.

TS 18234 Part 11: Location referencing container
This standard establishes the method of signalling the specific location referencing used by all TPEG applications that require detailed location information to be delivered to client devices such as TPEG-RTM, TPEG-PTI, TPEG-TEC or TPEG-PKI. The TPEG-location referencing container (TPEG-LRC) is described as well as how it is used to signal which specific location referencing method is in use for a particular TPEG Message. It is able to handle location referencing methods that are external to the present ISO series and the internal location referencing method (TPEG1-LOC) defined in Part 6 of this series.

It should be noted that system developers are not encouraged to use these standards, rather to use the more recent TPEG2 standards described later.

4.6.3 TPEG XML versions

During development of the binary versions of the standards, coding of applications in XML was beginning to be developed and it was agreed that all the binary applications should also be developed using XML. Hence, the development of the ISO/CEN 24530 series. Again, due to synchronisation issues between CEN and ISO, not all parts were published in both standardisation bodies. Additionally, synchronisation of part numbers were not carried through; for instance, RTM which was Part 4 in the binary version became Part 3 in the ML version and so on.
TS 24530 Part 1: Introduction, common data types.

This standard describes the TPEG application and the tpegML as the top-level 'containers' for TPEG messages in XML and the common data types that are used by tpegML applications (e.g. tpeg-ptiML). Inherently, tpegML is designed to 'map' the TPEG binary (ISO/TS 18234 series). However, additional tags are provided to create a message and message set structure to facilitate internet file delivery.

4.6.3.1 TPEGML applications

TS 24530 Part 3
TPEG-RTMML established the XML encoding of the method of the RTM application.

TS24530 Part 4

This established the XML encoding of the method of the Public Transport Information application. is intended to cover all modes of public (i.e. collective) transport as well as inter-urban and intra-urban travel. The application itself was designed to allow the efficient and language-independent transmission of public transport information either directly to an end-user, be it the public or another service provider, such as broadcasters, service operators or other information disseminating points, or to centres for onward transmission.

There were other parts of the TPEG-ML developed but they were never taken forward to the publication stage. It should be noted that system developers are not encouraged to use these standards, rather to use the more recent TPEG2 standards described later.

4.6.4 Location referencing

One of the main reasons for developing TPEG was to move the technology beyond the pre-coded location tables employed by RDS-TMC towards describing actual event locations. An 'on the fly' method was required that had greater locational resolution (not just junctions) and did not require synchronisation between the receiver and the service provider.

TPEG-LOC was developed to overcome these issues. TPEG-LOC is a combination of WGS84 location referencing plus a short description of the location. The short description is needed to confirm directionality and also to differentiate between roads in very close proximity. It is not that WG84 is inaccurate, rather that different maps vary in relation to an absolute WGS84 point. Hence, if the service provider is using a map from one provider which is different to the map in the receiver, then errors can occur.

During development of the original TPEG, a number of other location referencing methodologies were proposed leading to the concept of a location referencing container in ISO/CEN TS 18234 part 11 described earlier.

4.6.5 TPEG2

4.6.5.1 Background

The problems of synchronising standards in separate series for binary and for XML versions became a real burden for the development teams. This led to the decision in the mid-2000s to redefine all the applications using the UML [9] in a development tool *Enterprise Architect* software suite and translate from the UML to a binary and a XML implementation; this decision notable allowed non-coding experts, but more particularly TTI experts to help develop the applications without coding expertise constraints. Additionally, it was agreed that TPEG2 would only be standardised by ISO to reduce the ISO/CEN synchronisation issues. Each standard now has a UML part and then a translation into the binary and XML parts.

The standardisation is in the ISO 21219 series. Currently there are 20 TPEG2 parts published, or close to being published', as TS or full ISO Standards. It is the eventual aim that all TPEG 2 standards become International standards.

The Traveller Information Services Association (TISA) [16], a market-driven membership association with worldwide scope, was established as a non-profit company and now coordinate TPEG2 activities. The organisation is focussed on proactive implementation of TTI services and products based on existing standards, including primarily RDS-TMC and TPEG technologies.

4.6.5.2 TPEG 2 toolkit

Although there are 26 parts of TPEG2 declared, in reality a number will never be published as their contents are covered in other standards such as ISO 17572-2 and

3. The first five parts of the ISO 21219 standards provide definitions and an essential toolkit for the development of TPEG 2 applications.

The first non-TTI application parts are introduced below:

Part 1: Introduction, numbering and versions.
(TPEG2-INV) defines an index to the complete set of TPEG Generation 2 toolkit components and applications. New applications are enumerated with an AID as they are added to the TPEG applications family.

This standard will be updated when such developments occur, to indicate the latest status and the inter-working of the various TPEG specifications. It will be issued as a new editorial version every time a new issue of any other specification is issued. Preliminary AIDs are allocated and managed by TISA and are listed on the TISA homepage [16].

Part 2: UML modelling rules: TPEG2-UMR
This standard specifies rules for the creation and extending of TPEG application UML models. The rules are intended to ensure that TPEG application UML models can be interpreted unambiguously for conversion to physical format representations. TPEG application UML models that are defined according to these rules can be used for automatic generation of TPEG standards and for automatic generation of TPEG application physical format descriptions. This document also specifies the preferred structure of TPEG application specifications. The TPEG abstract data types and the set of TPEG tables of common use are specified in the annexes.

Part 3: UML to binary conversion rules (TPEG2-UBCR)
TPEG applications are modelled in UML to provide an application description that is independent of a physical format representation. By separating semantics from application description, applications can easily be developed at a functional level. Different physical format representations can be generated following a well-defined set of rules on how to convert UML classes to different physical formats.

This standard specifies the rules for converting UML models of TPEG application to the TPEG binary format. It contains the binary format definition of the abstract data types defined in ISO 21219-2. Rules for converting compound data types are also defined.

Part 4: UML-to-XML conversion rules
This standard specifies the rules for converting TPEG application UML models to the tpegML format description. It contains the XML format definition of the abstract data types defined in ISO 21219-2. Rules for converting compound data types are also defined.

4.6.5.3 TPEG 2 ancillary standards

Part 5: Service framework (TPEG2-SFW)
This standard establishes a method of conveying data for a wide range of applications that require the efficient transmission of point to multi-point data over potentially unreliable broadcast channels. It is also suitable for point-to-point and multicast applications and may easily be encapsulated in Internet Protocol.

This standard describes the basic capabilities of the second generation TPEG (TPEG2) for providing a multiplex of TPEG Services and applications. Together with the definitions of the general TPEG UML modelling rules and the particular physical TPEG representations for TPEG-binary streams (TISA: TPEG UML Conversion Rules) and tpegML files (TISA Specification: TPEG UML Conversion Rules), it replaces the former documents TPEG-INV and TPEG-SSF.

Part 6: Message management container (TPEG2-MMC)
This standard adds a basic toolkit definition to the ISO 21219 series specifying the message management container (MMC), which is used by all TPEG applications to provide information about the handling of messages on the TPEG client side. The MMC holds administrative information allowing a decoder to handle the message appropriately. This information is not aimed at the end-user. The MMC is a toolkit and not a stand-alone application but is included by TPEG applications.

Part 7: Location referencing container (TPEG2-LRC)
This standard establishes the method of signalling the specific location referencing used by all TPEG2 applications that require detailed location information to be delivered to client devices such as TPEG2-TEC. The TPEG2-location referencing container (TPEG2-LRC) is described and shows how it is used to signal which specific location referencing method is in use for a particular TPEG message. It is able to handle location referencing methods that are external to the present ISO series and the internal location referencing methods defined as parts of this series.

Part 9: Service and network information (TPEG2-SNI)
This standard establishes the method of delivering service and network information within a TPEG service. The TPEG-SNI application is designed to allow the efficient and language-independent delivery of information about the availability of the same service on another bearer channel or similar service data from another service provider, directly from service provider to end-users.

A number of tables of information are described, which provide comprehensive options for describing services, their timing, content, geographical coverage, etc. In all TPEG streams, it is mandatory to deliver to so-called GST (guide-to-service tables). Additionally, it is possible to signal the linkage of content between different bearers and services.

Part 10: Conditional access information (TPEG2-CAI)
This standard defines the TPEG conditional access information (CAI) application. It allows protection of the content of a TPEG service from unauthorised access. It further supports the management of subscriber information (e.g. Control Words and entitlement control messages (ECM)) on the client devices in order to set-up, prolong or revoke a subscription on a given client device.

The application defines

- the logical channel, for the transmission of the additional CAI;
- how the CAI is linked and synchronised to the scrambled content.

The basic concept behind the CAI application is to transport CAI in separate TPEG service components of a dedicated application type and to define an SNI table (see above) that contains the link between the scrambled content and the related CAI.

Part 24: Light encryption (TPEG2-LTE)
This standard defines the LighT Encryption (LTE) mechanism for TPEG Service Data. The objective is to provide a simple to use, yet effective conditional access mechanism for TPEG including encryption for use with both broadcast and/or point-to-point delivery.

It has been specifically designed for use in business-to-business (B2B) applications where, for both service providers and device manufacturers, a standardised conditional access mechanism avoids a proliferation of proprietary methods with multiplied implementation effort and lead times.

4.6.5.4 TPEG 2 applications

The 'application' parts of TPEG2 which provide the TTI to the end-user are:

Part 14: Parking information application (TPEG2-PKI)
This standard specifies the TPEG Parking Information application which has been designed to deliver parking information to a variety of receivers using a number of different channels, foremost being digital broadcasting and Internet services. Parking information may be presented to the user in many different ways, including text, voice, and graphics.

Some traffic congestion, particularly in urban areas, is attributed to drivers searching for parking spaces. Therefore, timely provision of parking information could help ease this situation. Furthermore, parking information would be valuable for visitors, particularly when it could be used to signal where a temporary parking facility is established for a special occasion.

Part 15: Traffic event compact (TPEG2-TEC)
This standard specifies the TPEG application TEC. The TEC application has been designed to support information about traffic events (e.g. road works, traffic jams). Critically, it mimics RDS-TMC event-codes so has a compact or small capability rather than a full encompassing event capability of a stand-alone TPEG application. A specific form of traffic event is local hazard warning which, being a safety-related message, is high priority to warn a driver that they may encounter dangerous situations (e.g. black-ice, accident beyond curves, obstacles on road, etc.).

Generally, the TEC application is designed to allow receivers to:

- ensure travel safety for the driver;
- enable the calculation of alternative routes;
- avoid delays (e.g. traffic jams);
- warn the driver of obstructions on route;
- provide the driver with information on infrastructural problems (e.g. closed fuel stations, non-functioning emergency telephones).

Part 16: Fuel price information and availability (TPEG2-FPI)

ISO/TS 21219-16:2016 specifies the TPEG application Fuel Price Information and availability (FPI). The FPI application has been designed to support information on fuel stations, their location, fuel types offered and fuel pricing and availability.

Part 18: Traffic flow and prediction application (TPEG2-TFP)
This document specifies the TPEG application traffic flow and prediction (TFP). It has been designed to provide information to a variety of receivers using different channels, including in the first instance via digital broadcasting and Internet. Traffic flow and prediction messages are intended for in-car applications and can also be presented directly to the user by textual, voice and graphical output devices.

TFP is status oriented, i.e. the transmitted information continuously updates the receiver's knowledge for a dedicated road network. In particular, the traffic states are delivered any time and for all road sections of the network, even when there are no abnormal traffic situations.

TFP aims to fulfil the following functions:

- provides dynamic navigation systems with up-to-date traffic state information;
- ensures travel safety for the driver;
- enables the calculation of alternative routes;
- avoids delays (e.g. traffic jams);
- lowers traffic load on over-saturated parts of the network;
- keeps the driver informed about current and upcoming traffic;
- provides compact and efficient coding of the traffic information.

Part 19: Weather information (TPEG2-WEA)
This standard defines the TPEG Weather (WEA) application for reporting weather information. It provides general weather-related information to all travellers and is not limited to a specific mode of transportation. The WEA application provides weather-related forecasts and status information over multiple time periods and for multiple, possibly linked, geographical areas. As with several other applications, presentation of the information is dependent on the specific capabilities of the receiving device. This standard therefore, does not define any prerequisites for the human–machine interaction of the device.

This application does not provide specific weather-related safety warnings to drivers; these are provided as safety-related messages as part of the TPEG2-TEC application.

Part 21: Geographic location referencing (TPEG-GLR)
This standard defines a method of using geographic location referencing (GLR) that can be used by relevant TPEG applications. The GLR type is used for defining geographic location references (points, polylines, and geographical areas). The GLR method is intended to be one of the methods that can be transported inside a TPEG-location referencing container (TPEG-LRC) for those TPEG applications providing information (e.g. weather) for primarily geographical locations.

The GLR specification is made basic and compact on purpose, such that it can also be employed advantageously in non-navigation devices for simple TPEG services such as weather information, safety alerts, etc. As such, the GLR location referencing method is intended to be complementary to map-related location referencing methods, where the focus is on referencing of man-made artefacts such as roads and highways. The scope of GLR is limited to geographic locations on the Earth's surface.

Part 22: OpenLR location referencing (TPEG2-OLR)
This standard specifies the logical data format of OpenLR™ location references and general requirements of the method. The toolkit is intended to be used in the TPEG location referencing container (TPEG2-LRC).

OpenLR™ has been designed for transferring traffic information from a centre to in-vehicle systems, built-in or used as an add-on (e.g. a personal navigation device or smart phone). The information transferred can consist of the current traffic situation at a certain location, a traffic forecast or special alerts. The corresponding locations are roads, a list of connected roads, points of interest, or areas.

In order to transmit location information from sender to receiver, the OpenLR™ method defines rules for generating map-independent location references; that is, the actual location references are generated dynamically and do not require use of pre-defined location references.

Part 23: Roads and multimodal routes (TPEG2-RMR)
This standard describes new mobility services like car sharing, car rental or park and ride, as well as the integration of different transport modes by multimodal or off-board navigation which are gaining increasing importance. Cooperative management of the transport infrastructure requires precise information and guidance on dedicated routes from a central knowledge base to a traveller's mobile device.

Such 'use cases' are addressed by the TPEG application defined in the document. The road and multimodal routes (RMR) application enables the service provision for road routes as well as multimodal routes including more than one transport mode and parking. For example, an optimal multimodal route may include a drive by car to a train station with parking facility, a train connection to a station nearby the destination and a local public transport ride from the train station to the traveller's destination.

The standardised delivery, via TPEG, of routing information has some potential benefits for the users of an RMR service, for instance:

- Enabling specialised routing services like scenic routing or eco routing;
- The best use of the overall transport network, i.e. not only the road network;
- Cost and time savings to travellers;
- Harmonisation of in-car navigation and traffic management, e.g. routing advices by variable message signs;
- Personalised service, i.e. information services considering the specific characteristics of a user.

Part 25: Electromobility charging infrastructure (TPEG2-EMI)
This standard defines the TPEG application ElectroMobility charging Infrastructure (EMI). It has been designed to support information about charging infrastructure for electric vehicles (not just cars). It includes the location of e-charging points and their suitability for different vehicle (e.g. connector type, charging modality). As electric vehicles typically occupy a 'charging space' for a substantial period of time, information on availability/waiting time and reservation options are also included so users may optimally plan their route/trip.

Standardised delivery, through TPEG, of information on charging infrastructures has the following benefits to an end-user of the service:

- Identifying suitable charging units, thus preventing unnecessary driving to find a compatible unit (which also has environmental benefits);
- Verifying the real-time availability of charging units;
- Being able to plan ahead and reserve a spot in a charging park and thus optimise the planning of a trip;
- Being able to select a financially attractive charging point in a charging park, the operator of which has billing agreements with the user's electromobility provider.

In addition to these end-user benefits, electromobility providers and charging park operators benefit from a standardised TPEG format as it allows an easier harmonisation of information between the management systems of electromobility providers and charge park operators and other third parties according specifications (e.g. Open Charge Alliance, e-Mobility ICT Interoperability Innovation, eMI3, etc.).

This TPEG application can work with other TPEG services to support a large numbers of charge parks with only modest additional bandwidth requirements.

Part 26: Vigilance location information (TPEG2-VLI)
This standard defines the application for Vigilance Location Information (VLI). Vigilance messages are intended for in-car applications to inform drivers when they should pay extra attention to their driving behaviour because of dangerous road stretches, traffic enforcement cameras or other locations that require increased driver vigilance. The warnings can be presented visually, audibly, or with the spoken voice, or as a combination of all three.

The objective of the service is to allows drivers to be more relaxed, in the knowledge that they will be warned when necessary. The situations where a vigilance message makes sense can be very different. For example, speed cameras are usually placed in areas where vigilance is required; the information about those locations promote safe driving and also more safety for other road users and outside traffic participants. Another example of areas requiring high driver attention are roads close by a school.

The information can be categorised into fixed or mobile:

Fixed or mobile locations:

- Fixed locations are relatively static such as an accident black-spot; and
- Mobile locations are transient in nature, such as the presence of a mobile speed camera.

Spot locations or zones:

- Spot locations refer to single points on a road network, with an indication of which direction of traffic is affected by the vigilance information; and,
- Zones refer to stretches of road network which represent a continuous area of warning affecting only one traffic direction.

The local regulations regarding the signalling of speed measurement systems, e.g. fixed speed cameras, or mobile speed radar locations can vary depending on the country or region. The signalling of speed measurement systems is encouraged by the local authorities in certain jurisdictions whereas it can be punishable by law in others.

4.6.6 Commercial exploitation

As at early 2020, TPEG2 is being implemented in 52 countries worldwide. The TISA organisation [16] is the project manager (along with ISO/TC204) for the development of the standard series and provides implementation support to its members.

Information on what TPEG applications are commercially provided is difficult to obtain, but as the TPEG applications are developed voluntarily by industry members of TISA it can be expected that there is a market for the applications.

According to some sources, at the time of writing there are 11 broadcast TPEG2-TEC, TPEG2-TFP and TPEG2-WEA services across 9 countries with a number of other in advanced trials. However, using mobile broadband (4G) carriers there are over 90 TPEG2-TEC, TPEG2-TFP and TPEG2-WEA services across 52 countries.

Currently there is no evidence that the more specialist TPEG2 services, like fuel price information, are available.

4.7 Outlook

Systems within vehicles are becoming more integrated, with data on the vehicles' position and status and the conditions surrounding the vehicle being automatically shared with the infrastructure and other travellers – this is called Cooperative Intelligent Transport Systems (C-ITS). In the future, TTI delivery to users may become part of this broader environment of shared information. Additionally, the sharing of information using applications, such as Waze, on smartphones to both collect and display traffic information is expected to increase. Thus, existing sources of TTI including Sat-Navs and broadcast radio may become less important over time.

The collation, exchange and transmission of traffic information will remain extremely important to support various business models and will involve both public and private organisations. The future is likely to see stand-alone systems becoming more integrated with developments involving social media and C-ITS services, but still using the backbones and message sets of DATEX 2, ALERT C and TPEG.

References

[1] UK Government. (n.d.). Retrieved from https://www.legislation.gov.uk/uksi/1989/1843/contents/made

[2] Wikipedia. (n.d.). https://en.wikipedia.org/wiki/Ceefax

[3] UNECE. (n.d.). *Introcuding UN/EDIFACT*. Retrieved from https://www.unece.org/cefact/edifact/welcome.html

[4] Hoffman, S. (1993). PLEIADES – Using transport technology to link Britain and France. *Vehicle Navigation and Information Systems (VNIS)* (pp. 92–95). Ottawa. Retrieved from https://ieeexplore.ieee.org/document/585592

[5] European Commission. (n.d.). *CENTRICO*. Retrieved from https://trimis.ec.europa.eu/project/central-european-region-transport-telematics-implementation-project
[6] European Commission. (n.d.). *Streetwise*. Retrieved from https://trimis.ec.europa.eu/project/seamless-travel-environment-efficient-transport-western-isles-europe
[7] European ITS Platform. (n.d.). *Easyway i-ii*. Retrieved from https://www.its-platform.eu/content/easyway-i-ii
[8] European Commission. (n.d.). *Datex2*. Retrieved from https://www.datex2.eu/
[9] OMG. (n.d.). *UML*. Retrieved from www.uml.org
[10] W3C. (n.d.). *XML Core Working Group Public Page*. Retrieved from https://www.w3.org/XML/Core/
[11] Open LR. (n.d.). Retrieved from http://www.openlr.org/
[12] DATEX 2 Forum. Retrieved from www.Datex2.eu
[13] European Commission. (n.d.). *RDS-ALERT*. Retrieved from https://cordis.europa.eu/project/id/V1029
[14] Wikipedia. (n.d.). *TPEG*. Retrieved from https://en.wikipedia.org/wiki/TPEG
[15] Wikipedia. (n.d.). *OSI*. Retrieved from https://en.wikipedia.org/wiki/OSI_model
[16] TISA. (n.d.). Retrieved from TISA: www.tisa.org

Chapter 5

European developments in traffic and travel information

Paul Burton[1], Bev Marks[2] and Danny Woolard[3]

5.1 Introduction

It is not possible to consider travel and traffic information (TTI) in the United Kingdom without taking into account the significant contribution of European research, development and implementation trials. Hence, this chapter is devoted to the story from individual European countries developing their own solutions to providing traffic and traveller information (TTI) to the realisation that to make the market viable, there needs to be Europe-wide solutions.

The start of the Europe-wide cooperation began in 1987 with the 2nd Framework (FP2) Research and Development DRIVE project, although there is evidence of cooperation between individual neighbouring countries on the European mainland, for instance, the Netherlands and Germany.

This chapter mainly deals with broadcast and distributed TTI; the development of other areas of TTI including variable-message signs (VMS) is discussed in other chapters of this book.

5.2 Background and context

5.2.1 The European Broadcasting Union

The European Broadcasting Union (EBU) [1] had been involved in the broadcasting of traffic content since the early 1980s; the aim was to give the European motorist better, relevant and timely spoken traffic information over the air. It promoted two initiatives: the *Broadcasts for Motorists Group* and the *RDS Forum* [2]. The *Broadcasts for Motorists Group* developed a message set (translated into many European languages) intended to be the basis of spoken traffic broadcasts. The *RDS Forum* promoted the development of the radio data

[1]CEN/TC278 WG4 and ISO/TC204 WG10 – Traffic and Traveller Information, Worcestershire, UK
[2]TISA (Traveller Information Services Association), Brussels, Belgium
[3]Chiltech Limited, Devon, UK

system (RDS) data stream on VHF/FM broadcasts and its associated facilities; this brought together transmission operators, broadcasters and equipment manufacturers to ensure a seamless service across Europe. Standards were developed (EN 50067) [3] that allowed for:

- Alternative frequency (AF) flag that gives a list of the frequencies that the network used in the area and its surroundings. This allows the receiver to retune easily and swiftly to a stronger signal on the same station.
- Traffic programme (TP) flag that signals which stations are likely to carry traffic information announcements. That emulated the earlier ARI signalling (see below) developed and used in German only.
- Traffic announcement (TA) flag which signals that the tuned station is currently broadcasting a traffic announcement; this allows an in-vehicle receiver to either un-mute the receiver or switch from an alternative audio source (such as a music player) to play the announcement.
- Extended other networks that signals which other stations are linked (from the same broadcaster) to the present station and hence enables access to traffic announcements on the linked stations. For example, a receiver tuned to BBC Radio 4 could automatically tune to BBC Hereford and Worcester to pick up a spoken traffic broadcast and then return to BBC Radio 4 afterwards.
- Traffic message channel (TMC) which supports digitally encoded traffic information. Digital coding is described in another chapter of this book.
- A number of other flags – CT (clock time and date), PI (programme identification), PS (programme service name), PTY (programme type) and RT (radio text and radio text plus).

5.2.2 Germany

In the 1990s, German broadcasters (led by their research institute, IRT) and a receiver manufacturer (Blaupunkt) developed the ARI (Autofahrer-Rundfunk-Information – driver radio information system). ARI signalled the presence of traffic information in FM broadcasts used by the German public network of FM radio stations using a 57 kHz subcarrier added to a station's FM signal. ARI was the forerunner of the RDS TA and TP flags and was replaced by RDS in 2005.

5.2.3 France

In the 1990s, France developed the Carminat system (part of the EU Eureka programme) in association with the then Philips Car Systems to develop a navigation and information programme which developed into the Renault and Tom Tom navigation and information system.

5.2.4 The United Kingdom

In the late 1970s, the BBC developed the world's first television teletext system called CEEFAX which sporadically broadcast traffic and travel information (TTI). The service started in 1974 and ended in 2012.

The BBC then developed the CARFAX system [4] using medium-frequency signalling in the 1970s and 1980s. CARFAX interrupted radio signals to switch equipped receivers to receive spoken messages. CARFAX was similar to ARI which was again superseded by RDS. CARFAX was never implemented beyond large area trials.

So, it can be seen that these examples that individual country's initiatives were all leading to European initiatives.

5.3 The European programmes

5.3.1 DRIVE research and development programme (1988–1991)

The original DRIVE Project (later known as DRIVE1) was part of the European Commission's 2nd Framework Programme (FP2). It came under the then DG XIII and was at the time called Road Transport and Traffic Telematics (RTTT). Many projects were launched under this programme, some horizontal such as CORD which was a coordination activity across all the other projects, but mainly vertical tasks such as RDS-TMC (described below) including VMS (VMS projects (described elsewhere).

5.3.1.1 RDS-ALERT

The RDS-ALERT project (RDS Advice and problem Location for European Road Traffic) [5]was part of the Second Framework Project (FP2) DRIVE 1 programme. It started in 1987 and completed in 1991. It was led by Castle Rock Consultancy (UK) and other participants were the BBC (UK), TRRL (UK), CCETT (FR), Philips Car Systems (NL) and Robert Bosch (DE).

The technical approach comprised:

- identification of traffic-related message content together with associated structures, protocol and message management techniques
- the carrying out of mobile RDS reception tests to gather data on error-statistics, especially in areas of severe multipath interference
- the computer evaluation of the error-statistics to establish an empirical model
- the establishment of a standard interface specification for the transfer of TMC data to other traffic-related technologies.

The project was subdivided into the following work packages:

- review and evaluate progress to date in achieving consensus
- define and establish liaison group structure
- liaise with interested parties
- develop, discuss and finalise location codes
- review message repertoire and propose input methods for broadcasts
- define assessment criteria and scenarios
- develop, test and evaluate software message decoding

- develop, test and evaluate message management strategies
- finalise the technical appraisal and optimise protocols
- prepare Interim Standards for RDS-TMC throughout Europe
- consensus building and revisions to Interim Standard
- prepare a proposal for a European standard.

During the project, among other facilities, two message protocols were progressed: a cause and effect protocol from France (ALERT A) and a straight message set which included cause and effect (ALERT B). At the completion of the project it was agreed that a combination of these two approaches would be used giving the ALERT-C protocol which was eventually standardised and is still in use at the time of writing this book.

The project was successful as a version of ALERT-C was finally produced in 1992 after agreement with of the EBU (Dietmar Kopitz) and European Conference of Ministers of Transport ECMT (Fritz Bolte – BAST). This was the precursor to standardisation.

In addition to ALERT-C, a more complex protocol ALERT Plus, was developed which allowed for more complex messaging. ALERT Plus was forwarded for standardisation in the European Committee for Standardisation (CEN) only but not developed further. It has subsequently been dropped from the CEN Work Programme. The French service provider MediaMobile introduced a service called Visionaute, part of car maker Renault's Carminat program. This was based on the extended version of the ALERT protocol called ALERT Plus and was available on an optional in-car device for the Renault Megane, but only available in the French Ile de France region.

5.3.2 DRIVE2 research and development programme (1992–1994)

Following the success of the original DRIVE programme, DRIVE 2 enjoyed a far greater take up by those organisations involved in what was then called 'RTTI' or 'Road Transport and Traffic Telematics' (which later became subsumed into the broader term 'Intelligent Transport Systems' or 'ITS').

5.3.2.1 ATT-ALERT

The ATT-ALERT project (Advanced Transport Telematics – Advice and problem Location for European Road Traffic) [6]was part of the FP3 DRIVE 2 Programme. The project was led by Castle Rock Consultancy but had many more members than RDS-ALERT. UK participants were Ford, the AA, BT and the BBC. TRL was associate partner.

The overall objectives of ATT-ALERT were to promote and standardise the current ALERT protocol and enhance its capabilities in providing a comprehensive driver information service. The project had two aspects; firstly it aimed to develop essential additions to the existing protocol and pursue standardisation on an international level and secondly, ATT-ALERT looked to a consistent application of the protocol on other media or bearers.

ATT-ALERT comprised three modules. The project aimed to provide key additions to the current ALERT protocol which were considered essential for DRIVE II field trials. The modules:

- studied the pre-standard and through links with other DRIVE II projects, and helped to evaluate field trials of their implementation
- developed an ALERT-COMMS module overseeing the development of protocols for three types of bearer channel: DAB (digital audio broadcasting), AMD (AM data systems) and DRC (digital radio communications systems)
- developed ALERT-OVER, the vital developments of ALERT radio communications protocols which were coordinated in an integrated manner through the provision of an efficient management and coordination structure. This module also provided consistent location coding principles for Europe.

ATT-ALERT did indeed interact with field trials, such as PLEIADES (see below), but as the trials were rather limited it was not possible to make a substantial contribution or to learn much from them.

ATT-ALERT provided a forum for development of the standard.

In terms of extending ALERT-C to other bearers, a series of bearer application protocols (BAPs) and bearer independent protocols (BIPs) were developed. These were never really progressed as there was no appetite to move into the, then new, DAB transmission domain. AM data systems and DRC were never a reality in the transport sector and so they remained solely an intellectual exercise.

5.3.3 The EU Third Framework Programme (FP3) (1992–1995)

As well as research and support projects in the DRIVE 2 programme such as ATT-ALERT, there were a series of field trials. In this context there were MELYSSA, PLEIADES, ARTIS and STORM field trials. One of these, PLEIADES, is described below.

5.3.3.1 PLEIADES

The PLEIADES project [7] considered a London to Paris and Brussels Corridor taking into account English Channel crossings (Eurotunnel opened in the latter stages of the project) and various river crossings.

The PLEIADES consortium was led by Wootton Jeffreys and had UK partners from BT, TRL, Siemens Plessey, Ford, Eurotunnel, the Department for Transport, and the AA.

The intention of the project was to implement and evaluate pilot systems for RDS-TMC, Trip-planning terminals, VHF/FM radio, Paging, Cellular phones, In-car navigation, Automatic traffic surveillance, VMS, Fog monitoring and Automatic vehicle identification. These systems were to operate in conjunction with an enhanced TTI and management service provided by interconnected traffic information centres and traffic control centres, both national and international.

Various technologies were trialled including RDS-TMC, but the size of the fleets and the technologies available (RDS-TMC decoders linked to early Psion Organisers) made meaningful results debateable. It did lay the groundwork for international cooperation soon to become the TMC Forum hosted by the ERTICO organisation in 1997–98 and various advanced pilots such as the RDS-TMC largescale pilot in the United Kingdom.

The swapping of cross boarder information laid the ground for more formal European funded projects such as CENTRICO who became a major player in the development of DATEX.

5.3.4 *The EU 4th Framework Programme (FP4) (1996–1999)*

5.3.4.1 FORCE/ECORTIS

It is not a straightforward task to separate the FORCE 1, FORCE 2, FORCE 3 and ECORTIS projects [8], save that FORCE 1 and FORCE 2 were promoted by the European Commission's DGXIII and FORCE 3 by DG VII. ECORTIS was part of the TEN-T Trans European Network implementation support programme from DG VII. The programme was always referred to as the FORCE/ECORTIS project, but it was never clear where one funding stream stopped and another began. These combined projects were led by the German Ministry of Transport (BMV) and managed by a combination of consultancies from Germany and the Netherlands (Michael Olberding and Maarten Bouwer, respectively) with assistance from David Bowerman at ERTICO.

The main objectives of the FORCE 1, FORCE 2 and FORCE 3 projects of the Telematics Applications Programme and the ECORTIS-TEN-T project were to enable the implementation of RDS-TMC services with European wide functionality, ensuring that those services are continuous, interoperable with any receiver from any manufacturer, achieve agreed quality levels and achieve a common functionality according to agreed definitions by:

- developing common guidelines for implementing and operating services
- encouraging common understanding of the European elements and basic services
- enforcing the co-ordination between service providers and all RDS-TMC actors
- increasing the exchange of traffic information between services
- consolidating standards and specifications
- fixing the framework for European activities.

During the FORCE/ECORTIS programme period there was support and guidance on the national trials taking place. Several audits of each trial were made by at least two independent experts from the project. These audits took two or three days and visited each organisation in the chain from data collection and collation to organising the carousels of messages, routing them to the relevant transmitters and broadcasting them. Overall project management, system design and results evaluation were also considered. In countries where receivers were developed the equipment manufacturers were also visited.

Much was learnt during this period, not just about the application of the technology but about interpretation of the standards and commercial imperatives. When RDS-TMC was conceived it was always considered that TTI would be free at the point of use and that public broadcasters would supply the transmission as part of their public service obligation. In the majority of European countries traffic data collection and collation was a public sector activity and so the concept of RDS-TMC being free at the point of use could be realised. That was certainly not the case in the United Kingdom where traffic information collection was undertaken by the AA, the RAC and Trafficmaster with a small amount from the BBC and the public traffic agencies; this made the concept of traffic information free at the point of use to be a difficult challenge.

The United Kingdom found that there was opposition from FORCE/ECORTIS projects to its concept of commercial services being enabled by the restriction of access to location codes provided by several commercial organisations.

5.3.5 The Trans European Network – Transport Programme (TEN-T) (1996–1999)

5.3.5.1 The UK RDS-TMC Trial

This section on the UK RDS-TMC trial is a good example of European funding really making a difference. It would not have been possible to make such a large-scale trial or to convince European players to engage without access to significant funding and the authority of the European Commission.

There was a degree of frustration following the lack of meaningful results from the third Framework field trials. This arose from the small numbers of receivers involved (due to budgetary constraints) leading to a lack of significant results. At that time, it seemed impossible to resolve the 'chicken and egg' situation – it was not possible to kick start services as there were no receivers and hence no users; and manufacturers could not be persuaded to make receivers available as there was no service.

In 1995, the Department of Transport let a feasibility study to the then Oscar Faber consultancy (led by Andy Graham) to propose a plan for a large-scale pilot where there would be sufficient receivers to get feedback from users and hence the value of a service. The report was available in early 1996 and proposed costs and timescales and possible collaborators.

In 1996, using the report from Oscar Faber, permission was sought from the Secretary of State for Transport, as he was a Signatory to resolutions both in ECMT and the Transport Council that encouraged the implementation of RDS-TMC. RDS-TMC was seen as a first transport telematics information system for Europe. The proposed project was closely related to the FORCE and ECORTIS projects. The Department needed to know whether or not there was a realistic prospect of such services being commercially viable and, if so what conditions need to pertain, for example, limited pump priming capital. The elements to be evaluated were:

- data input
- data collection

- data coding
- message management
- communications
- broadcasting
- receivers
- payment methodology
- user requirements and acceptance
- legal and institutional issues
- interoperability between services in the United Kingdom and the rest of Europe
- international data exchange.

Permission was given to apply for TEN-T funding in 1996. Over the two and a half years of the project the following grants were received; 1995 €450k, 1996 €300k, 1998 €400k, giving a total of €1150k. At the currency conversion rate prevailing at the time (£1 = €1.66), this gave a total of £693k. The TEN-T funding was for 50% of total costs for each non-governmental participant (European finance rules on 'Additionality' prevented any grant to publicly funded bodies.)

Organisations originally signed up to the project were the BBC, the AA, the RAC, the Highways Agency, the Scottish Office, the Welsh Office and the Department of the Environment in Northern Ireland. Oscar Faber were commissioned as the project managers.

As the project plans developed during 1996 and a Memorandum of Understanding was being developed between the parties, it became clear that the BBC was increasingly uncomfortable with the way the project was proceeding. Firstly, both the AA and RAC wished to develop their own location codes and not share them, hence they considered the location codes being a vehicle to charge users for any resultant service; the BBC felt that this might compromise their Charter giving them a public service remit. Secondly, the BBC were starting to consider digital audio broadcasting (DAB) as the future; as DAB transmits audio as digital data, spare bandwidth can also be used for transmitting other data alongside the audio; hence the BBC did not want to invest in a service based on what they considered to be a limited life technology soon to be replaced by DAB. At the time there was a growing desire to discontinue VHF FM Band II broadcasts in favour of the more efficient and better audio quality provided by DAB without the need to use alternate frequencies at each transmitter.

Despite many iterations of the Memorandum of Understanding it was increasing likely that the BBC would not be able to take part in the trial. The hunt was then on to find an alternative transmission operator. Help came in the guise of Bev Marks who had been working with a Milton Keynes based company, Communications and Measurement Technologies (C&MT) who were using the RDS sub-carrier of Classic FM to broadcast Differential GPS signals required pre 2000 to enhance the accuracy of GPS. See Chapter 6 for further information on the AS1 Licence. Bev introduced Danny Woolard, Founder and Managing Director of C&MT who was invited to attend a consortium meeting and it became clear that this was a solution, albeit in a more restricted sense than the BBC with its multi station network. So in late 1997 C&MT came on board.

Hence, the consortium going forward after signing the MoU were:

- The Department for the Environment, Transport and the Regions (DETR) – eventually to become DfT
- The Highways Agency
- The Welsh Office
- The Scottish Office
- The Department for the Environment in Northern Ireland (DoE NI)
- The Automobile Association (The AA)
- The Royal Automobile Club (The RAC)
- Communication and Measurement Technologies (C&MT).

By late 1997, all members were signed up to the consortium. The consortium then needed to look to which receivers could be used. There were three companies at the time with prototype receivers. Philips Car Systems with a Carminat full colour navigation system; Volvo with their full colour navigation system in the Volvo S80 and Bosch/Blaupunkt with the Viking 148 Receiver/CD player which was capable of speaking RDS-TMC messages with the location referencing held on an exchangeable smart card.

It was vital that there were sufficient receivers to get a valid idea of the value users would place on RDS-TMC, and what they would pay for a service. It was agreed with market research companies that 400–500 receivers would yield reliable statistics. Following long discussions with Bosch Blaupunkt and Philips car systems, 400 Blaupunkt Viking 148 receivers and up to 100 Carminat systems would be purchased using the TEN-T Funds. Volvo pledged that they would allow access to Volvo S80 customers.

The development of a UK service using two service providers (the AA and the RAC) on the same Classic FM data stream was an interesting challenge. It was agreed that the 65k locations (the RDS TMC maximum) be split into two with the AA having location codes from 0 to approximately 32k and the RAC from 32k upwards. Separate smartcards were produced for each service, yellow for the AA and blue for the RAC; in this way the two services could coexist using the same RDS-TMC data stream. Only those receivers with the appropriate card could pick up each service. This solution however did not go down too well with the FORCE/ECORTIS project auditors who did not believe the solution was viable, and could not, or would not understand that it was not the public authority providing the data and the public broadcaster distributing it. This led to a tense time, but eventually the objectors had to agree that although the services were not provided in the way that they envisaged, there was nothing in the standard that precluded it. This resistance to what was proposed continued throughout the project as mainland Europe could not understand that commercial services was the way that sustainable services would go. It should be noted that the value chain proposed for this project has now been adopted worldwide.

The production of the Viking 148 receivers was complex in that all the locations for each service had to be recorded and constructed in voice form. The voice files were recorded in a studio in Dusseldorf by a German speaking announcer; this caused much mirth trying to get a correct pronunciation of place names like Godalming and Leicester. The sets were produced on time and to specification. Sadly, the same could not be the same for the Carminat receivers; despite promises no receivers were forthcoming. Volvo kindly agreed to add RDS-TMC location references to their navigation system in the Volvo S80 saloon and a number of Volvo drivers were contacted and interviewed giving vital information to the project.

Initially Oscar Faber provided the market research advice, but it soon became clear that more specialist effort would be needed. A specialist IT marketing and survey company, Prodata Partners Ltd. were engaged after competitive tenders by the UK Department for Transport to undertake an in-depth study of user reactions and produce results and recommendations.

The next task would be to select 400 users to take part in the trial who were a representative sample of the motoring public. The AA and RAC agreed to ask a number of their members to volunteer to be part of a trial. Questions were asked to respondents including the annual and type of mileage, car make and model, their gender and age. From a pool of respondents, Prodata Partners Ltd. determined a statistically representative sample and wrote to each volunteer. It was important that users travelled in the broadcast area and on the roads that TMC covered. Successful volunteers were given a free Viking 148 TMC receiver with a compact disc (CD) radio which they could keep after the trial (see Figure 5.1). The change of ownership was a headache for the Department for Transport which had bought the receivers and so legally owned them, but the Department could not have any responsibility if the installation or use damaged the volunteers' vehicles. A 'contract' between the Department and the volunteers was provided that indemnified the Department and transferred ownership to the volunteers.

Figure 5.1 A Bosch Blaupunkt Viking 148 receiver

eBay was gaining momentum at the time and there was concern that if receivers were sent out to the volunteers that a number would simply be sold on, or not ever installed and others not working correctly. As RDS-TMC is a broadcast one-to-many one-way technology, it would not have been possible to determine which sets were currently receiving the TMC service at any time. The solution was to identify the nearest Halfords store to each respondent and to ship a set and the relevant AA or RAC card to that store. Volunteers would pay £20 to Halfords for the fitting of the radio to cover the cost of time and any wiring convertors; Halfords gained by the users probably purchasing Halfords items whilst waiting for their receiver to be fitted. Halfords were given a briefing sheet on the installation and testing of the TMC facility as well as explaining to the users. Halfords would also be seen as at the forefront of the latest in-vehicle technology, albeit to a small user base.

Oscar Faber undertook the daunting task of packaging and shipping the receivers and cards to Halfords stores to match up with the volunteers. This distribution was successful in the majority of cases, saving incidents such as when a small number of sets were stolen from a Halfords stockroom.

Meanwhile a helpline was set up to assist with queries from the volunteers. The response was steady but there were some 'interesting' questions that says a lot about the Great British public. A call came in, to the effect that the user could not insert the smartcard into the Viking 148 receiver. Establishing that the card was blue and had an RAC RDS-TMC logo on it and that the user had a Viking 148 receiver installed, the user was asked to identify the slot for the card. The dialogue continued 'can you now put the card into the slot on the radio – just above the orange sleeve?' – 'no it's too thick' – 'ok..... is the small writing on the card legible?' – 'no, not really' – 'ok... ... can you see the gold coloured connection patch on the card?' – 'yes' – 'squeeze it between your thumb and forefinger' – 'yes' – 'using your other hand hold the card between your thumb and forefinger' – 'what now?' – 'move your hands apart – you will then find that the card comes out of its protective sleeve and will now fit into the receiver' – 'oh......... .... thank you'... a wise bit of advice – experts never underestimate the stupidity of the human race! Some users missed the slot altogether and inserted the card between the base of the DIN receiver and the aperture for the radio, resulting in the receiver having to be removed to retrieve the card. At least the project knew it was dealing with a representative sample of motorists.

Receivers were not distributed and fitted until the TMC service was up and running and tested. Again some basic assumptions of the RDS-TMC 'experts' were not acted out in reality. RDS-TMC assumes an incident occurs at a point, then the effect and hence the disruption travels upstream to another point determined by the *extent* in a TMC message. However, the software writer, not unreasonably, assumed that where a traveller entered a queue was the incident and the *extent* was to where the queue finished, at the incident itself; a complete reversal of the design. It took many hours of analysis to uncover this fundamental misunderstanding – a cautionary tale to all implementers.

The traffic messages were generated at the RAC and the AA. The RAC used bespoke software to convert its traffic data into TMC messages and transmit them over private BT circuits to C&MT for insertion into TMC message carousels. The

AA used software provided by a German company GEWI who were also involved in the implementation of TMC in Germany.to create TMC messages to again transmit over a private BT circuit to C&MT.

Both the RAC and the AA created their own location codes for the test areas and supplied them to Bosch and Volvo for inclusion on their devices. As stated earlier, each organisation had its own range of location codes and the TMC messages contained only those location codes. So when a receiver with an AA card received a message with a location code in the RAC range, it simply ignored it and vice versa. There was a fair amount of iteration between the map-makers and the receiver manufacturers to get the map data in the right format.

C&MT performed the role of infrastructure provider, aggregator and data-broadcaster for both the AA and the RAC services. C&MT, as the operator of the Additional Services Licence, proposed that carrying the TMC service required additional RDS bandwidth and that this should be requested by the Department, to add high level credibility. C&MT received messages over private BT circuits from the two content providers and formed them into carousels that repeated every 2-5 minutes or so depending on the message volumes. Important messages were inserted more than once so that they were broadcast more often. C&MT developed an innovative system called 'intelligent polygons'. Instead of broadcasting every message at every transmitter, the C&MT system used the location in the message and the message type to create a virtual geographic polygon based on severity, road class and affected direction (if a dual carriageway/motorway). The polygon was overlaid over the service areas of transmitters to determine which transmitter received which message. Some messages were transmitted over more than one transmitter when the message was in border areas.

Once the carousels were formed they were delivered to the relevant transmitters by BT private circuit. All the main transmitters that transmitted Classic FM were used (7 in all). Secondary (more local) transmitters picked up the data stream over the air to retransmit. C&MT were responsible for the end-to-end integration.

Once all of the elements were in place and end-to-end testing complete the receivers were delivered to the volunteer users.

Prodata Partners Ltd. were responsible for all communications with the users, except the helpline which was operated by Oscar Faber. Users were surveyed by written questionnaire, by telephone and by focus group sessions. The analysed results were available at the end of the project and were focused on answering the questions posed at the commencement of the project.

The project was considered to have been a success with much learned and a positive conclusion. Prodata Partners Ltd. compiled a report that details all aspects of the service. The results can be summed up by the following highlights:

- A public–private partnership worked together to produce a complex system, to time, budget and specification; the consortium provided a fully featured TMC service to test users for the whole period of the demonstration
- The market take-up models suggest that by the end of 2004 the number of RDS-TMC subscribers is likely to be 3.1 million

- Nearly 87% of the participants felt that RDS-TMC had saved them time and stress
- Over 45% of participants said they had changed their travel plans (changed routes, modes, timing or cancelled journeys) at least once as a result of receiving RDS-TMC travel information
- The likely market for TMC is as an adjunct to navigation systems rather than as a stand-alone service in its own right
- 79% of navigation system users would be prepared to pay £80 per year for RDS-TMC services
- 50% of drivers using stand-alone RDS-TMC would be prepared to pay over £100 for an improved TMC service
- Given the positive user reaction and the willingness to pay for the service, RDS-TMC services have a considerable market potential.

All in all the research findings supported the conversion of the demonstration into full commercial services.

The overall project spend came in at just under the £2m allowing the contributions by the project partners and the costs to the Department for the Environment Transport and the Regions (DETR) on project management, the purchase of receivers and project evaluation.

So what happened next?

C&MT were acquired by ITIS at the end of the project which led to the formation of the first commercial service (see below).

Encouraging as the results were, and the fact that users appeared prepared to pay up to £80 per year, it was felt that they were probably a little over enthusiastic. Given that there was no built-in access control to TMC services, charging customers directly was going to be problematical. It was envisaged that a business-to-business service might be the way to go, with users not being charged directly but via a motoring club subscription or ownership of a particular receiver or vehicle.

The Automobile Association made a press announcement (16 October 2000) on the introduction of a commercial RDS-TMC service; in the end this did not happen.

In early 2001, Toyota was considering installing a navigation system from Asin Warner (TNS200) into the last year of the Avensis body shell to extend its life until the new model was ready. This navigation system had the facility to receive and display RDS-TMC. An agreement was made that ITIS would supply a TMC service where access would be via their own TMC location database (built by Oscar Faber) and stitched into the Navteq map used in the system. ITIS obtained a commercial AS1 licence on Classic FM's RDS stream. Over the following years significant numbers of vehicle and equipment manufacturers signed up for the ITIS (now INRIX) service.

Subsequently, TrafficMaster (with the RAC) started an RDS-TMC service. Between INRIX and Trafficmaster in 2011 it was estimated that there were over 5 million UK RDS-TMC satnav users. As most satnavs contain a Europe-wide map, in most cases RDS-TMC is available in the United Kingdom to foreign drivers in their own language and to UK drivers in mainland Europe. RDS-TMC is now a multi-million pound per annum industry using the business model developed in the UK trial. RDS-TMC is available in more than 45 countries worldwide.

None of this would have been possible without the TEN-T funding.

5.4 The International Transport Forum (ITF)

The International Transport Forum (ITF) is an inter-governmental organisation within the OECD (Organisation for Economic Co-operation and Development) system. It is the only global body with a mandate for all modes of transport. It acts as a think tank for transport policy issues and organises the annual global summit of transport ministers. The ITF's motto is 'Global dialogue for better transport'. Between 1953 and 2007, the organisation had existed for over fifty years as the European Conference of Ministers of Transport (ECMT).

As the ECMT, it made a major contribution alongside the EBU in shaping RDS-TMC by acting as a bridge between the DRIVE research and development projects and the 'Establishment'.

5.5 The TMC Forum (1999–2007)

As the FORCE/ECORTIS projects came to an end in 1999 and support for implementations, trials and developments was coming to an end, it was realised that a support organisation was needed going forward. The EC suggested that a forum be established to support RDS-TMC. The TMC-Forum was established as a non-profit organisation whose members included service providers, receiver manufacturers, car manufacturers, map vendors, broadcasters (public and private), automobile clubs, and public authorities. It was formed under the auspices of ERTICO whose members automatically became members and others could join for a modest fee. The TMC Forum was run by ERTICO staff members but had a Supervisory Board made up of representatives from each sector. The TMC Forum discussed traffic information related matters. It assisted in the maintenance of the TMC-Standard (ISO EN14819-series). As RDS-TMC services became more and more common, and followed the business model established in the United Kingdom, public authorities and public service broadcasters became less engaged in RDS-TMC. Commercial discussions about the licensing of non-standardised location methodologies like AGORA dogged the forum and created a tension with those who advocated the free at the point of use services and those who had a more commercial imperative.

In late 2007, the TMC-Forum and the TPEG-Forum merged into the Traveller Information Services Association (TISA). The TISA is now the industry association that is responsible for maintaining the TMC Standards as well as developing and maintaining todays TPEG2 Standards. Further details of commercial development and activity in TTI can be found in Chapter 6 of this book.

5.6 The EBU (B/TPEG) 1997–2007)

In 1996, several of the public broadcasters, including the BBC, were looking towards the development of broadcasting TTI on higher data rate bearers such as DAB. The EBU, which is an organisation representing the majority of public service broadcasters, commissioned a project called B/TPEG (mirroring JPEG and MPEG). It was agreed that TPEG would be bearer independent so that it would include all access control and security rather than relying on the facilities of a particular bearer.

The challenge taken up by B/TPEG was to get over the service/client synchronisation problems in RDS-TMC:

- The need to have the same location table at the receiver as well as the service provider. TPEG should create location references on-the-fly so that the receiver and service supplier did not have to have fixed predefined location sets in synchronicity. It was acknowledged on the fly location tables would take up far more bandwidth than the 16-bit references that RDS-TMC used.
- The need to get away from the fixed text event messages to a methodology that could provide information in a variable and detailed way but could also be interpreted by a simple receiver.

Additionally:

- is unidirectional – it is broadcast one-to many
- supports 'thin' and 'thick' clients (thin clients access central data resources whereas thick clients have most resources downloaded to and resident on the end-user computer system)
- can have encryption in its own right
- can be filtered by an intelligent receiver
- is language independent
- has adaptation layers for different bearers.

To achieve this, experts were invited to monthly two day B/TPEG meetings in Geneva. As work progressed and services started being envisaged, the TPEG Forum was formed (in a similar way to the RDS Forum and the DAB Forum sponsored by the EBU).

The first instantiations (practical examples) of TPEG were designed to use a binary communications stream. A number of applications were described based on parts of the standard including the service and networks layers. The first applications were a road traffic message (RTM) which provided a rich description of traffic situations, and a location referencing application that provided an on-the-fly method that did not require synchronisation between the receiver and the service provider. Various other applications were developed such as public transport information and parking information. These applications were developed and submitted as CEN an ISO Technical Specifications (TS) (the CEN ISO TS 18234 series). Unfortunately due to the adoption rules in CEN and ISO some standards were developed in parallel and some in CEN and some in ISO only. It became clear that the RTM application was considered too complex at the time for a navigation system to decode and attach to the route guidance applications, so a 'lighter' version (TPEG-TEC) was proposed and produced.

By the late 1990s, XML applications were becoming more common and it was agreed in the TPEG Forum that TPEG-ML applications should be developed in parallel and submitted to CEN and ISO as TS (ISO CEN TS 24530 series). Again, due to the adoption rules in CEN and ISO some standards were developed in parallel and some in CEN and some in ISO only; this can be seen in the lists of standards.

Worse was to come, in that there was no consistency in part numbers, for instance Location Referencing was part 6 in the Binary version but part 4 in the ML version. It was also becoming difficult to synchronise developments between the binary and ML versions.

In the early 2002, it was agreed to develop TPEG2 where applications were modelled in UML and then automatically transcribed into binary and ML. This ensured that there was always synchronicity between the two instantiations. The TPEG2 standards are in the ISO TS 21219-series.

It was agreed that, due to earlier synchronisation issues, TPEG2 would be standardised in ISO only. At the time of writing, some 20 TS have been completed with some being moved to International Standards.

In 2007, after joint activities between the TMC and TPEG Forums and the German Mobile.Info group, it was proposed that the TPEG Forum should merge with the RDS Forum to form the Traveller Information Services Association (TISA).

5.7 Traveller Information Services Association – TISA (2007–Present)

The TISA [9] was formed in 2007 by bringing together the TPEG Forum and the TMC Forum and the German Mobile.Info project. It is a non-profit organisation under the Belgian law; the executive office of the TISA is hosted by ERTICO, in Brussels. The TISA has members outside Europe so can be considered a worldwide organisation, not just European.

TISA is a market-driven membership association with worldwide scope focussed on proactive implementation of TTI services and products based on existing standards, including primarily RDS-TMC and TPEG technologies.

TISA supports the maintenance and development of standards that provide elements or a framework for services and products leading to economic implementation and rapid market acceptance across a wide range of travel information services and products. In addition to road traffic information, for example, public transport, points of interest, weather and environmental data will continue to be in the TISA focus of important topic areas.

TISA's mission is to develop and promote open standards and policies that:

- facilitate a timely and cost-effective deployment of TTI services and products that save end users time and money, increase traffic safety, and minimise environmental impact

- improve the quality and minimise the cost of such services and products by maximising interoperability worldwide

5.8 Standardisation (1990–Present)

Standardisation of TTI services is undertaken in CEN (Committee Europeen de Normalisation) and ISO (International Standards Organization). Some ancillary communications standards such as RDS and DAB are standardised in ETSI (European Telecommunications Standards Institute) CENELEC (Committee Europeen de Normalisation pour Electrotechnical) and ITU (International Telecommunication Union).

Standards committees are currently organised into Technical Committees that look after an agreed scope of work. For TTI, the Technical Committee is CEN/TC278 – Intelligent Transport Systems [10]. The scope of CEN/TC278 is: 'Standardization in the field of telematics to be applied to road traffic and transport, including those elements that need technical harmonization for intermodal operation in the case of other means of transport'.

It supports:

- vehicle, container, swap body and goods wagon identification
- communication between vehicles and road infrastructure communication between vehicles
- vehicle man–machine interfacing as far as telematics is concerned
- traffic and parking management
- user fee collection
- public transport management
- user information.

Each Technical committee is subdivided into Working Groups under a Convenor. For TTI, the majority of the work is progressed by Working Group 4. Some standards such as DATEX 2 are progressed under Working Group 8 – Road Traffic Data and recently some under WG17 – Urban ITS. Cooperative ITS is progressed in Working Group 16.

Some standards are progressed in IS/TC204 – Intelligent Transport Systems [11]. Its scope is: 'Standardization of information, communication and control systems in the field of urban and rural surface transportation, including intermodal and multimodal aspects thereof, traveller information, traffic management, public transport, commercial transport, emergency services and commercial services in the intelligent transport systems (ITS) field'.

The relevant Working Groups in ISO are: Working Group 10-Traveller Information Systems; Working Group 9 – Integrated transport information, management and control; Working Group 19 – Mobility Integration and Working Group 18 – Cooperative Systems.

Both the Committees have been active since the 1990s and can work in parallel under the Vienna Convention. In most cases of TTI the parallel work is under CEN lead.

Standardisation comprises a series of steps from initial working drafts, though committee drafts (CDs) to technical reports (TR) and TS to full European Standards (EN) or International Standards (ISO IS)

Work undertaken solely in CEN/TC278 but withdrawn

- ENV 12313-4:2000: traffic and traveller information (TTI) – TTI Messages via Traffic Message Coding – Part 4: Coding Protocol for RDS – traffic message channel (RDS-TMC) – RDS-TMC using ALERT Plus with ALERT C – Soon to be withdrawn;
- ENV 12315-2:1996 TTI – TTI Messages via Dedicated Short-Range Communication – Part 2: Data specification – Uplink (vehicle to roadside).

Work undertaken in CEN/TC278 and ISO /TC204 but withdrawn

- CEN/TS 14821:2003 TTI – TTI messages via cellular networks – Parts 1 to 8

Standards in CEN and ISO

- EN ISO14819-1:2003 Intelligent transport systems – TTI messages via traffic message coding – Part 1: Coding protocol for RDS – traffic message channel (RDS-TMC) using ALERT-C
- EN ISO14819-2:2003 Intelligent transport systems – TTI messages via traffic message coding – Part 2: Event and information codes for RDS – traffic message channel (RDS-TMC) using ALERT-C
- EN ISO14819-3:2003 Intelligent transport systems – TTI messages via traffic message coding – Part 3: Location referencing for radio data system – traffic message channel (RDS-TMC) using ALERT-C
- EN ISO14819-6:2003 TTI – TTI messages via traffic message coding – Part 6: Encryption and conditional access for the RDS – traffic message channel ALERT C coding
- EN ISO 14823:2017 Intelligent transport systems – graphic data dictionary
- ISO/TR 14823-2:2019 Intelligent transport systems – graphic data dictionary – Part 2 Examples
- CEN ISO/TS 18234-1:2013 Intelligent transport systems – TTI via transport protocol experts group, generation 1 (TPEG1) binary data format – Part 1: Introduction, numbering and versions (TPEG1-INV)
- ISO/TS 18234-2:2013 Intelligent transport systems – TTI via transport protocol experts group, generation 1 (TPEG1) binary data format – Part 2: Syntax, semantics and framing structure (TPEG1-SSF)
- CEN ISO/TS 18234-3:2013 Intelligent transport systems – TTI via transport protocol experts group, generation 1 (TPEG1) binary data format – Part 3: Service and network information (TPEG1-SNI)
- ISO/TS 18234-4:2006 Intelligent transport systems – TTI via transport protocol experts group, generation 1 (TPEG1) binary data format – Part 4: Road traffic message (RTM) application

- ISO/TS 18234-5:2006 Intelligent transport systems – TTI via transport protocol experts group, generation 1 (TPEG1) binary data format – Part 5: Public transport information (PTI) application
- ISO/TS 18234-6:2006 Intelligent transport systems – TTI via transport protocol experts group, generation 1 (TPEG1) binary data format – Part 6: Location referencing applications (LOC)
- CEN ISO/TS 18234-7:2013 Intelligent transport systems – TTI via transport protocol experts group, generation 1 (TPEG1) binary data format – Part 7: Parking information (TPEG1-PKI)
- CEN ISO/TS 18234-9:2013 Intelligent transport systems – TTI via transport protocol experts group, generation 1 (TPEG1) binary data format – Part 9: Traffic event compact (TPEG1-TEC)
- CEN ISO/TS 18234-10:2013 Intelligent transport systems – TTI via transport protocol experts group, generation 1 (TPEG1) binary data format – Part 10: Conditional access information (TPEG1-CAI)
- CEN ISO/TS 18234-11:2013 Intelligent transport systems – TTI (TTI) via transport protocol experts group, generation 1 (TPEG1) binary data format – Part 11: Location referencing container (TPEG1-LRC)
- CEN ISO/TS 24530-1:2006 TTI – TTI via Transport Protocol Experts Group (TPEG) Extensible Markup Language (XML) – Part 1: Introduction, common data types and tpegML
- CEN ISO/TS 24530-2:2006 TTI – TTI via Transport Protocol Experts Group (TPEG) Extensible Markup Language (XML) – Part 2: tpeg-locML
- CEN ISO/TS 24530-3:2006 TTI – TTI via Transport Protocol Experts Group (TPEG) Extensible Markup Language (XML) – Part 3: tpeg-rtmML
- CEN ISO/TS 24530-4:2006 TTI – TTI via Transport Protocol Experts Group (TPEG) Extensible Markup Language (XML) – Part 4: tpeg-ptiML

Standards in CEN only

- EN 16157-1:2018 Intelligent transport systems – DATEX II data exchange specifications for traffic management and information – Part 1: Context and framework
- EN 16157-2:2019 Intelligent transport systems – DATEX II data exchange specifications for traffic management and information – Part 2: Location referencing
- EN 16157-3:2018 Intelligent transport systems – DATEX II data exchange specifications for traffic management and information – Part 3: Situation publication
- CEN/TS 16157-4:2014 Intelligent transport systems – DATEX II data exchange specifications for traffic management and information – Part 4: VMS publications
- CEN/TS 16157-5:2014 Intelligent transport systems – DATEX II data exchange specifications for traffic management and information – Part 5: Measured and elaborated data publication

- CEN/TS 16157-6:2015 Intelligent transport systems – DATEX II data exchange specifications for traffic management and information – Part 6: Parking publications
- EN 16157-7:2018 Intelligent transport systems – DATEX II data exchange specifications for traffic management and information – Part 7: Common data elements
- FprCEN/TS 16157-8 Intelligent transport systems – DATEX II data exchange specifications for traffic management and information – Part 8: Traffic management publications and extensions dedicated to the urban environment
- prCEN/TS 16157-9 Intelligent transport systems – DATEX II data exchange specifications for traffic management and information – Part 9: Traffic signal management publications dedicated to the urban environment

Standards in ISO only

- ISO/TS 18234-8:2012 Intelligent transport systems – TTI via transport protocol experts group, generation 1 (TPEG1) binary data format – Part 8: Congestion and travel time application (TPEG1-CTT)
- Standards in ISO only:
- ISO/TS 18234-8:2012 Intelligent transport systems – TTI via transport protocol experts group, generation 1 (TPEG1) binary data format – Part 8: Congestion and travel time application (TPEG1-CTT)
- ISO/TS 21219-1:2016 Intelligent transport systems – TTI via transport protocol experts group, generation 2 (TPEG2) – Part 1: Introduction, numbering and versions (TPEG2-INV)
- ISO 21219-2:2019 Intelligent transport systems – TTI via transport protocol experts group, generation 2 (TPEG2) – Part 2: UML modelling rules (TPEG2-UMR)
- ISO 21219-3:2019 Intelligent transport systems – TTI via transport protocol experts group, generation 2 (TPEG2) – Part 3: UML to binary conversion rules (TPEG2-UBCR)
- ISO 21219-4:2019 Intelligent transport systems – TTI via transport protocol experts group, generation 2 (TPEG2) – Part 4: UML-to-XML conversion rules
- ISO 21219-5:2019 Intelligent transport systems – TTI via transport protocol experts group, generation 2 (TPEG2) – Part 5: Service framework (TPEG2-SFW)
- ISO 21219-6:2019 Intelligent transport systems – TTI via transport protocol experts group, generation 2 (TPEG2) – Part 6: Message management container (TPEG2-MMC)
- ISO/TS 21219-7:2017 Intelligent transport systems – TTI via transport protocol experts group, generation 2 (TPEG2) – Part 7: Location referencing container (TPEG2-LRC)

- ISO/TS 21219-9:2016 Intelligent transport systems – TTI via transport protocol experts group, generation 2 (TPEG2) – Part 9: Service and network information (TPEG2-SNI)
- ISO/TS 21219-10:2016 Intelligent transport systems – TTI via transport protocol experts group, generation 2 (TPEG2) – Part 10: Conditional access information (TPEG2-CAI)
- ISO/PWI 21219-13 Intelligent transport systems – TTI via transport protocol experts group, generation 2 (TPEG2) – Part 13: Public transport information (TPEG2-PTS)
- ISO/TS 21219-14:2016 Intelligent transport systems – TTI via transport protocol experts group, generation 2 (TPEG2) – Part 14: Parking information application (TPEG2-PKI)
- ISO/TS 21219-15:2016 Intelligent transport systems – TTI via transport protocol experts group, generation 2 (TPEG2) – Part 15: Traffic event compact (TPEG2-TEC)
- ISO/TS 21219-16:2016 Intelligent transport systems – Traffic and travel information via transport protocol exports group, generation 2 (TPEG2) – Part 16: Fuel price information and availability (TPEG2-FPI)
- ISO/PWI 21219-17 Intelligent transport systems – Traffic and travel information via transport protocol experts group, generation 2 (TPEG2) – Part 17: speed information (TPEG2-SPI)
- ISO 21219-18:2019 Intelligent transport systems – Traffic and travel information (TTI) via transport protocol experts group, generation 2 (TPEG2) – Part 18: traffic flow and prediction application (TPEG2-TFP)
- ISO/TS 21219-19:2016 Intelligent transport systems – Traffic and travel information (TTI) via transport protocol experts group, generation 2 (TPEG2) – Part 19: Weather information (TPEG2-WEA)
- ISO/TS 21219-21:2018 Intelligent transport systems – Traffic and travel information via transport protocol experts group, generation 2 (TPEG2) – Part 21: Geographic location referencing (TPEG-GLR)
- ISO/TS 21219-22:2017 Intelligent transport systems – Traffic and travel information (TTI) via transport protocol experts group, generation 2 (TPEG2) – Part 22: OpenLR location referencing (TPEG2-OLR)
- ISO/TS 21219-23:2016 Intelligent transport systems – Traffic and travel information (TTI) via transport protocol experts group, generation 2 (TPEG2) – Part 23: Roads and multimodal routes (TPEG2-RMR)
- ISO/TS 21219-24:2017 Intelligent transport systems – Traffic and travel information (TTI) via transport protocol experts group, generation 2 (TPEG2) – Part 24: Light encryption (TPEG2-LTE)
- ISO/TS 21219-25:2017 Intelligent transport systems – Traffic and travel information (TTI) via transport protocol experts group, generation 2 (TPEG2) – Part 25: Electromobility charging infrastructure (TPEG2-EMI)
- ISO/TS 21219-26:2018 Intelligent transport systems – Traffic and travel information via transport protocol experts group, generation 2 (TPEG2) – Part 26: Vigilance location information (TPEG2-VLI)

5.9 Outlook

5.9.1 Contribution of Europe to TTI development more generally

As will be seen from the previous paragraphs, the development of the TTI owes everything to the contribution of European Initiatives. The need to have language independent standardised systems in Europe has produced standardised systems that have a worldwide appeal. Because of the language independency, localised systems have not got a foothold so the standards like RDS-TMC have had a long lifetime and continue to develop whist retaining backward compatibility.

5.9.2 Cooperative ITS (C-ITS)

It is likely that the rise of Cooperative ITS will mean that bespoke applications will gradually be superseded by information built into an integrated offering. Using the message sets and many of the location referencing facilities of existing standardised systems allows a coexistence of legacy and emerging services leading to gradual integration whilst the market penetration of C-ITS systems grow.

5.9.3 Connected cooperative automated mobility (CCAM)

Connected cooperative automated mobility (CCAM) will require many services to make it fully operational. For instance the management of traffic regulation and signing databases will require standardised message sets of signs and road markings; this is where existing worldwide standards like ISO 14823:2017 – Intelligent transport systems – Graphic data dictionary can make a significant contribution.

5.9.4 Social media-driven services

Social media services in traffic and travel information are being developed and deployed rapidly. It is vital that to ensure compatibility with existing and future systems, they are not developed in a silo and the data they use and create are formatted in such a way that a variety of services can be offered. The development of social media in relation to TTI is described in other chapters of this book.

5.9.5 National access points

The Commission Delegated Regulation (EU) 2017/1926 requires that Member States facilitate the easy exchange and reuse of data for the provision of comprehensive travel information services. Transport authorities, transport operators, infrastructure managers or transport on demand service providers, as appropriate, should make the static data, corresponding metadata and information on the quality of the data accessible to users through a national or common access point. This regulation will impact the formatting of traffic and travel data so as to allow easy exchange of data using DATEX II and the message sets developed previously in Europe.

5.9.6 Standards

The present way of producing and maintaining standards will have to change due to the rapid development, changes to scope and use of standards. Presently standards are considered to be paper based in editions even though they are generally distributed in electronic form; standards like the Graphic Data Dictionary (ISO 14823) are in a constant state of change as new signing is developed and requirements advancing. This will be a challenge for the standardisation bodies.

References

[1] European Broadcasting Union. [Online] ebu.ch.

[2] European Broadcasting Union. RDS-FORM. [Online] www.rds.org.uk.

[3] Dietmar Kopitz, Bev Marks. *RDS:Radio Data System.* Artech House, London, UK. ISBN: 9780890067444.

[4] BBC Research Department. *The 'CARFAX' road Traffic Information system.* s.l.: British Broadcasting corporation, 1984.

[5] European Commission. RDS-ALERT. [Online] https://cordis.europa.eu/project/id/V1029.

[6] European Commission. ATT-ALERT. [Online] 1992–1994. https://cordis.europa.eu/project/id/V2028.

[7] European Commission. PLEIADES Paris–London Corridor. [Online] 1999. https://cordis.europa.eu/project/id/V2047.

[8] European Commission. FORCE1, 2 and 3 and ECORTIS. [Online] https://trimis.ec.europa.eu/project/radio-data-systemtraffic-message-channel-european-interoperability.

[9] TISA. Traveller Information Services Association. [Online] www.TISA.org.

[10] CEN. http://itsstandards.eu. [Online] https://itsstandards.eu.

[11] ISO/TC204. ISO/TC204 Intelligent Transport Systems. [Online] https://www.iso.org/committee/54706.html.

Chapter 6

The role of the commercial sector in developing road traffic information in the UK and Europe

Danny Woolard[1]

6.1 Introduction

The real 'key' to the development of commercial traffic services was a proposed 'Demonstration' trial project to broadcast digital traffic information by radio data systems (RDS) sponsored by the UK Department for Transport (at that time called the Department for Environment, Transport and the Regions, or DETR). This is described in another chapter of this book. However, one big challenge remained: with the BBC not able to commit to the Demonstration project because of issues with its Charter and its decision to put resources in other areas, it meant that there was a big question mark over the availability of RDS capacity to actually undertake a trial.

The first section below describes various organisations and developments that have all played a part in making the trial a reality. Subsequent sections describe how the Demonstration was turned into a national commercial service and how that service has developed up to the present day. Technical details and references concerning RDS and the traffic message channel (RDS-TMC) can be found in other chapters of this book. This chapter concentrates on the people, organisations and circumstances that contributed to commercial road traffic information development.

6.2 Demonstration and early development

6.2.1 Communications & Measurement Technologies Ltd. (CMT)

This part of the story really did come from one of those rare chance meetings and discussions (of course in a pub), that resulted in Communications & Measurement Technologies Limited (CMT), entering into discussions with the DETR about becoming involved in the Demonstration. But first some background to CMT and its role in the UK services.

[1]Chiltech Limited, Devon, UK

CMT was a small privately owned company whose principle business was GPS systems integration technology. The company designed and manufactured a number of standard products as well as undertaking bespoke development and building of integrated GPS navigation and wireless data communications systems, principally for the oil and gas survey markets.

The company was established in 1988 and the founders all had a strong background in the marine survey and navigation market. This may seem miles away from the main topic of this book but, at the time, the commercial market for GPS navigation was 'exploding' and, due to the US Department of Defense imposed downgrading of GPS accuracy, known as 'Selective Availability', or SA, techniques were developed which involved communicating the observed measurements computed at a known fixed (reference) location, and transmitting the calculated error to a mobile GPS receiver, typically via terrestrial radio data technology, such as that manufactured by CMT. This technique was known as 'Differential GPS' or dGPS. Typically, the accuracy of a stand-alone GPS receiver with SA was about 100 m, but applying differential corrections would typically allow accuracies of <3 m, depending on the technique.

Throughout the later part of the 1980s and early 1990s, availability of the expanding GPS satellite constellation, developing GPS technology and reducing cost of GPS receivers started to create many new uses for GPS. It also created a number of commercially available differential GPS services for professional survey use. One such application was the use of GPS in precision farming, using GPS to precisely navigate agricultural vehicles/implements such as harvesters and fertilizer application tractors. This, of course, needed high accuracy and therefore the application of a dGPS technique.

A number of methods of 'communicating' dGPS corrections were used, from terrestrial short- range radio telemetry to satellite communications, of which the latter was used extensively in offshore marine survey applications. However, the cost of such transmissions did not lend itself to the emerging new markets. CMT became aware of a US company, Differential Corrections Inc., (DCI), that had developed a compression method to enable dGPS corrections to be broadcast over a low bandwidth carrier, specifically RDS.

DCI were the Licensee for part of the RDS sub-carrier on Classic FM in the UK (known as the AS-1 Licence) and appointed CMT as a sub-licensee for its technology and operator of the AS-1. CMT created a division of its systems business known as Focus FM to operate a national dGPS service. This became a successful operation with many users across a broad range of industries, but the most successful was the Precision Agriculture market after signing an agreement with Massey Ferguson Harvesters.

As mentioned above, CMT became aware of DETR's proposed trial and were asked to present to the DETR how they could potentially assist in providing capacity on the AS-1. During the early discussions, however, it became clear that if the dGPS service was not to be impacted in terms of its quality of service and the requirement to maintain a specific RDS update rate, then additional capacity would be required. Therefore, before any commitment, both by the DETR and indeed

CMT to assist in the trial, it was agreed to approach Ofcom (then the Radio Authority) to seek a short-term increase in RDS capacity to allow TMC messages to be broadcast. This, of course, required the main broadcaster, Classic FM, to agree to this increase. It was further agreed that it should be the DETR that should approach Ofcom given that they were the sponsors of the trial.

After what seemed like years, Ofcom finally agreed to allowing an additional signal bandwidth (in the form of an additional RDS 'group') to be used for the purposes of conducting the proposed trial. The result was that CMT agreed to fulfil the role of both provider of broadcast capacity and to build the infrastructure to support the trial.

One of the many challenges encountered in developing and building the network was that the RDS infrastructure installed at each transmission site was owned and maintained by the independent broadcast company NTL, now Arqiva. This added another layer of complexity into the development as there was no direct connectivity to the RDS encoders at each transmission site, therefore individual land lines were required to be leased and installed. Readers should be aware that this was all 'pre-internet' and far more challenging than would be the case nowadays.

It was agreed that six out of Classic FM's 20 primary transmission sites should be used, selected to ensure maximum coverage of the country's strategic road network managed by the Highways Agency.

Once the supporting network was developed and tested, CMT as part of the UK consortium, fulfilled the role of infrastructure provider and systems integrator. Traffic information compiled by the two providers, the AA and the RAC, was formatted in the ALERT-C protocol and sent to CMT. A system developed by CMT then undertook the network control and distribution of the messages to the most relevant transmission sites. This included 'Geoscoping' each traffic message based on its severity and geographical relevance.

Oh, and what became of dGPS? In 2000, the US Department of Defense under the Clinton administration ended the use of Selective Availability. A number of services ended, including that of Focus FM.

6.2.2 ITIS Holdings

During the UK trial of RDS traffic information, CMT recognised the opportunity for commercial traffic services but further realised that this was outside of its core GPS systems business and potentially a distraction. In view of this, the company decided to evaluate the commercial opportunity and to seek additional funding for its business. A number of potential partners were approached. Naturally, discussions were held with both the AA and RAC given their involvement in the trial and as two key providers of traffic information in the UK. Both also had business divisions providing TTI services, AA Roadwatch and RAC Traffic and Travel. It has to be said that neither party saw enough commercial potential at the time to consider investing further after the UK Demonstration. Discussions were also had with both TrafficMaster and a small emerging telematics and traffic business, ITIS.

ITIS had an interesting business concept – it aimed to create and develop the vehicle telematics business, providing a number of driver and concierge services and in parallel use the location data from in-car devices to create road traffic speed data, potentially detecting congestion. Of course, most readers would recognise this today as 'floating car data', or as ITIS termed it 'floating vehicle data or FVD' (not just cars). There is more about this in another chapter of this book. So it has to be recognised that ITIS were very much ahead of the game in recognising the real potential of FVD.

In May 2000, following a period of discussions and in parallel to the UK Demonstration, CMT agreed to sell its 'fledgling' traffic business and technology to ITIS. The first priorities were then to ensure that the broadcast licence was secured and that planning for rebidding of the AS-1 licence (see below) was in place.

It should also be recognised that ITIS invested significantly and supported the development of what became the first true commercial TMC service, despite the risks and uncertainty over whether vehicle manufactures and navigation systems manufacturers would support private services. After all, broadcasting traffic information over RDS was very much developed with public services in mind.

6.2.3 The Additional Services 1 (AS-1) Licence

Reference has been made to the AS-1 in the previous section and it is worth explaining the background to this specific licence and the relationship with the host broadcaster, Classic FM.

Outside of the BBC Radio environment, the first and only national independent FM broadcaster at that time was Classic FM, occupying the frequencies 99.9 to 101.9 MHz. Classic FM were awarded their licence (known as INR1) and commenced in 1992. However, the Radio Authority (Ofcom) at the time retained part of the RDS capacity within the radio spectrum which it proposed to licence separately by auction. This licence was known as the Additional Services or AS-1 licence. (And, yes, there was a proposed AS-2 which was never issued!)

The AS-1 licence was auctioned for a period of eight years in 2004 and finally awarded to DCI Inc. who proposed the dGPS service as the main application. As previously discussed, CMT, as the UK partner for DCI, developed and operated the AS-1 licence for the primary purpose of its dGPS service.

An acknowledgement should be made at this point to Classic FM and Arqiva whose engineering teams were very supportive during the development of the AS-1 services.

6.3 Development of a national service

Other chapters in this book discuss the aims and outcomes of the Demonstration. The following sections are focussed on how the UK's RDS-TMC service 'went commercial' and the challenges that had to be overcome.

6.3.1 Commercial models

The first real challenge was creating a business model. There were few, if any, examples of viable business models except perhaps the limited service operated by Mediamobile in France which used a variant of TMC protocols known as ALERT Plus. However, given that most of the vehicle manufacturers and their navigation system providers were focussed on 'Free to Air' TMC, a model had to be created based on licensing the content to the car manufacturer directly. At the time, there was a working group within the TMC Forum (a group of public and private sector organisations formed to discuss and promote TMC) developing encryption techniques, known as Conditional Access. The aim of this was to create an extension to the ALERT-C protocols (ISO 14819-6) to standardise an encryption method that could be adopted by all commercial services, providing a means of individual user subscription to a traffic service.

However, ITIS were keen to launch a commercial service prior to the above becoming standardised, developed and integrated into receivers and indeed services.

6.3.1.1 Free to air commercial model

ITIS proposed an alternative method, one which would not require any specific changes to the technology. This was based on the fact that the UK did not have any public TMC location table (a look-up giving a unique code to all relevant locations). This was a result of a decision made during the Demonstration. The two service providers each shared a single location table, being split by address locations. However, as ITIS proposed a new service it was decided to develop a unique Location Table and work with the two digital map providers Navtech (now HERE) and Tele-Atlas (now TomTom). The integration of Location Tables and digital maps are discussed later.

There were many 'Chicken and Egg' situations overcome during the development of services in the UK. The steps (also the 'risks'), taken by ITIS to create a service and commence commercial services was a real catalyst to creating a viable business for traffic services.

Having commenced commercial broadcasting in 2002, early adopter discussions opened with Toyota GB. Toyota was planning a 'model run-out' marketing campaign with their Avensis family car. This vehicle was available with a 'Turn by Turn' (TbT) navigation system provided by the supplier Aisin AW. Toyota GB wanted to add standard features to enhance the appeal of the vehicle and to standardise on traffic enabled TbT navigation was a first in the UK. Toyota even gave the feature a specific name, ETA, Electronic Traffic Avoidance.

Working closely with Toyota, Aisin and Navtech (which undertook the integration of the ITIS Location Table), ITIS learnt many lessons, not least that developing services in close alliance with all players was essential. ITIS felt that they were pioneering in this area. After all, there were no other services to emulate.

Two other customers then adopted the ITIS TMC Service, premium brand BMW and Siemens VDO (The original Phillips CARIN System). These early three clients proved the commercial business model of providing traffic content as a service, with a specific licence element being the availability and use of a commercial TMC Location Table.

6.3.1.2 Encrypted services

As previously mentioned, the launch of the Free to Air commercial service in the UK was driven by timing. However, the work in the TMC Forum's Encryption Standards Task Force continued to develop technology and method that could be standardised and deployed in any commercial service and in many countries other commercial services were also being planned and developed.

The principal of any commercial model is the need to cover the cost of data collection and dissemination by charging either end users or intermediaries to receive and use the data. A method was developed to encrypt the data with the key to decrypt the data being made available upon payment of a subscription fee.

It was fundamental that the addition of encryption had to conform to the RDS and RDS-TMC specifications combined; it had to be backward compatible; i.e. it had to require no changes to existing devices, it had to support both lifetime and term based subscriptions, and it had to ensure minimal data capacity overhead.

A number of candidate proposals were put forward; however, a submission by Deutsche Telekom was best judged to meet the above criteria. The method proposed and developed encrypts the 16 bits that form the location code (pointer to a geographic location within a location table) in each TMC message which is subsequently decrypted by two 'Keys'. The first is provided under non-disclosure by the service provider to the terminal manufacturer (of an in-vehicle device), the second 'key' is broadcast in the Encryption Administration Group' which is broadcast as a variant of an 8A TMC message.

The full specification of the adopted and implemented system is defined is ISO Standard 14819 Part 6. Encryption and Conditional Access for the Radio Data System – Traffic Message Channel.

It has transpired that no 'single end user' subscription services have been deployed or supported and all commercial services that are in use today are transacted on a Business to Business contractual basis between service providers and vehicle manufacturers or terminal manufacturers. Recovery of costs are typically embedded in the cost of the vehicle option or in-vehicle device, such as a Personal Navigation Devices.

6.3.2 Digital maps and location coding

This section describes a top-level description and overview of digital mapping and the use of TMC location coding. It further acknowledges additional location referencing methods that are, or have replaced, TMC location codes. It is not the purpose of this chapter to go into full technical detail of digital mapping technology and standards.

Beyond the early development of text to speech receivers using RDS-TMC, the primary 'use case' for RDS-TMC was in turn-by-turn navigation systems or full screen, map based, navigation receivers. The objective was integration of real-time traffic to provide dynamic routing and guidance; that is, calculating a route to a destination based on the user's current GPS-derived location and adapting that route based on up to date traffic situations.

The fundamental component of a navigation system's architecture is the digital map: a machine-readable digital representation of the geometry of the navigable road network. Two principle commercial businesses at the time, developed digital maps. One was Navtech Inc., a business which has its roots in Phillips Car Systems. This company has gone through various different ownerships and branding but is now the foundation of the company HERE. The second company was Tele Atlas, whose history tracks back to Sony and was later sold to TomTom. Digital maps databases are the resulting products and are available globally from both providers. There are, of course, other providers of digital maps, including Google and Apple but the focus on in-car navigation development was very much down to Navtech and Tele-Atlas.

Both companies' map data is similar in structure, defining 'link-and-node' geometry referenced to a location of the navigable roadway. There are many other attributes assigned to each map element; these can include road class, road name, point of interest and physical address information. Creating and updating digital maps is a very costly exercise, requiring much of the network being physically driven as part of the data collection process. Many techniques have been developed by each of the companies, including licensing of national mapping agency content, and satellite imagery. Much of the data collection now uses video recognition for road assets such as speed signs and both companies employ techniques based on user-generated data to update map features as road geometry.

6.3.3 Location coding

The focus of this section is the background to the development of the UK RDS-TMC service and is mainly centred on the development and use of TMC Location Codes. It has to be stated, though, that the use of Location Codes as defined in the TMC Standards is now entering its 'twilight years', being superseded by other proprietary location methods and open standards such as Open LR and other 'on-the-fly' location referencing methods. These are described in some detail in another chapter of this book.

However, TMC location codes have been, and are, a very effective and efficient method of defining a precise location of a road attribute in just 16 bits of data. The restriction at the time of definition was due to the RDS group structured protocol which has very limited bandwidth.

The technique basically assigns a unique 16-bit number (i.e. in the range 0–65,535) to a point (node), segment, or area of a geographical network. The coding rules are defined in the ALERT-C Standards. Each TMC code is assigned to an element coded within the digital map. Hence, the requirement to integrate the TMC location database of a public or commercial traffic service with the digital map. The transmitted and decoded location is then referenced to the digital map for routing and display purposes.

6.3.4 In-car navigation, personal navigation and other traffic receivers

I'm sure many have recollection of James Bond in his Aston Martin DB5 with a map display and flashing light showing his position on a map, thinking 'One day . . . .'.

With the advent and rapid development of GPS, this all became a reality in a very short space of time. As discussed earlier, the principal business of CMT was GPS and, specifically, differential GPS systems. It has been staggering to see high end receivers, which cost many tens of thousands of dollars and requiring the need of a differential GPS technique to increase accuracy to better than 100 m, reduce to a few dollars for a GPS receiver on a chip, or in a low cost navigation device, or phone. Unexpected also has been the demise of one dGPS business seeing the opportunity to emerge as another.

The development of in-car navigation leading onto Personal Navigation Devices and now onto Apps on Smartphones has, as mentioned, been very rapid. Both Phillips and Bosch were very much leaders in the early development of turn by turn navigation. That is not to say that the Japanese industry and the likes of Pioneer, Panasonic, Sony and Alpine have not been as advanced. However, Bosch and Phillips were also part of the original group of organisations including the BBC which ultimately developed the early ALERT-C TMC standards under the early European DRIVE research and development program.

The early in-car navigation systems integrated a GPS receiver, a digital map data base (typically stored on a Compact Disk drive) and an in-car map display. Map displays had low resolution and limitations in processing power resulted in slow route calculation. However, they were the foundation for all future generations of Personal Navigation Devices.

In the early years of UK traffic services, the general view was that navigation was limited to high end vehicle brands such as BMW and Mercedes. The reality, however, was that as the cost of GPS and processing power fell, most of the mainstream brands started including navigation as a standard fit or as a low-cost option. Within the space of five years following the launch of RDS-TMC in the UK, all brands, such as Ford, GM Vauxhall, Nissan, Citroen and Renault offered TMC in most of their vehicle lines. All of these supported the ITIS TMC Service or, as it emerged, the competitive service from TrafficMaster.

6.3.5 Personal Navigation Devices and mobile/smartphone navigation

The ending of selective availability on the GPS signals marked a major turning point in the development of personal navigation and, indeed, that of the traffic industry. A number of companies created consumer software applications, notably Garmin, TomTom, Navman and Magellan. TomTom released their first navigation software which ran on Windows CE, designed for a PDA (personal digital assistant) in 2002. Whilst the PDA itself did not include a GPS receiver, a small 'GPS Dongle' was supplied connected via Bluetooth communications. TomTom's first real Personal Navigation Devices, the TomTom GO, was launched in early 2004 followed by a number of other manufacturers keen to exploit this very new market. In 2004 over 100,000 devices were sold in Europe alone. The Personal Navigation Devices market exploded, with millions of devices being sold annually by 2008. The integration of traffic into these devices led to a number of challenges however. Early devices, such as those made by Garmin and TomTom, used external FM RDS receivers and antennas. The biggest challenge was installation of the antennas, often with poor sensitivity, coupled with the very low level of RDS sub-carrier injection (i.e. signal level) which meant that performance in an 'in-car' environment was somewhat compromised. For one who was very involved in the integration of traffic and navigation, there was an undeniable feeling of frustration given it was always perceived that it was the traffic service that was at fault!

Whilst the Personal Navigation Devices market continues, it is really only Garmin and TomTom that continue to manufacture devices in any substantial quantity. It is, however, interesting to mention the chosen path to deliver traffic services to these devices.

Garmin continue to provide broadcast traffic services using digital audio broadcast (DAB – see below) from a number of global service providers as the carrier for their 'HD' traffic services. This has led to a number of DAB TPEG services (see below) being created to support Garmin as a customer. TomTom, on the other hand, pursued integration of a SIM-enabled mobile modem, the cost of connectivity being supported by the cost of the device itself and a range of subscriber services. TomTom has therefore not been limited in terms of where their traffic enabled devices are sold. Of course, TomTom themselves are one of the major traffic service providers globally, whereas Garmin continue to contract with third party traffic service providers.

The real 'gamechanger,' in terms of consumer navigation, came with the advent of the GPS-enabled mobile phone, and in particular the Smartphone. It was 2008 when Apple announced its iPhone and App store to global acclaim. Whilst initially the use of the GPS function was restricted for developers of third-party Apps, this soon changed and mobile navigation quickly became the most popular App and was, of course, supported by traffic services.

One aspect of this new market had a knock-on effect to the traffic service providers, particularly those who had developed and integrated the use of Floating Vehicle Data (FVD) as a primary source of road speed detection. The rapid growth of navigation-enabled smartphone Apps meant a significant rise in the level of 'FVD Probes'. Of course, the real beneficiaries of this were Google and Apple (teamed with TomTom) in having the native mapping Apps on the mobiles. In 2009, Google announced its 'free' navigation software on Android devices followed by Nokia, in 2010. This significantly changed the markets for both the mapping companies and the traffic service providers.

The mobile navigation market continues to grow, with the integration of in-car screens becoming the norm. Navigation services are mostly supplied free to air and this has meant that real time traffic has truly become widely acceptable and consumer-driven.

So where next? With the majority of in-car, portable devices and Apps supporting connected services, including traffic, RDS-TMC is now nearing the end of its 'Technology Life'. With the development and roll out of 3G, 4G and now 5G wireless networks, these are becoming the 'norm' in terms of connecting services to the driver. However, it has to be acknowledged that RDS-TMC has been a huge success in creating services and adding real value to the in-car navigation experience, and it really should be seen as the foundation for all of the services that we see today.

6.4 The development and expansion of RDS-TMC services

6.4.1 Europe

With growing interest from more automotive brands to include navigation on their 'mainstream' volume models and with the advent of Personal Navigation Devices that supported TMC, either as standard or as an option, created the momentum for the expansion of services, both in Europe and in other global markets. The period between 2003 and 2006 saw new public services and new emerging commercial traffic services. The 'Chicken and Egg' problem, or service–receiver availability issue had very much been solved.

Despite public services being developed and expanding in Germany and France, the vehicle manufacturer (OEM) focus was very much on commercial services to fill the growing need for traffic information supported by high quality traffic data. The public services were very much seen as providers of journalistic information, i.e. road incidents and planned events such as roadworks. The information coming from road agencies had little or no focus on the needs of the dynamic navigation receiver market, where the real requirement was for 'Level of Service' traffic information in the form of road speeds and congestion data. It is this data that is needed to support dynamic routing.

Along with the growth of the UK service provided by ITIS, commercial RDS-TMC services began in both Germany and France. The German service was developed and operated by T-Systems Traffic, a division of T-Mobile. This service went live in 2005 under the brand TMC-Pro, broadcast via commercial radio networks. The service used the now standardised Conditional Access encryption (ISO 14819-6), with the business contractual relationships between T-Systems Traffic and the automotive manufacturer on a business-to-business (b2b) basis. In 2009, T-Systems Traffic was acquired by Navteq (now HERE).

In France, whilst there was a free service covering the Péage (Autoroutes) broadcast on a specific FM channel 107.7 (discontinued in 2012), two main commercial services were launched, the largest of which was developed and rolled out by Mediamobile, known as V-Traffic. This service was a natural progression and development of the original TMC+ service operated by Mediamobile but, again, used the now standardised conditional access encryption. Mediamobile was a subsidiary of TDF, the French Broadcaster, but with additional shareholders Renault and Vinci. In 2019, Mediamobile was acquired by Be-Mobile, the Belgian content services provider, part owned by Proximus, the Belgian mobile network operator. The other commercial TMC service launched in France was established by Via Michelin, a company owned by Michelin and Carte Blanche Conseil.

In recognition of the part played by the TMC Forum, a lot of cooperation occurred between the early service providers and their customers, particularly OEMs such as BMW, Renault, Toyota and Mercedes. I would use the word 'Pioneering' to describe the organisations involved in these developments. The service providers learned from each other and the OEMs were more than just customers, they really were partners in the growth of the global TMC market.

6.4.2 Beyond Europe

From 2006 onwards, most EU countries were developing either public or commercial services (or both), and beyond Europe, the US market was developing with services provided by Navteq using Satellite radio as a carrier (XM Sirius) and also by Total Traffic Networks established by the broadcaster, Clear Channel Inc., (now IHeart Media) and TeleAtlas. This used conventional FM RDS across Clear Channel's extensive FM network.

The service in Australia was developed and continues to be operated by Intelematics, a wholly owned subsidiary of the Royal Automobile Club of Victoria. This service is branded as SUNA Traffic and has a full national network serving all state capitals and major metropolitan cities.

A number of Asian countries have, or continue to operate, services as does Latin America. These have been operated with varying degrees of success from a commercial perspective.

6.5 From RDS-TMC to DAB broadcast

This section considers what the industry's aim was in developing future services using an alternative broadcast technology, namely DAB. As discussed in other chapters of this book, the focus was on preparing and developing standards for traffic and traveller information (TTI) over DAB by the TPEG Forum. This forum's focus was to develop standards primarily driven by the needs of the broadcasters, and whilst there were developments in the area of location referencing, this was undertaken in isolation from the automotive and navigation systems industry.

6.5.1 Mobile.Info and RTIG

It was becoming clear, therefore, that the automotive industry needed technology to support its next generation services. In Germany a Federal project was established called Mobile.Info. This was led by BMW and focussed on developing initial TPEG services for broadcast over DAB.

In parallel, industry discussions between the TMC and TPEG Forum were in progress which led to a joint initiative called RTIG, ('Road Traffic Information Group') jointly chaired by the Chairmen of the respective Forums (Danny Woolard TMC and Bev Marks TPEG). The purpose of the group was to both address the technology, standards and needs driven by the Mobile.Info project and also to merge the two groups; this second aim led to the formation of the Travel Information Services Association.

6.5.2 DAB TTI services

It was believed (by many), that DAB would provide a natural progression as a broadcast bearer of TTI using the TPEG Standards. However, the roll out of DAB Digital Radio was fraught with problems and challenges such as the base broadcast protocol (DAB – MP2 vs. DAB + AAC), spectrum allocation and interoperation between EU member states. The UK had already implemented DAB services on both public (BBC) and commercial multiplexes. Given the early lead by the UK, and the fact that the BBC were part of the original development and supporting consortia, the networks were supporting DAB (MP2). As time passed, DAB+ was developed which offers greater channel capacity and more efficient coding and, therefore, most countries implementing DAB later than the pioneers, were electing to develop DAB+ networks. This was leading to further stalling (politically) and uncertainty in terms of roll out.

Ultimately this led to concerns across the traffic information industry, including the automotive makers, and to questions as to the viability of using DAB as the next generation bearer. Efforts then started on developing the needs of TPEG over IP (internet) and using mobile network communications, such as 3G, as the main communications bearer.

Whilst DAB TPEG broadcast services have been developed, for example in Germany (ARD) and the UK (INRIX), they are not widely supported by receiver and device manufacturers. One notable exception has been Garmin who do support DAB-TPEG broadcast TTI.

6.6 Conclusions and outlook

In summary, the global adoption of RDS-TMC traffic services and the support by the automotive industry, their Tier-1 navigation system providers, Personal Navigation Device suppliers and a number of other supporting technology/software providers earns RDS-TMC the position as the foundation of dynamic navigation systems.

Most countries continue to support services and sales of vehicles and compatible devices also continue. However, at the time of writing (Spring 2020), it should be noted that there is wide agreement, particularly with the advent of connected services, that RDS-TMC has a finite life. Most service providers are contractually obliged to maintain service for at least 3 years following a vehicle's sale, but in the author's estimation, it is likely that the commercial service providers will indicate that they will cease service in the 2025–2027 timeframe. This is primarily due to the economics of maintaining the broadcast network.

For the future, it looks as though TTI will develop using mobile digital communications and wireless and internet delivery. There may always be a requirement for public service broadcasting but there will also be a market for commercial service providers of TTI.

Chapter 7

Dynamic road traffic signage

John Mitchell[1]

7.1 Introduction

The use of variable-message signs (VMS) to deliver road safety and other information to road users is a well-established and widely adopted intelligent transport system (ITS) tool. In addition to promoting road safety, VMS is a highly efficient way of implementing active traffic management, such as improving traffic flow and by reducing traffic congestion. VMS is a highly flexible type of road traffic sign and can serve a number of different functions on the road, for example, providing drivers with regulatory warning, or advisory traffic information on a sign, often in various display formats (i.e. text only, pictograms or both pictogram and text).

This chapter describes the development and implementation of VMS, which provides dynamic traffic information to road users. The main focus of this chapter is related to information delivery and display options by VMS on the UK's Strategic Road Network (SRN), but it also considers the provision of traffic information on non-strategic roads, primarily in urban areas.

The first section of this chapter summarises the historical background of the evolution of VMS use on UK motorways and trunk roads. Within this background section, it describes the development of pictograms (i.e. pictorial symbols) shown on VMS.

The next section outlines established UK and European guidance on the use of VMS, including underlying principles of VMS message design.

Following on from this, consideration is given to the different types of VMS currently used on the UK's SRN, with particular emphasis on the display features of the different types of VMS and how display technologies can place limitations on what can be displayed on the sign.

The following section, related to the control of VMS on the SRN, describes how message sets are generated for VMS, identifying the bodies responsible for undertaking this task, both in England and in other parts of the UK.

A detailed description is then provided of typical applications of VMS (both fixed and portable) on the SRN. Typical applications of fixed VMS include: showing drivers

[1]TRL Ltd., TRL, Wokingham, UK

tactical information (primarily for incident management), strategic information, travel time information and non-traffic message information. Examples are also provided of traffic message provision for drivers travelling on traffic management schemes, such as motorway incident detection and automatic signalling (MIDAS), Smart motorways and for also for road works traffic management.

The use of portable VMS is considered, both for traffic information provision in urban areas and providing safety related information at road works.

This chapter also describes the relationship between VMS and other technologies and services providing information for road users, such as MIDAS, Smart motorways and Urban Traffic Management and Control (UTMC) systems.

The final part of this chapter considers anticipated future developments in the use of dynamic traffic information systems. Specific areas include connected and autonomous vehicles and the provision of real-time information in the development of integrated transport strategies.

7.2 Background and context

A review of motorway signals in use in the 1990s was undertaken by Rutley [1]. When motorways were first introduced, it was generally considered that their superior geometric design over established roads, their absence of steep gradients, sharp bends, roundabouts, traffic signals, the prohibition of cyclists and pedestrians and the existence of a hard-shoulder to avoid broken down vehicles having to stop on the main carriageway, would make unnecessary the use of additional traffic control measures other than normal lane road markings, directional signs and a few information type signs.

Some major accidents on motorways caused by poor weather conditions, e.g. fog, resulted in 1965 in the introduction of 'Motorwarn' signals. These signals consisted of two amber fog lamps placed one above the other. Signals were mounted on posts located at one mile intervals. The battery-operated amber lamps warned of the danger ahead, for example, fog, accident, or the risk of skidding. Motorwarn signals were activated by a microwave radar gun from a passing police car. A modern version of Motorwarn signals is in still in use; the amber lanterns flash alternatively and when activated, display an advisory 30 mph speed limit.

It was soon apparent that more sophisticated warning devices were required, so in November 1966, the whole of the 1,000 mile motorway network at that time was equipped with a light-emitting diode (LED) signalling system that was remote controlled. This warning system was the now familiar motorway matrix signal (MS1), which could display a recommended maximum advisory speed limit, or the indication of one or more closed lanes, see Figure 7.1. The MS1 matrix warning system was first introduced in 1967 on the M4 at the Severn Bridge.

Before the introduction of the MS1 matrix signals, two trial systems had been used, both of which were centrally controlled. In 1964, large matrix warning signs were installed on a 38 km section of the M5. These warning signs could display one of three alternative messages: 'SKID RISK', 'ACCIDENT' or 'FOG', each of these

Figure 7.1 Early version post-mounted matrix signal (MS1)

messages being accompanied by the word 'SLOW'. At a similar time, remotely controlled matrix signs were installed on the London end of the M4 motorway from Chiswick to London Airport, which enabled the police to divert traffic from the M4 onto the parallel A4 trunk road in case of an emergency.

The Motorway Signal Mk2 (MS2) was introduced in the early 1990s on trunk roads and motorways. Initially, several MS2s were installed on a trial basis on the M25 London Orbital motorway. The MS2 comprised of an enhanced message sign (EMS), which consisted of letters, numerals and other alpha-numeric characters, these being displayed as two rows of 12 characters (2 × 12), or alternatively two rows of 16 characters (2 × 16) and an enhanced matrix indicator (EMI). The EMI, which was mounted on a display board comprised of amber and red lanterns, was dimensionally much larger than the MS1 matrix indicator. There were several variations in the type of MS2, i.e. mounted on a cantilever arm, or installed on a portal gantry. Most MS2s were comprised of a 320 mm character height EMS with amber lanterns, although there were variations on this (i.e. one type of MS2 included an EMI, and a 400 mm character height was used on two different types of MS2).

The MS2 was replaced in 1999 by the motorway signal mark 3 (MS3), see Figure 7.2. The two-line MS3 is capable of displaying either two rows of 16 characters, with a nominal character height of 400 mm, or alternatively two rows of 12 characters, combined with an EMI. Three-line MS3s can display three rows of 18 characters, with a nominal 400 mm character height. The cantilever-mounted MS3 is still a standard motorway signal which is used to display strategic traffic messages.

In 2000, the Highways Agency initiated a programme to develop a new VMS called a motorway signal mark 4 (MS4), with the main objective of replacing the life-expired post-mounted versions of the matrix signs (MS1) that were widely installed in the central reservations of inter-urban motorways.

Figure 7.2 Motorway signal mark 3 (MS3)

The motorway signal Mark 4 comprises a high definition, dual-colour display that can show pictograms, as well as aspects (such as advisory speed limits, or lane closure 'wickets') that are shown on EMIs.

Prior to the introduction of the MS4 onto the SRN, the Highways Agency commissioned the Transport Research Laboratory (TRL) to conduct extensive off-road research trials to investigate the viability of the MS4 concept. A summary of these off-road research trials, which took place between 2000 and 2001, is contained in TRL's MS4 off-road research summary report Cooper and Freeman [2]. These trials were conducted, in both daylight and darkness conditions, using a prototype version on the MS4 sign to determine the legibility distances and the comprehension of traffic messages in text only format, pictogram only format, or a pictogram combined with text. Dynamic testing at normal motorway speeds (70 mph) investigated the comprehension and reading times of pictograms shown on the new MS4 sign, see Figure 7.3. Static trials determined the legibility distance measurements for text only messages constructed from several different alpha-numeric fonts and in addition for a set of pictograms shown as tactical warning messages, such as 'Accident SLOW DOWN'.

Following the successful conclusion of the off-road research trials, key findings from these trials were fed into an on-road trial of the MS4 in 2002, which was carried out on the M4 motorway between Junctions 12 and 14. Following the successful completion of the on-road trial, the planned roll-out of the MS4 onto the SRN commenced in 2003.

Figure 7.3 Prototype motorway signal mark 4 (MS4)

7.3 Development of pictograms for VMS on the UK's SRN

7.3.1 MS1 pictogram development

A review of motorway signals Rutley [1] described trials of potential improvements to the motorway signal mark 1 (MS1) matrix signal, which took place in the 1970s. The rationale behind conducting these trials was that it was considered that drivers would be better prepared for identified hazards on the road ahead and would be more likely to respond in a more positive way to advisory speed limits shown on motorway signals if the signals themselves showed the nature of the hazard ahead and, where appropriate, the distance to the hazard ahead.

For the on-road trial, 12 pictograms were designed to cover a wide range of reasons for which specific motorway signal was to be used, these are shown in Figure 7.4. In 1979, ten of the existing MS1 signals on the northbound M1 motorway in Bedfordshire were replaced by experimental signals which were controlled by a typewriter terminal in the Police Motorway Control Office.

Analysis of speed data obtained during this on-road trial, which ran between November 1979 and April 1981, showed an average speed reduction of approximately 10 per cent for short length vehicles when the 'Accident' pictogram was displayed and approximately 5.1 per cent for road works when the 'Roadworks' pictogram was shown; for the all other pictograms, the speed reduction was 8.5 per cent. These results showed that the additional information provided by the motorway signals was of considerable benefit in causing a positive response from drivers, with all of the inherit safety benefits.

Figure 7.4 Experimental pictograms shown on MS1 trial on the M1

7.3.2 MS4 pictogram development

During the development stages of the MS4 motorway signal, TRL undertook a short research study of proposed pictograms for the MS4 off-road trial to establish whether the pictogram designs were adequately comprehensible, or whether improvements were required Cooper *et al.* [3].

Seven different pictograms, of two levels of resolution (i.e. 'low resolution' and high resolution') were tested, with each of these pictograms showing hazard warning signs (i.e. the appropriate symbol contained within a red warning triangle). Examples of the proposed pictograms are shown in Figure 7.5. A small number of this set of pictograms had previously been tested as parts of the European TROPIC project TROPIC [4]. The proposed pictograms were tested by a small survey of members of the general public.

7.3.3 Development of MS4 displays for Smart motorways

Prior to the introduction of Smart motorways, with a permanent or temporary conversion of the hard shoulder to normal traffic lane, research was conducted by Blakeman *et al.* [5] to investigate the provision of on-road driver information, specifically lane control,

Pictogram	Design		Findings
	Low-resolution image	High-resolution image	
Accident			Participants had some problems in comprehending both the high and low resolution versions
Danger of ice			Participants had some difficulties in comprehending the low resolution image. The high resolution image was well understood
Slippery road			Participants had significant problems in comprehending both the low and high resolution versions

Figure 7.5 Examples of proposed MS4 pictograms for off-road trial

using only cantilever-mounted MS4 signs that were installed in the verge. This new concept was a departure from the then (2011) standards for Managed Motorway schemes, which required the use of advanced motorway indicator (AMI) to be installed on gantries, with one AMI per lane as lane control signals, in addition to gantry-mounted VMS, to provide supporting information for the lane control signals.

TRL's research, using DigiSim, their advanced driving simulator, together with desk top simulation methods, was undertaken to establish if drivers could comprehend the tactical message shown including lane control information (i.e. advisory/mandatory speed limit or a lane closure 'wicket' sign containing a single red 'X') would add to the complexity of the information. Two sets of images were shown to drivers participating in the simulation trials, one set showing lane control signals and VMS (MS4) installed on a motorway gantry and the other set showing the corresponding lane control signalling and VMS (MS4) displayed on a cantilever-mounted MS4, see Figure 7.6.

Key findings from this research established that the comprehension of safety critical information, such as that provided by a lane closure pictogram, shown on a single-verge-mounted MS4, was equal to or higher than the same information shown on gantry-mounted AMI signals plus a gantry-mounted MS4.

7.3.4 Development of graphical information display panels

In 2002, the Highways Agency investigated the potential use of graphical congestion display panels (GCDPs) to show both real-time traffic congestion and journey time information on sections of the UK's trunk road network. The Highways Agency commissioned the Transportation Research Group of the University of Southampton, in

Figure 7.6 Gantry-mounted AMI signals plus MS4 (left) Cantilever-mounted MS4 (right)

Association with Atkins Transport Systems, to undertake an initial research study to investigate the use of GCDPs. This study by Richards, *et al.* [6] investigated the comprehension of traffic congestion and journey time information that can be displayed on GCDPs, with the aim of establishing some prototype design guidelines. As part of this research study, a simulator study tested several designs of both traffic congestion and journey time information shown on GCDPs. This research concluded that further research should be undertaken to address specific areas, such as the harmonisation of similar display graphical information display panels in European countries. In addition to this, it was recommended that research should be undertaken into specific features of the display panel, such as text size and colour, appropriate colour coding of the routes and a suitable symbol to show a blocked route.

A further programme of research and development into the use of graphical information display panels to display traffic congestion information on the SRN was commissioned by the Highways Agency, in 2009. As part of this programme, a research study conducted by Shalloe *et al.* [7] investigated the effect of increased information shown on a VMS (MS4 type) displaying graphical information in terms of both driver behaviour and attitude. This study developed both draft designs and design guidelines for graphical route information panels (GRIPS) to be used to display traffic congestion information on the UK's motorway network. The main objective of this research was to develop an understanding of the effect of increased information shown to drivers, in addition to the effects of a pictorial representation of congestion shown on the motorway network, on both driver attitude and on driver behaviour. Basic GRIPS designs were developed using the findings of targeted literature reviews and driver/industry expert focus groups. As part of this research, a scenario study investigated people's comprehension of a number of different GRIPS Designs and to establish potential effects of drivers' behaviour related to route choice. Findings from this research study promoted a number of guidance principles for the design of road-based graphical displays. Guidance principles included methods of improving driver comprehension and acceptance, such as providing information about road curvature and relative distance. The choice of colour representing a road closure was another factor, with red, as illustrated below, being identified as the most appropriate colour, with either a red bar or red cross symbol used to depict the closure, see Figure 7.7.

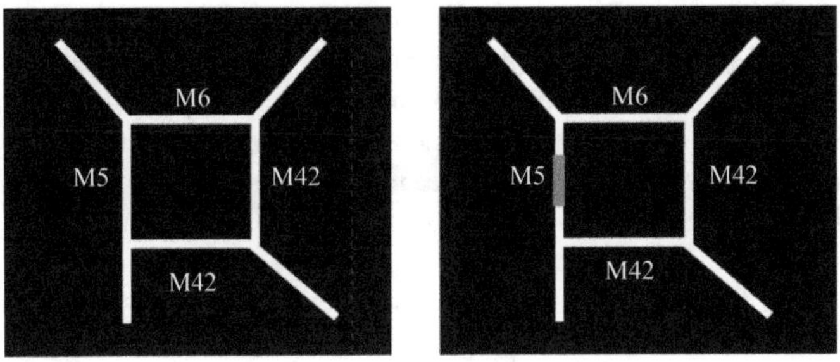

Figure 7.7 Experimental GRIPS display of the Birmingham Box

7.4 Established guidance on the use of VMS, including VMS message design

7.4.1 Highways England/Highways Agency guidance

Highways England (previously the Highways Agency) has published several guidance documents in the form of either technical standards or technical advice notes, these being related to the different types of motorway signalling (VMS) available at that time of issue and a guide to the use of VMS for strategic traffic management on trunk roads, including motorways. Two technical standards and advice notes were published in 2005, these being, TD 46/05 – Motorway Signalling, Highways Agency [8] and TA 83/05 – Guide to the use of VMS for strategic traffic management on trunk roads and trunk road motorways, Highways Agency [9]. Both technical standards are contained within the Design Manual for Roads and Bridges, Highways England [10].

TD 46/05 describes the different types of motorway signals, including VMS, present at the date of publication (2005) and describes the criteria required for their provision. This standard provides an overview of the types of motorway signal and VMS in use at that time, i.e. MS1, MS2, MS3 and gantry-mounted lane signals. Signalling standards are provided, for example for 2-lane motorways, high flow motorways.

TA 83/05 provides guidance for the design, implementation and evaluation of Strategic Traffic Management Systems using VMS. Strategic Traffic Management provides motorists with information and advice on the road conditions ahead. Advice related to Strategic VMS types includes VMS technologies and types and applications of Strategic VMS, such as providing information at strategic junctions, diversion routing and local traffic management.

In 2008, the Highways Agency published an Interim Advice Note 109/08, Highways Agency [11], relating to the MS4. This Advice Note gives guidance on the use of cantilever-mounted MS4 motorway signals including the display options available, including EMI aspects, pictograms and text messages which include the use of mixed case text. In addition, guidance is provided on the siting of MS4s, to minimise sign obscuration issues.

In 2017, Highways England produced a detailed policy document on the use of variable signs and signals (VSS), Highways England [12]. The term 'VSS' encompasses VMS and electronic light-emitting matrix signals, the latter being required for lane control. Highways England's current operation of VSS is contained within this policy document, which is divided into three main sections: [1] Signs and signals policy guidance, [2] Tactical signs and signals policy guidance and [3] Strategic signs policy guidance.

The section related to signs and signals policy guidance describes the relationship between tactical and strategic signs and signals, the location of messages on VMS, the use of junction numbers and place names, diversions, height and width restrictions, flashing and scrolling VMS messages, the design principles of VMS messages, including upper or mixed case text, pictograms and on the overuse of VMS messages.

The following section, related to tactical signs and signals policy guidance, provides information on lane closures, advisory and mandatory speed limits, incident management and unconfirmed reports, use of the word 'END', police pursuits, severe weather, up and over signing, MIDAS signalling and tactical road works information.

The final section of the policy document, related to strategic signs policy requirements, provides guidance on future notification on road works, dual messages, travel time messages, GRIP messages, signing for delays, now open and time to reopen messages, multimodal messages, toll roads, campaigns, special events, postponed or cancelled events, crisis management, area wide severe weather warnings and trapped traffic.

7.4.2 European guidance

European guidance on the principles of VMS Message design is contained within the document: 'Variable Message Signs Harmonisation – Principles of VMS Message Design', ITS-Platform [13].

This guidance document focuses on the general principles of VMS message design, which is synthesised through its 32 stated principles. One of the key aims of this guidance document is for any VMS operator in Europe, who follows the recommendations contained in these stated principles when composing VMS messages, to contribute to greater harmonisation of signing on the Trans European Road Network.

The general principles of VMS message design contained within these guidelines can be applied to a number of different categories of road/traffic scenarios for which VMS messages can be displayed: traffic congestion, traffic flow information, unplanned events, road works, dynamic traffic management, weather related information, pre–announcements, co-mobility and campaign messages.

The guidance document recognises that there is diversity in VMS technology. Design principles described in this document are applicable to all of the current types of VMS used, i.e. warning/regulatory signs used in lane control systems and text only VMS:

- Normally 2 or 3 lines of text;
- Text VMS with pictogram area on the left side;
- Text VMS with 2 pictogram areas on the left;
- Text VMS with pictogram area on both sides; and
- Fully matrix panels which can display bitmaps.

7.5 VMS currently in use on the UK's SRN

The most basic type of VMS currently used on the UK's SRN is the Motorway Signal Mark 1 AMI signal, see Figure 7.8. This signal is an upgraded version of the MS1 introduced back in the 1990s. The current type of MS1 can display a number of different legends (either pictorial or text, including temporary maximum advisory speed limit, risk of fog warning ahead, and lane closure wicket (for a closure for up to three traffic lanes). This motorway signal is installed in the central reserve of a motorway and can also be found located at motorway to motorway interchanges, entry slip roads to sections of motorway controlled by Motorway Signal Mark 3s (MS3) VMS, or gantry-mounted signals.

EMSs are installed on portal gantries where they are used as tactical VMS. This type of tactical VMS (i.e. 2 rows of 12 characters) is used to support the AMI matrix lane control signals installed on the gantry. The characters used on this particular type of VMS have a reduced character height, which is nominally 320 mm.

TD 46/05 describes the message form that can be displayed on the Motorway Signal Mark 3 (MS3). This can comprise of 2 rows of 16 characters, with a nominal character height of 400 mm in EMS only mode, or alternatively, 2 rows of 12 characters where the EMS is combined with an enhanced message indicator (EMI), which can display, for example, a lane closure wicket aspect. MS3s can be mounted on portal gantries as a tactical VMS, as described previously.

Figure 7.8 MS1 displaying a 50-mph maximum advisory speed limit

Table 7.1 MS3 – variations in display format

Character format (400 mm character height)			
Rows	**Characters**	**Mode**	**Mounting arrangement**
2	16	EMS	Portal gantry
2	12	EMS + EMI	Portal gantry
3	18	EMS	Cantilever arm
3	12	EMS + EMI	Cantilever arm

A second type of MS3 is mounted on a cantilever arm. This type of MS3 combines an EMS and EMI with two variants: MS3 (2 rows of 16 characters in EMS only mode, or 2 rows of 12 characters, plus an EMI), or alternatively, 3 rows of 18 characters (with a nominal character height of 400 m) in EMS mode, or 3 rows of 12 characters, plus an EMI in combined mode. The variations in MS3 display format are shown in Table 7.1.

The MS4 can either be mounted on a cantilever arm, or installed on portal gantries, with the cantilever arrangement being significantly more common. The MS4 can display a wide range of red/white/yellow bitmap images, but it is only allowed to display messages (text, or pictograms and EMI aspects, for example, mandatory or advisory speed limits) that are prescribed in the Traffic Signs Regulations and General Directions, 2016, Statutory Instruments [14]. The top two thirds of the MS4 display panel can display EMI aspects and warning messages, including those containing a red warning triangle. Examples of pictorial messages that can be displayed on an MS4 are shown in Figure 7.9. Text only messages can be displayed in capitals only (uppercase) or as mixed case. A specific font (Transport Medium, in either upper case or mixed case format) is used for text messages. Due to the nature of the MS4s display panel, an array of LEDs with 20 mm spacing permits a variable character height.

7.5.1 Control of VMS messages in England

Historically, VMS messages had previously been set by police officers in police control offices. As the Highways Agency started to introduce Regional Control Centres (RCCs) across the United Kingdom, the responsibility for setting VMS on motorways was progressively transferred to Highways Agency operators.

Highways England currently maintains 7 RCCs throughout England. RCCs are responsible for setting tactical* messages on VMS. Tactical legends include confirmed and unconfirmed incidents, tactical diversion messages, site specific authorised

*Tactical messages are defined as: *any messages set to assist with local incident/hazard management, ordinarily in support of signals and or messages set by automated systems for similar purposes.* (Source: Policy and Procedures for the use of variable message signs by Regional Control Centres Issue 3 Highways Agency, 2007.)

Figure 7.9 Examples of messages shown on MS4

messages, link messages, network messages, strategic diversion messages and approved national campaign messages.

Highways England's National Traffic Operations Centre (NTOC) (previously called the National Traffic Control Centre) remains responsible for setting strategic[†] messages). Strategic messages are only applicable to VMS that can support 3×16, or 3×18 message format. NTOC may also set Driver Information Messages. NTOC also has the capability of setting any messages on behalf of the police, or for the RCC. Another chapter of this book – Smart Motorway Information Development – describes UK motorway management in more detail.

London Roads are managed by Transport for London (i.e. London Streets Traffic Control Centre).

7.5.2 Control of VMS messages in other parts of the UK

Welsh roads are managed by the Traffic Wales. Tactical and strategic messages that are shown on VMS on motorways are controlled by one of two Traffic Management Centres depending on geographical location of the road, i.e. North Wales Traffic Management Centre, or the South Wales Traffic Management Centre.

Scottish Roads are managed by Traffic Scotland. In Scotland, tactical and strategic messages shown on VMS are controlled by Traffic Scotland Control Centre.

7.6 Limitations to the use of VMS

Despite its many advantages over fixed plate traffic signs, there are potential limitations and some disadvantages to the use of VMS.

VMS is often found on high speed roads, such as motorways, where traffic speeds are generally much higher compared with those found on urban roads. When travelling at high speeds on motorways, drivers often have a very short time, typically just a few seconds, to read a message shown on a VMS. This may be further reduced due to obscuration caused by other vehicles. This very short time window restricts the amount of information that drivers can read on the VMS sign. Displaying too much information,

[†]Strategic messages are defined as: 'a wide range of advisory diversion signing system which can provide specific information to accord with agreed diversionary options'.

which drivers would extremely difficult to read in the time available could also result in information overload and may also increase the possibility of driver distraction; both could have safety implications.

Being an active traffic sign, VMS brightness is another factor related to drivers' ability to correctly read a message shown on the sign, especially if the sign's ambient light level sensing system is not functioning correctly. This may result in a VMS message being too bright in darkness, or not bright enough in daylight, for drivers to be able to correctly read the message shown on the sign.

There is also the issue of a blank VMS. If a VMS is blank, it is not possible to tell if the VMS is functioning correctly and whether traffic conditions on the road ahead are normal. A blank VMS is likely to be interpreted by drivers that traffic is running as 'normal' on the road ahead, with no incidents or other events causing a delay. However, unwanted situations occasionally occur, for example, a communication systems failure, or a power failure, the latter being more likely to occur on battery powered portable VMS. On Welsh and Scottish motorways, real-time information is shown on one small section of the VMS display panel, which gives drivers reassurance that the VMS is functioning correctly.

Technology restrictions related to display format apply to some types of VMS. In the UK, the type of VMS which is currently used to display strategic traffic information has a fixed message length format, this being in terms of the number of lines available for a message to be displayed and the number of characters that can be shown on each line. This technological constraint limits the length of a traffic message and in addition, does not allow pictograms to be shown on the sign.

7.7 Typical VMS applications on the UK's SRN

Typical applications described in this section are only applicable to permanent VMS, i.e. VMS installed on portal gantries, or mounted on cantilever arms. Guidance on the different applications of VMS on the SRN is contained within the Highways England policy document on the use of VSS, Highways England [12]. Sections 7.7.1–7.7.10 summarise the guidance contained within this policy document for commonly shown messages on motorway VMS.

7.7.1 Tactical messages

Tactical messages shown on VMS provide drivers with early warning of incidents, accidents, or adverse weather conditions on the SRN. When viewing these particular messages, drivers may be required to take some degree of action as an active response to the message shown on the VMS, for example, reduce their speed, change their lane of travel, or in some extreme circumstances, stop.

Examples of commonly used tactical incident management and weather messages are shown below:

- 'Accident slow down';
- 'Congestion slow down';

- 'Report of debris';
- 'Incident slow down';
- 'Lane closed on slip road';
- 'Queue on slip road';
- 'Report of obstruction';
- 'Salt spreading';
- 'Fog slow down'.

7.7.2 Strategic messages

Strategic messages are shown on VMS to provide drivers with information in advance of any incidents or events that are highly likely to have an impact, either on the road that a driver is currently travelling on, or alternatively on other specific routes which may have an impact on a driver's journey. There are two main types of strategic message – driver information (link) message and driver information (network) message.

7.7.3 Driver information (link) messages

A link message is the simplest type of driver information message. This type of message shows the cause of a delay and its location. Examples of commonly used link messages (see Figure 7.10) include:

- 'Accident after J29';
- 'Congestion after J29';
- 'Long delays after J29'.

7.7.4 Driver information (network) messages

Driver information (network) messages are generally more complex. Network messages provide drivers with additional information, such as the specific location of an incident. Figure 7.11 below shows a typical example of the type of a driver information network messages often seen on motorways.

Figure 7.10 Example of a driver information (link) message

Figure 7.11 Example of a driver information (network) message

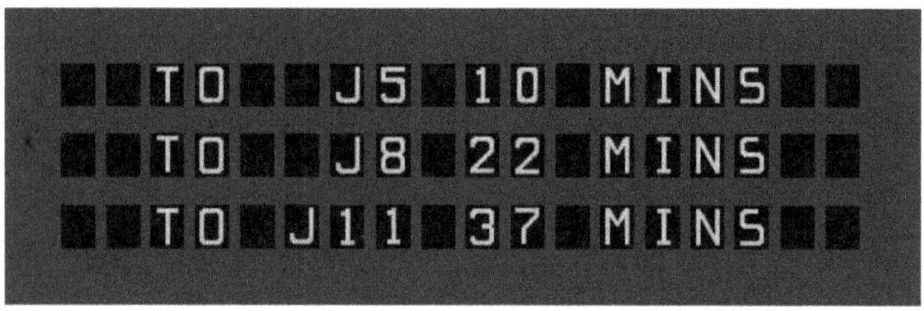

Figure 7.12 Example of a travel time message

7.7.5 Travel time messages

Travel time messages provide drivers with travel time information, i.e. average journey times, this normally being in whole minutes, see Figure 7.12. Travel time is usually to a particular destination, or junction (or junctions) on the same road that a driver is currently travelling. Travel time messages can also provide drivers with information on the time taken to reach a particular destination or junction via different roads, but this is much less commonly seen.

Travel times are obtained from near real-time information, which is calculated every minute, and historical travel time information.

7.7.6 Motorway incident detection and automatic signalling (MIDAS)

MIDAS systems have been installed on motorways in England for a number of years. Smart motorways, where the hard shoulder has been converted to a normal traffic lane, were first introduced in 2015. Messages shown on VMS to support both MIDAS and Smart motorways are set by RCCs.

A set of messages have been developed that are exclusively used with MIDAS systems:

- 'Queue caution';
- 'Queue ahead';

- 'Queue after Jct';
- 'Queue on slip road'.

A further set of messages have been developed that are exclusively used with Smart motorways:

- 'Congestion stay in lane';
- 'Congestion caution';
- 'Congestion after Junction'.

7.7.7 Graphical route information messages

GRIPS have been developed for motorways in a simple two-colour graphical display format. The panel displays traffic conditions on the road ahead and uses a second colour, i.e. red, to highlight areas of the motorway where delays are occurring.

Due to its display requirements GRIPS messages can only be displayed on MS4s, as only this type of VMS used on motorways has a fully programmable two-colour display.

7.7.8 Diversion messages

Diversion messages can be displayed on VMS.

Strategic diversion messages are used to advise drivers to use an alternative route due to an incident, accident, or an event on the road ahead, for example, 'M6 ACCIDENT J2-J3 FOR M5 USE J1'. Strategic diversion messages are set by the NTOC.

Tactical diversion route messages are set by one of the 7 RCCs. A commonly shown tactical diversion route message includes a simple geometric shape as part of the message. This type of tactical message format can be displayed on an MS4: 'DIVERSION FOLLOW circle/square/triangle, diamond'.

7.7.9 Non-traffic information messages

A review of the use of non-traffic messages was undertaken in 2002 by Cooper and Mitchell [15]. Non-traffic messages were shown on VMS located on English motorways for a number of short-term safety schemes in the 1990s. Schemes included three National Courtesy Days in 1995, 1996 and 1998. Two surveys were conducted on the M25 in 1993 and 1994, following the installation of new VMS signs and in addition two further schemes were initiated in 1997: the Cheshire Driver Behaviour Initiative and the Kent Speedwatch scheme. All of these schemes were initiated to investigate the effects of the VMS messages on driver attitudes and/or behaviour. In Scotland, VMS can be used continuously to display safety messages in the absence of traffic messages.

For the National Courtesy Days, a number of different courtesy message were shown on VMS. These included: 'Courtesy day – don't cut in', 'National Courtesy day', 'Keep calm – keep your distance', 'Courtesy day – don't hog the middle lane'.

For the Cheshire Driver Behaviour Initiative, which was part of a highly published local safety campaign related to speeding and other desirable driver behaviour, two different messages were show on VMS: 'Watch your speed', or 'Keep your distance'.

The Kent Speedwatch scheme showed four different messages on VMS located on the M2 and M20 motorways: 'Watch your speed', 'Don't hog the middle lane', 'Keep your distance' and 'Tiredness kills – take a break'.

At present, road safety campaign messages can be displayed on VMS. These messages can be shown either when they are required, or alternatively as part of a planned campaign, such as promoting the safety of more vulnerable road users, such as motorcyclists. Common examples of this type of campaign message are:

- 'Red X means lane closed';
- 'Think bike, think biker';
- 'Don't hog the middle lane'.

VMS can also display a further set of road safety campaign messages which are used to support substantial associated road safety activities, such as mobile phone use whilst driving. Common examples of this type of message are 'Don't drink and drive', 'Don't drive tired', 'Watch your speed' and 'Don't phone while driving'.

Special events messages can also be displayed on VMS, with the provision that previous occurrences of a particular event have disrupted traffic on parts of the SRN in the vicinity of the event. There is a fixed VMS message format which informs drivers about a future event:

- Location of an event and its description (e.g. 'Twickenham rugby event');
- Date(s) of the event (e.g. 'Saturday 12 March');
- 'Expect delays'.

7.7.10 Road works information

Permanent VMS is used to display tactical road works messages and future notification of road works messages.

Tactical road works messages are commonly used when road workers are setting out temporary traffic management, such as fixed plate traffic signs and cones. The most commonly used tactical roads works message is 'Workforce in road – slow'.

With regard to future notification of road works, VMS can be used to warn drivers of planned works, which are likely to result in delays or increased journey times, see Figure 7.13. Highways England has specified that for planned works on

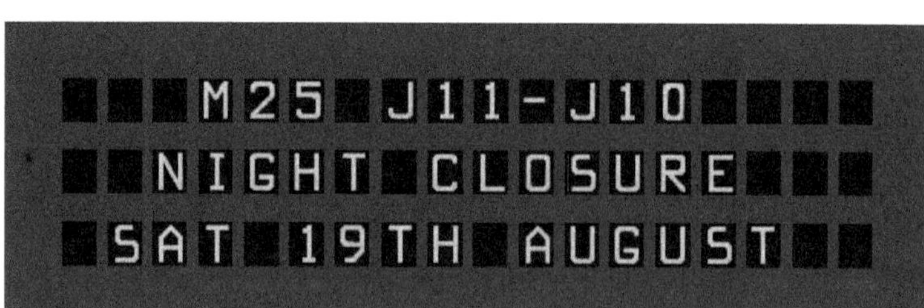

Figure 7.13 Example of a future notification of road works message

the SRN, it is only permitted to show these messages one week before the road works commences. In addition, VMS should only display future roads work information in a specific format:

- Place/location of road works;
- Date(s) of road closure;
- Time(s) of road closure.

7.8 Urban traffic information provision by VMS

VMS is used to provide drivers with information about local traffic conditions, road closures, car-parking availability and for specific safety issues, for example, warning the drivers of high sided vehicles of the presence of a low railway bridge on their route.

For local traffic conditions, VMS designed specifically for urban use can display appropriate text messages and may also have the capability of showing warning pictograms, for example, ‘congestion’ or ‘accident’. These symbolic messages, shown in conjunction with an appropriate text message, can either re-enforce a pictogram’s meaning or give drivers a warning instruction such as ‘Slow’. VMS may also be able to display regulatory road traffic information, such as mandatory speed limits.

For road closures in urban areas, which can be as the result of a planned event, or road works, VMS can provide drivers with diversionary route information which may be linked to UTMC systems. Car parking availability can also be UTMC compliant or be stand alone.

The use of VMS signing has proven to be a major benefit for parking systems. VMS directs motorists to the most appropriate parking location to minimise their journey time when looking for a parking space, see Figure 7.14. VMS also causes a reduction in circulating traffic looking for a parking place. Another advantage of the use of VMS is that it reduces environmental pollution, this being particularly important at sensitive heritage sites and in areas of architectural importance. Most VMS used for car parking displays the total number of spaces available in a particular car park, together with information such as ‘Spaces’, ‘Full’ or ‘Closed’.

VMS can also be used as a safety measure, such as an aid in the prevention of railway bridge strikes. The use of VMS for this purpose occurs in both urban and rural areas but is more commonly used in the urban environment. The use of VMS as a low bridge warning aid is often seen as a last resort due to the high level of cost to local authorities, so VMS is normally only used at low railway bridges where there is a history of bridge strikes. VMS can be installed in close vicinity to a low railway bridge, or alternatively on its approaches, providing a warning for high sided vehicles, or giving the drivers of such vehicles alternative route information.

Full-colour VMS can be used in urban areas to display warning information to drivers. Where urban road works (referred to as street works) are present, VMS can be used to show appropriate warning messages, such as the ‘road works ahead’ warning pictogram sign plus the appropriate distance (shown in yards) to the start

Figure 7.14 Example of car-parking availability shown on VMS

of the works. On multi-lane urban roads, the 'wicket' pictogram sign, showing both open and closed lane(s), can be displayed. Where road works are occurring in urban areas, it is likely that appropriate warning information will be displayed on portable VMS.

7.9 European perspective on the use of VMS

VMS is widely used for traffic management and for providing traffic related information in many different European countries.

In the Netherlands, there is extensive use of VMS on motorways, with over 5,600 VMS locations, Beck, *et al.* [16]. There are also 430 dynamic route information panels (i.e. graphical congestion displays) installed on Dutch motorways. Dynamic route information panels were first introduced in the Netherlands in 2005. The modern type of dynamic route information panel has the facility of changing the colour of discrete sections of the sign message according to the level of congestion on the motorway, i.e. the black default colour setting is changed to yellow or to red (which is used to indicate delays) depending on the congestion levels. Hard shoulder running, where a motorway hard shoulder is temporarily converted into a normal traffic lane, has been used in the Netherlands since the 1990s. Active signage, in the form of lane control signals, in addition to VMS, tells drivers if they are able to use the hard shoulder as a normal traffic lane. The Netherlands is also a partner in a Cooperative ITS Corridor project, where cooperative services will be provided on the Rotterdam – Venlo (Austria) motorway network. One target identified for this project is for In-Vehicle Signage, where information displayed above the road on VMS is sent to passing vehicles with the aim of increasing the level of road safety.

On controlled sections of motorway in Germany, VMS is used to manage traffic and optimise traffic flow by displaying mandatory variable-speed limits (VSLs), lane control signals, or pictograms, which provide either mandatory or warning information. The use of VMS on controlled sections of German motorway is described by Busch [17] Control is either traffic dependent (i.e. incident or congestion warning, or specifically related to entry/exit slip roads), environment dependent (i.e. slippery road/ rain/fog warning) or special functions, such as the prohibition of lorry overtaking. VMS showing graphical representations of section of the motorway network is commonly used on German motorways. This type of graphical sign shows the level of congestion, such that the black default setting on a graphical display of the section of motorway network in question can be changed to red according to the level of congestion. Graphical information panels can either be mounted on overhead gantries, or alternatively be ground mounted. There are several other types of VMS commonly found on German motorways. VMS can be incorporated into fixed plate directional signs, where the VMS message can be changed according to the traffic conditions ahead, for example, a route diversion message can be shown where severe congestion is present several junctions ahead. VMS can also be incorporated into a fixed plate directional type sign, which shows the destinations for several junctions ahead. The inclusion of small VMS panels enables travel time information to be displayed for each of the respective destinations shown on the fixed plate directional sign.

7.10 International perspective on the use of VMS

The use of VMS for both traffic control and for the provision of traffic information is found in many developed countries outside of Europe. The use of VMS for traffic control and the provision of traffic information is widespread in countries such as the United States, Australia and Japan.

In the United States, VMS (referred to as 'Changeable Messages Signs') is used to display traffic, operational, regulatory, warning and guidance information. In the United States, there are many applications of the use of VMS, which include incident management, adverse weather condition warning, lane control, travel times, warning scenarios, traffic regulations, speed control, destination guidance, route diversions and toll lanes. In addition, individual States and local highways agencies are permitted to display safety messages, transport related messages, emergency homeland security messages, or AMBER (America's Missing: broadcast emergency response) alert messages. Fixed location VMS is widely used on freeways and portable VMS is commonly used at temporary traffic management control zones (referred to as 'works zones').

In Australia, VMS has been installed on many of the individual State freeway systems and on major roads serving these important network links. There are several different types of VMS used in Australia, ranging from one or two-line message VMS to VMS with graphical displays. In Western Australia, VMS has been used as part of the Graham Farmer Freeway Tunnel system, with VMS installed on approach roads to the tunnel and at the entrance points, where they are used to control traffic entering or leaving the tunnel. Portable VMS is used at appropriate locations on many different types of road, with applications such as traffic calming, advance warning, and associated with speed radar.

In Japan, active traffic information is provided on highways by VMS and graphic information boards. VMS is used to display pictograms and bi-lingual text messages (in both English and Japanese). Journey times (bi-lingual – English/Japanese) are displayed on VMS travel time information boards and graphical information boards (which have been used in Japan since the 1980s), to show traffic conditions on the road ahead. The black default colour of the network representation on these graphical display panels changes to amber and then to red as the level of congestion increases.

7.11 Lane control signals

7.11.1 Advisory and mandatory VSLs

In the United Kingdom, advisory speed limits can be displayed as lane control signals, with a maximum advisory speed limit of 60 mph being shown, down to a minimum advisory speed limit of 20 mph, in 10 mph decrements. Advisory speed limit reductions are not normally greater than 20 mph at a time.

For multi-lane motorways that have portal gantries where matrix signals – AMIs have been installed, advisory speed limit signals taken as a whole set will either be off (i.e. blank, with no speed limit shown), or switched on, with the same advisory speed limit shown over all live traffic lanes. However, there is one exception to this particular rule in situations where a gantry is located at the parallel diverge of motorway, where different speed limits can be displayed on groups of signals, for example, one group of a signals displaying 40 mph and another group displaying 60 mph. In situations where lane control signalling is not provided on gantries, such as on some Smart motorways, advisory speed limits applying to all traffic lanes can be displayed on a cantilever-mounted MS4. In the past, advisory speed limits were also shown on some types of MS2 or MS3 containing an Enhance Motorway Indicator (EMI).

Similarly to advisory speed limits, mandatory speed limits, can be displayed as lane control signals, but these signals can display the national speed limit sign (70 mph on motorways), or a mandatory speed limit of 60 mph with 10 mph decrements down to 10 mph. Where portal gantries are present containing AMI signals, the mandatory speed limit is displayed over all the live traffic lanes. The mandatory speed limit sign (speed limit roundel) is located on the upper right-hand area of an MS4 display panel.

Figure 7.15 Example of a lane closure sign displayed on an MS4

When the reason, or reasons, why a speed restriction no longer applies, the national speed limit sign together with the message 'End' is displayed over each of the live traffic lanes where portal gantries are present, or a single national speed limit sign plus 'End' is displayed on a cantilever-mounted MS4.

7.11.2 Other lane control signals

Lane control arrows and lane closure red Xs can also be displayed using AMIs where portal gantries containing these signals are present. In situations where gantries are not present, for example on some Smart motorways, cantilever-mounted MS4s can display lane control applying to all traffic lanes shown as a lane closure pictogram, this containing a red X with the lanes open depicted by vertical arrows, see Figure 7.15.

7.12 Intelligent road studs

Intelligent (active) road studs are now becoming commonplace on many types of road in the United Kingdom. The use of solar powered LED road studs have been proven to be a major safety benefit in darkness conditions, as they can be visible to drivers at a distance of up to 1,000 m, this being a significant improvement over the visibility provided by traditional retro-reflective road studs. Intelligent road studs are well suited to complement centre line road markings and slip road markings on different types of road. There many other applications of the use of active road studs, including dynamic lane marking, tunnel marking, all-weather lane marking and the marking of railway level crossings.

Intelligent road studs are current being installed by Highways England at one of the busiest junctions on the motorway network where two motorways (M57 and M58) join with three A roads. The LED road studs are linked to traffic signals at the motorway junction so that the road studs are illuminated when the traffic signals turn green, enabling drivers to see clearly the lane they should be following. Highways England has previously used intelligent road studs at the Hindhead Tunnel on the A3 in Surrey so that drivers can be guided through the tunnel.

A European funded research project undertaken and, coordinated by TRL [18], investigated the use of intelligent road studs. A driving simulator study conducted by IFSTTAR in Paris investigated the behaviour and attention levels for different driving scenarios on curved sections of road with different lighting scenarios (i.e. no lighting, illuminated by conventional street lights and illuminated with active road studs). Results from this study showed that drivers kept to their lanes and had a higher standard of driving when guided by intelligent road studs. In addition, field testing took place on off-road test facilities in Austria and Spain and on-road testing took place on the A6 motorway between Vienna, Austria and Bratislava, Slovakia.

7.13 Portable VMS

7.13.1 Road works applications

Portable VMS is becoming more commonly used on motorways for providing drivers with road works information. VMS generally provides a higher visual impact for drivers compared to fixed plate signs, and at road works, it is often used in conjunction with fixed plate road works signs. Portable VMS is often located close to fixed plate advance road work warning signs, where queuing is common, or between 200 yards and 800 yards of a lane closure in place of fixed plate lane closure wicket signs.

Portable VMS can be used to show a message giving the travel time through road works, an example is shown in Figure 7.16. When road works are being set out, real-time traffic information is often not available, this being due to either systems being switched off (for example, MIDAS loops are typically disabled during road works).

Figure 7.16 Portable VMS showing the delay through road works

Portable VMS can be used to show an appropriate travel time message, with real-time accurate information on the time taken (or the delay experienced) to drive though the road works.

7.13.2 Urban applications

Portable VMS is sometimes used in urban locations for non-road works traffic information, such as providing drivers with information on local traffic conditions, local road closures, or advising drivers of future planned events.

7.14 Outlook

7.14.1 Relationships with other technologies and services

The use of VMS is closely linked with other technologies and services. On motorways VMS is an integral part of the MIDAS system, Smart motorways and in urban areas, UTMC.

MIDAS uses sensor information (historically from induction loops, but now additionally adopting radar detection) for setting mandatory VSLs and for queue protection information in the form of appropriate VMS messages.

Some Smart motorways have the hard shoulder converted permanently to a normal traffic lane. On other Smart motorways, the hard shoulder is opened as a normal live traffic lane at peak times of the day. On Smart motorways VSLs are often set at peak times. If an incident occurs, these mandatory VSLs can be set with a red X (showing the lane closed ahead). In these situations an appropriate tactical message is shown on a VMS, which may be located on a gantry, or alternatively on a cantilever-mounted MS4.

UTMC systems allow many of the different applications used within modern traffic management systems to communicate with each other and to share information. Data from multiple sources, such as traffic signals, Automatic Number Plate Recognition (ANPR) cameras, VMS, public car parks and air quality/weather monitoring stations to be incorporated into a central control system, or database.

7.14.2 Anticipated future developments

Future developments are expected to be in the fast-emerging areas of connected and autonomous vehicles, this including traffic sign recognition systems. With the future deployment of connected services on the SRN, In-Vehicle devices displaying traffic information shown on VMS is likely to become more commonplace on high speed roads and on other trunk roads. Depending on the system used, this may require a change in traffic sign architecture with the use of wireless communications technologies.

With advances in the wireless communications, it is foreseeable that, in future digital traffic sign posts will be implemented which would transmit traffic sign information shown on VMS wirelessly to vehicles, with the signs displayed automatically on in-vehicle displays. This information may change over time, reflecting real-time traffic conditions.

Recognition of conventional fixed plate traffic signs linked to in-vehicle displays is already a well-established technology, with several major car manufactures offering this feature in some of their top of the range models. However, it has been established that VMS can be difficult to read with digital camera sensors installed on vehicles due to these sensors using control systems and technologies that are designed for the human eye, EuroRAP [19].

Future developments are also expected to be in real-time information provision as part of the development of integrated transport strategies.

References

[1] Rutley K. *A review of motorway signals.* Transport and Road Research Laboratory, Department of Transport Research Report 339 1992

[2] Cooper B., Freeman M., and Mitchell J. *MS4 off-road research summary report.* TRL Report TRL 604 TRL Limited 2002

[3] Cooper B., Burton S., and Mitchell J. *Comprehension of proposed MS4 pictograms.* TRL Unpublished Project Report PR/TT/115/99, Transport Research Laboratory, 1999

[4] TROPIC. *Traffic optimisation by Integration and Control.* Transport Research, European Commission Fourth Framework Programme 1999 https://cordis.europa.eu/project/rcn/105469/factsheet/en

[5] Blakeman P., Boddington K., Diels C., and Weekley J. *Managed Motorways 2 Concept Development Tasks 2, 3 4 and 5 Key Findings Report.* CPR 1062, TRL Limited, 2011 https://highwaysengland.co.uk/knowledge-compendium/knowledge/publications/managed-motorways-2-concept-development/index.html

[6] Richards A., McDonald M., Fisher G., and Brackstone M. *Investigation of Driver Comprehension of Traffic Information on Graphical Congestion Display Panels using a Driving Simulator*. EJTIR, 4, no. 4 (2004, pp. 417–435) 2005

[7] Shalloe S., Sharples S., Burnett G., Crundall, D., Meekums R., and Morris D. *Developing a Graphical Route Information Panel (GRIP) for use on the UK motorway network. The first steps.* Transportation Research Part F 27 (113–149) 2014

[8] Highways Agency. *TD 46/05 Motorway signalling.* Design Manual for Roads and Bridges, Volume 9, Section 1 2005

[9] Highways Agency. *TA83/05 Guide to the use of variable message signs for strategic traffic management on trunk roads and trunk road motorways.* Design Manual for Roads and Bridges, Volume 9, Section 4 2005

[10] Highways England. *Design Manual for Roads and Bridges* (DMRB), 2018

[11] Highways Agency. *Interim Advice Note 109/08. Advice Regarding the Motorway Signal Mark* 4. 2008 http://www.standardsforhighways.co.uk/ha/standards/ians/pdfs/ian109.pdf

[12] Highways England. *Highways England Policy for the use of Variable signs and Signals*. Version 2.2 2017 https://www.artsm.org.uk/content/uploads/2018/06/Highways-England-policy-for-the-use-of-Variable-Signs-and-Signals-VSS-3.0.pdf
[13] ITS Platform. *Variable Message Signs Harmonisation Principles of VMS Messages Design*. Supporting Guideline VMS-DG-01 Version 02-02-00 2015
[14] Statutory Instruments. *The Traffic Signs Regulations and General Directions*. The Stationary Office Limited 2016
[15] Cooper B. and Mitchell J. *Safety and effectiveness of the wider use of VMS – Final Report*. TRL Report 526, TRL Limited 2002
[16] Beck M., Schol E., Tierolf J.W., *et al. ITS in the Netherland Progress Report 2014-2017*. Ministry of Infrastructure and the Environment in cooperation with Connekt/ITS Netherlands, 2017
[17] Busch F. *Intelligent Transport Systems for Interurban and Regional Traffic Control – German approach and research samples*. Traffic Engineering and Control, Technische Universität, München 2010
[18] TRL. *Intelligent Renewable Optical Advisory Systems (INROADS)*. 2015 https://cordis.europa.eu/project/rcn/102011/reporting/en https://www.inroad.eu/
[19] EuroRAP. *Roads that cars can read – A Quality Standard for Road Markings and Traffic Signs on Major Rural Roads*. 2013

Chapter 8
Smart motorway information development

Neil Hoose[1]

8.1 Introduction

Driver information technology can be used to provide dynamic, 'tactical' management and control for the immediate road ahead. In these situations, the signs and signals display 'control' information, that is to say, legends and messages that the driver must respond to and abide by. Failure to comply with the control instruction is an offence and the driver can be prosecuted and penalised accordingly. The intention is to have a direct and immediate impact on the traffic in the vicinity of the sign. Alongside the mandatory instructions, supporting information can be provided to give context to the instruction and improve compliance by drivers.

The most prominent use of these techniques is in motorway control and the implementation of 'smart' motorway schemes by Highways England. Very similar arrangements have been designed and installed throughout the world [1]. This chapter describes the approach adopted by Highways England as it can be used to illustrate the evolution of the 'smart' motorway design from basic queue detection through control of speeds to provision of additional physical capacity by part-time use of the emergency lane to removal of emergency lane altogether. This chapter describes management of the network principally from the perspective of the road operator. However, a key part of that management is providing advisory and mandatory information to the road users.

Dynamic motorway control is used on those parts of the network that are subject to high levels of traffic demand and where disruption can cause significant economic costs in terms of delay and the consequences of road traffic collisions. These systems are installed to support the broader, key objectives of the highway operator, Highways England. Their key objectives include safety of road users and the workforce engaged in maintenance and construction, providing acceptable and predictable levels of service to their customers, i.e. vehicle drivers and passengers, supporting the national and local economy and mitigating the environmental impacts.

In England, the motorway network is a vital national infrastructure. In 2018, it carried 21% of the traffic despite only comprising 1% of the road network length [2].

[1]Bittern Consulting Ltd and Centre for Transport Studies, Department of Civil and Environmental Engineering, Imperial College London, UK

Hence maintaining a high level of service to road users is of national importance. The motorway network serves both long-distance traffic and the needs of local traffic, particularly for commuting. It links major population centres and strategic transport locations such as ports and airports. Customers clearly require safe journeys but also want acceptable and, possibly more important, predictable journey times [3,4]. Reducing the amount and duration of congestion, and removing external causes of it, are therefore a primary operational objective. Dynamic, tactical control systems are a key part of the toolkit that delivers that aim.

This chapter describes the development of the use of tactical control and information from its origins in the queue and incident detection systems of the 1980s and 1990s to the smart motorway systems of the 2010s. The focus is on the local use of information infrastructure to deliver tactical traffic outcomes as opposed to the wider strategic use of information for journey planning and guidance. The descriptions centre around the operational regimes and how they are implemented and avoids repeating technical descriptions of how individual technology components function.

8.2 Background and context

The current advanced motorway control systems, sometimes referred to as 'Smart' motorways, are the result of an evolution of technology, applications and organisational change over some four or five decades. During that time there have been significant changes in the challenges faced by road operators and advances in technology and techniques to address those issues.

Highway operators have three imperatives that they must keep in mind in everything they do from outline design through construction, operation and maintenance to demolition and removal. They are:

- To maintain and improve the safety of network users and those working on the network.
- To reduce and manage congestion and provide predictable and reliable journey times.
- To mitigate environmental impacts, notably water pollution, air quality and noise.

These are all significant challenges in a stable situation but, of course, the world continues to change. Traffic growth makes it more difficult to achieve targets in all areas. Available capacity is swamped by excess demand and congestion increases rapidly. Measures to reduce incidents can reduce the number of such events but the level of traffic demand means that the congestion caused by each incident still increases. Providing additional capacity is the obvious solution to increasing demand but in densely populated areas, this is very difficult. Additional land may not be available and, in any case, there is a lengthy planning and approval stage to be gone through. This can last many years and that delay will reduce the benefits of the additional capacity because traffic growth in the intervening period will have absorbed some of the available increase.

Actively managing traffic can alleviate some of the effects of traffic growth. It can be used to smooth flow so that the practical capacity is maintained, and the occurrence of flow breakdown is reduced or delayed. It can be used to reduce the time to clear incidents, thereby reducing the duration of the associated congestion, and to reduce the likelihood of further incidents at the tail end of any queue. It can also enable the operation of more lanes of traffic within the existing highway layout by allowing the use of emergency lanes by running traffic at times of peak demand, or even complete removal of the emergency lane.

The tactic of using speed limits displayed over each traffic lane in order to smooth traffic flow was first implemented on a dual, four-lane stretch of the M25 to the West of London during the 1990s [5]. No change was made to the alignment and a full-length emergency lane was in place. The technique was made possible by three evolutions of technology. Firstly, traffic speed and flow in each lane could be measured in real-time, processed locally at the roadside and key data communicated to a central processing system in real-time. This enabled dynamic signal settings. Signal technology, specifically light-emitting diodes had matured to the level where they could be deployed in roadside conditions in a configuration that allowed mandatory information to be displayed. This meant that compliance with the signal over the lane was required and failure to comply was an offence. Finally, camera technology and data processing systems had been developed to the point where information collected automatically could be submitted to a court and successful enforcement action taken.

At the start of the twenty-first century, the need to provide additional capacity on some sections of England's motorway network was becoming more acute. Although controlling speed limits on existing lanes had produced measurable benefits on dual four lane sections the technique did not seem to be as effective for dual three lane sections. Unfortunately, this alignment comprised the largest proportion of the network. To widen to a full four lanes with emergency lane in each direction was both technically difficult and prohibitively expensive in many cases. Experience with speed control sections indicated that the technology was available to allow the existing emergency lane on a dual three lane section to be opened to running traffic at times of peak demand. The additional technology needed was real-time video for 100% of the emergency lane to ensure there were no obstructions prior to opening to traffic and additional sequencing rules to control the signals over all the lanes, including signals positioned over the emergency lane. The systems were enhanced by the increased capability of 'matrix' based electronic message signs. These were now capable of a greater range of text messages and could also display standard traffic sign graphics in the correct colours. This capability provided context information alongside the lane signals to improve drivers' confidence in the signalling system and their compliance with mandatory signals.

The 'active traffic management' approach of dynamically opening the emergency lane to cater for peak demand and manage congestion resulting from incidents was piloted on the M42 motorway to the East of Birmingham in the English midlands. After a successful implementation that showed a range of worthwhile benefits [6] the system was rolled out to other locations.

A number of issues arose that led to a further change in design and resulted in the 'all-lane running' (ALR) layout. Opening and closing the emergency lane requires manual control and takes a significant amount of resource in the control room. Maintaining all the on-road equipment is a considerable undertaking and the costs were higher than anticipated further exacerbated by the delay costs to motorists if the emergency lane could not be safely opened due to a technical fault. Finally, a small number of drivers incorrectly interpret the situation and drive in the emergency lane when it is not open to traffic. This is clearly undesirable and increases the risk of a collision for those who are stopped because of a genuine emergency and to maintenance staff who use the emergency lane as a starting point for setting out temporary traffic management.

To address these issues, the operational logic is reversed. Instead of adding a lane when demand is high, all lanes are provided all of the time and there is no emergency lane. Should a situation arise where a lane that is no longer safe for traffic, then the signalling is used to close that lane quickly and appropriate teams to deal with the situation are despatched. This approach removes the need for the technology and control room resources to check that the emergency lane is clear before opening to traffic. Advances in electronic message matrix signs mean that mandatory signs and context information can be displayed simultaneously on the same panel, as can graphics showing information relevant to individual lanes. These can be mounted on a cantilever instead of a gantry as shown in the images in the chapter on Dynamic Road Traffic Signage. This reduces the quantity of signals that need to be maintained.

8.3 Technologies available

8.3.1 Technologies for measuring traffic and road conditions

Smart motorways make use of data collected from the road to support a variety of functions such as queue detection, speed/flow state, presence of stationary vehicles in a live lane and to provide operators in a control room with visual information. Some data is used directly by the control systems to set electronic signals whilst some is presented to the control room staff who then use their operating procedures to initiate any response actions.

Three types of traffic data are usually collected; traffic flow in terms of number of vehicles per unit time, traffic speed averaged over a unit time and ‘detector occupancy’. The last one is a measure of how long a detector in an individual lane is in a ‘detect’ state.

Traffic flow and speed are usually measured for each traffic lane. The values are then aggregated over a single minute by counting the number of vehicles detected and averaging their measured speed respectively. Flow can then be summed over all lanes to give the flow in the traffic direction per minute. Individual lane average speeds are then further averaged across all lanes to give the directional average per minute.

Detector Occupancy depends on both the speed and size of the vehicle. The longer the detector occupancy the slower the vehicle is travelling. This is a more reliable measure for detecting queuing traffic than the average speed which has two drawbacks. At low speeds measurement is less reliable because drivers may vary their speed within the detection zone area, particularly if the traffic has become unstable and is in 'start-stop' mode. Also, the data has to be averaged over a minute so there is a delay between the individual measurements and the aggregate result being available to the next level of the system. Therefore, using detector occupancy will enable a queue detection algorithm to react more quickly by looking for individual extended occupancy. For example, the algorithm used by Highways England, HIgh OCCupancy (HIOCC), triggers an alert if a continuous detector occupancy of 4 s or more is measured in any lane. However, the algorithm can be over-sensitive and create false alerts so a second step of checking that the average speed is also lower than a pre-determined threshold has been added to try and address this issue. Typical time to detect a queue is around one minute but this can extend to several minutes if the traffic volume is low and it takes longer for the queue to grow across the detection site.

The three quantities described above are measured at individual detection sites spaced regularly along the motorway. Typical spacing is 400 m $\pm$ 50 m between detection sites determined by gantry spacing [7]. Longer spacings reduce the effectiveness of the queue detection algorithm and could allow significant discontinuities in traffic behaviour to go undetected. Reducing the spacing has not be shown to provide additional benefits that outweigh the capital and maintenance costs [8]. Measurement was initially made using inductive loop sensors in pairs buried in each lane creating a detection zone of around 6 m in length, based on 2 no. $\times$ 2 m square loops spaced 2 m apart. These are largely being replaced by side-fire-mounted radar devices. These are mounted at a height of approximately 6 m on masts installed 4–6 m back from the edge of the carriageway. The devices project a pair of radar beams across the carriageway, orthogonal to the traffic's direction of travel. The detection zone is thus less than 0.5 m along the carriageway. Data processing of the signal allows data for each lane across two, four-lane carriageways to be measured by a single device as shown in Figure 8.1.

The presence of stationary vehicles in live lanes is clearly a hazard, particularly under low-flow and high-speed conditions when no queue will form. A 'scanning' radar is being deployed by Highways England to provide alerts to the control room operators. The device was originally developed to track ground movements in airports but has been adapted to roadside operations. It scans a radar beam horizontally through 360° at 5Hz. This enables it to detect and track radar echoes. By limiting the area of interest to the road part of the scan the device can identify group echoes that are potentially an individual vehicle. These groups can then be tracked and any that move slowly and then come to a halt are used to determine the presence of an isolated stationary vehicle. The device has an effective range of 250 m radius. They are mounted about 4 m high on a pole or gantry bracket at the side of the carriageway. Device spacing is usually in the range 300–500 m to ensure better than 95% coverage but spacing is highly influenced by the horizontal and vertical curvature of the road and the presence of structures that can create a 'radar shadow' where detection is not possible.

Figure 8.1 Image of side-fire radar detector used to measure traffic on M42 near Birmingham, UK

Closed-circuit television (CCTV) is a key tool for control rooms that are overseeing Smart Motorway sections. High specification pan-tilt-zoom (PTZ) cameras are installed in sufficient numbers and at close enough spacing to provide 95% coverage. Note that this is not the instantaneous coverage i.e. 95% of the motorway section cannot be viewed by looking at all the cameras simultaneously. It means that 95% of the motorway comes within the field of view of one or more PTZ camera, when pointed in the correct direction. Control room operators can scan through CCTV views to try and identify any issues on the road, but this is generally an inefficient use of their time. Instead they use the cameras to confirm the nature of alerts received from phone calls, on-road patrols and automated systems, notably the stopped vehicle detection system. They can then then decide on the appropriate response, check adjacent cameras and monitor the situation using the CCTV as their eyes on the road.

Those sections of Smart Motorway on the Highways England network where the emergency lane is opened and closed in response to traffic demand are equipped with an additional CCTV system dedicated to monitoring that lane immediately prior and during opening. This comprises fixed view CCTV cameras mounted on posts or short cantilevers at frequent intervals along the edge of the emergency lane, an example of which is shown in Figure 8.2. The spacing and cameras' field of view is such that at least 100% coverage is achieved and that any car-sized object would be clearly visible in the image presented to the control room operator. This enables the control room operator to check the emergency lane is clear of obstructions prior to setting the signals that allow traffic to use that lane as a running lane. By stepping through the camera views in strict sequence an operator can be confident they have not missed any obstructions or other potential hazards.

Figure 8.2 Fixed CCTV cameras for monitoring the emergency lane prior to opening to traffic

8.3.2 Technologies for providing information

Smart motorways require a variety of information to be passed on to drivers. On one hand we have fixed information, such as direction signing or the start and end of variable speed limit control, which can be provided via static, painted signs. On the other hand we have mandatory instructions as to the maximum speed limit and the closure of lanes. In between these two are the dynamic advice about hazards ahead, the need to slow down or limit lane changing and additional context to back up mandatory instructions. Also in the mix, is the use of classic red-amber-green traffic lights to control the in-flow into the motorway at junctions; so called 'ramp metering' or 'access control'.

There is a finite amount of space at the roadside to mount equipment, and their supporting structures, so any dynamic display device needs to support multiple functions. It was only once the type of display devices described in Section 7.3 of Chapter 7 on Dynamic Road Traffic Signage, namely MS4 and AMI, became feasible and a competitive supply chain established that the complexity of dynamic signalling required to make smart motorways a practical proposition became available.

In the case where there is no emergency lane along the carriageway the initial setting out of temporary traffic management for maintenance becomes more hazardous. It was also deemed necessary to install specialised electronic signs that mimic the fixed plate signs used in the approach to temporary traffic management. Remotely operated temporary traffic management signals [9] are installed to prevent the need to have roadworkers carrying signs along or across carriageways when they cannot start from a safe area. They can be switched via wired or wireless communications from the temporary traffic management installation team whilst they are in a safe location.

Smart motorways make use of various combinations of AMI and MS4 signs to provide the necessary legends to inform drivers and to manage traffic in a variety of conditions.

8.4 Managing capacity

8.4.1 Controlled motorway

The first step in managing the capacity on motorways in a dynamic fashion was to directly influence the maximum speed and reduce the speed difference between lanes. The intention behind the former is to create stable flow at a speed that can maximise the use of the available capacity. By forcing speeds down earlier than might occur by the traffic naturally impeding itself you can increase the probability of achieving the maximum practical flow that is predicted by the Greenshield speed-flow diagram. What is more you may even manage to push the achieved maximum capacity more towards the theoretical maximum value [10].

The reduction of speed difference between lanes decreases the incentive for drivers to change lanes. When a driver changes lane, particularly if they force a gap in the lane that they are moving to, this can lead to the formation of shock waves. Such shock waves may be damped or amplified by surrounding vehicles depending on the distance between them and the braking behaviour of drivers. If the wave is amplified this creates a breakdown of flow and a shift to the lower side of the Greenshield curve before the maximum capacity is reached. Once this has happened, traffic speeds will remain slow, perhaps less than 20 mph, and flow will remain constrained until the traffic demand has reduced significantly. Figure 8.3 illustrates these points.

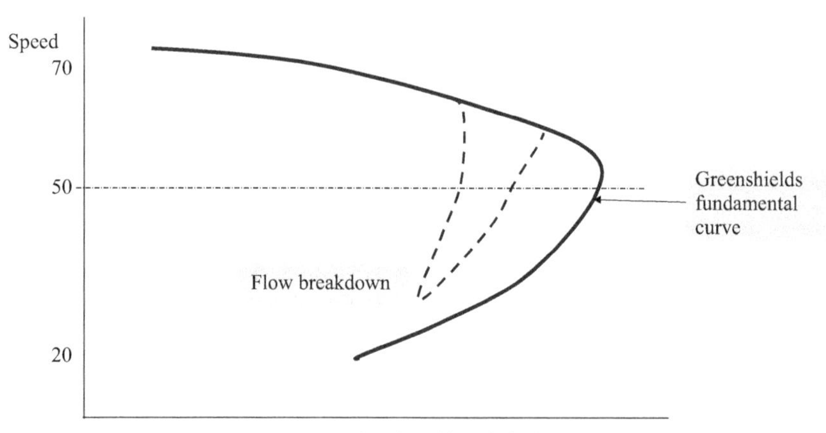

Figure 8.3 Sketch of speed-flow (Greenshields) curve showing how curve changes under flow breakdown

Figure 8.4 Controlled motorway section of M25, London, UK showing 60 mph speed limit

Speed control is achieved by displaying legally enforceable speed limits on electronic signs above each lane as seen in Figure 8.4. In the United Kingdom, the speed limit is enforceable because the format of the display is the same as for a fixed plate speed limit sign, i.e. a number inside a red ring. A specific piece of regulation is also enacted by way of a statutory instrument (SI) that allows the speed limit to be varied in time and space on the designated section of motorway without the need to publish a temporary traffic restriction order. This means the speeds can legally be varied by the road operator and contraventions by drivers enforced through the courts. Fixed plate signs are installed at the start and end of the section that the SI applies to informing drivers that they are entering or leaving a section of motorway where the speed limit may be varied.

Speed limit signs, referred to as lane control signals or, more recently, advanced motorway indicators (AMI), are mounted on a portal gantry. Typically, gantries are spaced at 800–1,000 m. The actual spacing depends on the alignment of the road with the aim to ensure that the next set of signals are visible to drivers very shortly after passing under a gantry. There may also be constraints caused by issues related to the foundations of the gantry and the proximity of fixed plate directional signing.

In order to provide drivers with additional, contextual information that gives some explanation as to why lower speed limits are being displayed, and to encourage compliance with them, it is usual to install some electronic variable message signs (VMS) within a controlled motorway section. These may be text only (MS2) or mixed text and graphics (MS4) as described in the chapter on Dynamic Road Traffic Signage. These will provide additional information by way of simple legends such as 'QUEUE AHEAD', 'CONGESTION – STAY IN LANE' or 'CONGESTION AFTER JNC', see Figure 8.5. There is a difference in meaning between 'QUEUE',

Figure 8.5 Controlled motorway with MS2 providing additional context information

which means that traffic is virtually stationary at some point, and 'CONGESTION' which indicates dense and slow moving, but not stationary, traffic conditions. Whether the general public are aware of this distinction is not clear, nor how it is intended to change their response.

Speed restrictions and supporting information are provided to drivers in three situations; queue warning, congestion management and incident management.

8.4.1.1 Queue warning

The objective of queue warning is to reduce the likelihood of collisions caused by fast moving vehicles running into the rear of stationary, or very slow moving, vehicles that are at the tail end of a queue. By imposing a speed restriction upstream, the longitudinal speed difference is reduced and the braking distance shortened. Additionally, the presence of the speed limits and the additional information via VMS should raise the awareness of drivers of a potential hazard ahead.

AMI will be set on the three gantries upstream of the tail end of the queue, through the length of the queue and on the gantry immediately downstream of the head end of the queue. Downstream will be set to show the national speed limit indication; in the United Kingdom, this is a white circle with a black diagonal bar. This tells drivers that normal motorway speed limits now apply. This set of signals is of increasing importance as increasing numbers of vehicles are fitted with systems capable of reading speed limit signs automatically. The speed limit is displayed to the driver and may be linked to a speed limit warning system. Without a specific sign to show the end of the variable speed limit such systems would not be able to work properly on these sections of motorway.

Along the length of the queue a '40' mph limit is displayed. There is a school of thought that the AMI should be blank as the traffic speed is constrained anyway, usually to a speed lower than 40 mph. However, this gives an ambiguous message to drivers. It might encourage them to try to speed up and this might lead to accidents within the queue. Also, it is not clear that the speed limit last displayed still applies because no other speed limit has been shown. Some drivers may interpret blank signals as indicating the road operator is not aware of the traffic condition and that the situation is not being managed.

The 40 mph aspect is set on the gantry immediately upstream of the tail end of the queue and '60' set on the two AMI gantry locations prior to this. As the end of the queue moves this sequence will move along the gantries with it. Clearly this is hard to achieve through manual operation so there is a direct link to the queue detection system in order to know where the end of the queue is. The AMI gantry set to 40 is regarded as the 'primary' signal and the two gantries to be set to 60 mph are the associated 'secondary' signals. Within the signal control system data every gantry has software links to the secondary gantries to be triggered if it is set as a primary, including AMI positioned on slip roads.

The VMS (MS2 or MS4) will also automatically be set with a legend that includes the text queue, e.g. 'QUEUE AHEAD' or 'QUEUE AFTER JNC'. This is intended to provide the contextual information to tell drivers to anticipate finding very slow or stationary vehicles ahead and to be prepared to stop.

The primary function of the advance signalling is to raise awareness with drivers of a hazard ahead and reduce the potential for collisions at the back of the queue. Furthermore, by slowing drivers down earlier the arrival rate at the queue can be reduced and hence the speed at which the queue grows slowed. The extent that this latter effect occurs depends on a combination of local factors such as the cause of the queue, the amount of road capacity lost, the density of the upstream traffic and the behaviour of drivers in response to the posted speed limits. If the traffic flow is already comparatively high then the effect will be quite limited, particularly if the cause of the queue has resulted in one or more traffic lanes being blocked and there is a significant loss of capacity for traffic at the queue seedpoint location.

8.4.1.2 Congestion management

Alongside the detection of queues using the HIOCC algorithm, the traffic measurement system also records speed and flow in each lane for every minute. This data can be used to identify approximately where the current traffic situation is on the speed flow curve. Rather than try and position itself on a line the speed-flow chart is divided into a set of regions. Each region has a target speed limit. When the current situation changes between regions the signals are adjusted accordingly. The data from an individual detector site is used to set the signals immediately upstream of the detector location [11,12] as shown in Figures 8.6 and 8.7.

When the traffic data indicates a transition from a region in the chart where the national speed limits apply to one where 60 mph should be applied two signal sets are changed [11]. Upstream of the detector site will be set to the 60 mph speed limit

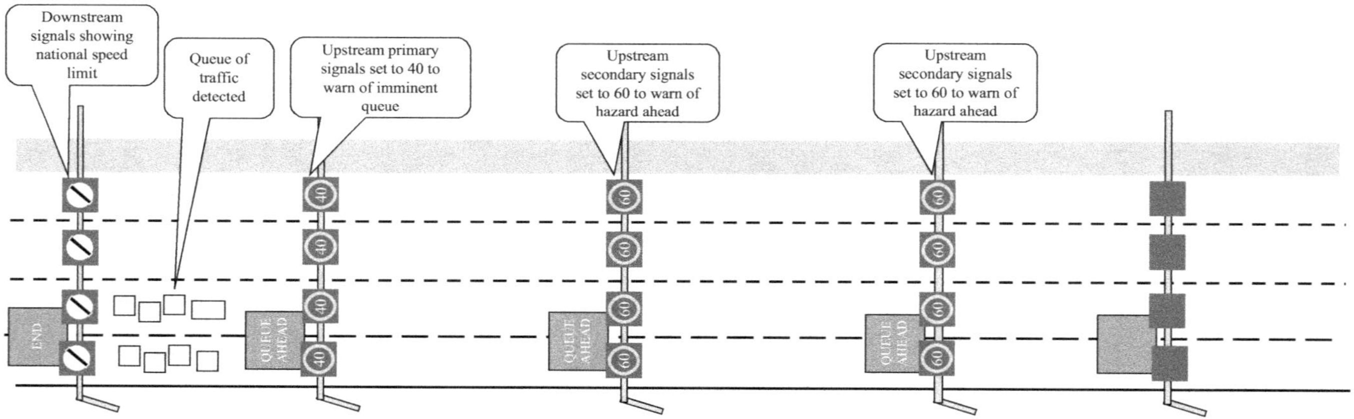

Figure 8.6 Pattern of lane signalling when a queue has been detected

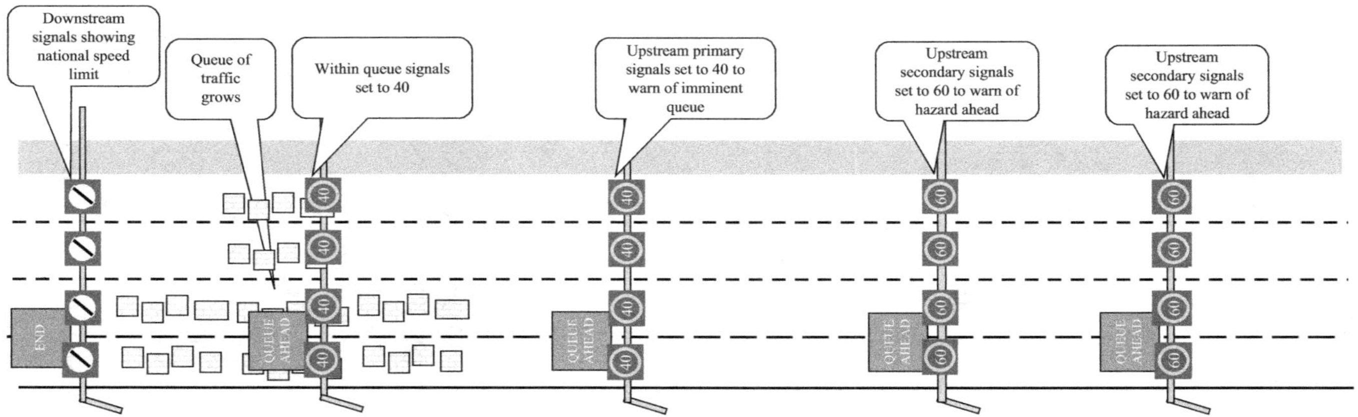

Figure 8.7 Pattern of lane signalling as queue grows

on all lanes whilst, at the same time, the next downstream signal will be set to show the symbol for national speed limits. This is to explicitly indicate that the reduced speed limit has come to an end.

As the traffic detection systems determines the extent of any spread of congestion additional signals will be switched on. If the congestion is spreading upstream, as is normal, then additional upstream signal gantries will have their signals illuminated. Under some circumstances, congestion can spread downstream as well. In this case the signals showing the national speed limit symbol will change to a speed limit whilst the national speed limit symbol moves to the next downstream signal set.

If the traffic data from a detection site indicates that traffic conditions have moved to a speed-flow region with a different speed limit then the signals at the gantry upstream of the detector site will be switched to the next limit. The system usually only varies between 60 mph and 50 mph as 40 mph is reserved for queue protection and incident management. Also, depending on the precise local road geometry and driver behaviour, 40 mph is either at or slightly below the point of maximum flow. Therefore, applying such a limit is unlikely to be beneficial. Successive signal gantries will vary independently between 60 mph and 50 mph depending on the traffic conditions. If traffic does slow down and falls into the lower part of the curve then generally some vehicles will come very close to, or actually, stop. At that time the queue protection algorithm will take over and signals will be set as described in the previous sub-section.

There is a precedence between the speed limits so that a lower speed limit will be set in preference to a higher one. This means that the speed limit on a signal gantry will be set in response to whichever algorithm, queue protection or congestion management, requires the lowest limit. This means that within a given section where both algorithms are operating a mix of 40, 50 and 60 mph speed limits will be encountered by the drivers.

VMS signs will also be set to indicate to drivers that there is congestion. Typical legends are 'CONGESTION STAY IN LANE' or 'CONGESTION AHEAD'. Signs upstream of the most upstream gantry may also be set to give advance warning to drivers, e.g. 'CONGESTION AFTER JNC'.

8.4.1.3 Incident management

When traffic lanes have to be closed this is usually the consequence of an incident. The signs and signalling can be used to help manage traffic and keep drivers informed. Where a lane has been closed, this can be shown immediately upstream by displaying a solid Red 'X' in the AMI with flashing red lights at the corners of the signal. This shows the lane ahead is closed and drivers should not pass under the sign; in fact, it is an offence to do so. Drivers are warned of the need to change lanes by the display of 'lane divert' (↘) arrows on the secondary AMI gantries associated with the gantry displaying the Red X. In some cases, the road operator may also choose to further restrict the speed limit to 20 mph in open lanes past the incident scene. Setting such lower speed limits is done manually from the control centre as is setting the Red X on the closed lanes. The lane divert arrows will be set automatically by the signal control system once the Red X has been selected.

Drivers can be given additional context information via the VMS. The type of information displayed depends on the signs' capability. Legends may range from short text messages such as 'ACCIDENT AHEAD', 'SLOW DOWN, INCIDENT', or 'LANE CLOSED AHEAD' through simple graphics such as 'wicket' graphics showing which lanes are closed (|TT) or full colour graphic signs alongside one of the text messages.

Thought needs to be given to how far upstream explicit warnings that show which lanes closed should be displayed. There needs to be enough advance warning to allow drivers to safely move into the open lanes. However, if the signing is too far upstream this can result in a very long queue in those lanes and a large expanse of road that is empty. Some drivers, those with less patience, will not join the queue but instead will drive alongside the queue until they spot a gap or force their way in once the reach the head of the tailback. This may upset drivers who have waited patiently and result in road rage incidents. Further back in the queue some drivers who had joined may observe the vehicles passing them, lose patience and decide to pull out into an apparently empty lane but which has other queue 'jumpers' approaching at relatively high speed. Clearly such a situation is potentially unsafe.

Signalling speed limits too far upstream can also have an unintended effect, although less hazardous. After responding to the first speed limit they encounter, if drivers do not encounter a visible reason for that limit within a couple of minutes, they will increase their speed again. Their awareness has been raised so the risk of them not seeing a hazard ahead is reduced but the credibility of the signals has been undermined and the beneficial effects of slowing traffic ahead of the problem may well have been lost. Also, the next time they encounter a set of speed limits drivers may be less inclined to pay sufficient attention.

A similar problem can be encountered when there is a report of an incident but no confirmation that anything has occurred at the reported location. The prudent approach is to introduce an initial speed limit, typically 50 mph, until such time as the existence and nature of the incident can be confirmed. However, if the incident does not exist or has cleared of its own accord very quickly, this undermines the credibility of the signalling and information systems in the eye of the public and can lead to poor compliance when the information is correct. One strategy to address this is to use the legend 'REPORTED INCIDENT' on the VMS to convey the unconfirmed nature of the situation at that time.

A driver's confidence in the signalling system can be undermined by signals being left on for too long after the need for them has dissipated. Again, there is a trade-off to ensure that signals are not being removed and then re-introduced at too fast a rate. The automated queue protection system used by Highways England automatically starts to remove signals through a timed sequence of increasing speed limits subject to a minimum duration on the primary signal. Where signals have been set manually there is a more significant risk of speed limits being left in place until someone in the control room notices that they are still set, or a complaint is received from a member of the public or a road patrol.

8.4.2 Dynamic hard shoulder

In the previous section, we looked at how AMI and MS4 could be used to improve safety by warning of queuing traffic ahead and to improve traffic flow by limiting speed to try and even out the traffic. The same mix of technology can be used to manage traffic past an incident by signalling lane closures and speed in combination. Whilst this signalling strategy can be used to try and get best use of the available capacity it does not of itself provide additional capacity on a reliable and predictable basis.

The obvious way to increase capacity is to provide additional traffic lanes. This can be very expensive as it will often require the acquisition of additional land and will certainly require existing bridge and viaduct structures to be widened and strengthened. A more cost-effective alternative is to make use of the emergency lane (the 'hard shoulder' in UK parlance) as a running lane at times of peak demand or to offset the loss of capacity from lanes closed by an incident. Referred to as 'dynamic hard shoulder' (DHS) running this can make use of the techniques described in the previous section supplemented by signing to support the opening and closure to traffic of the emergency lane.

Some physical changes are needed to implement DHS smart motorways. Additional refuge lay-bys are required to provide a safe location for vehicles unable to travel to the next junction when the emergency lane is open to traffic. Fixed signing is also required to make it clear to drivers they are entering a section where the speeds and number of lanes may vary, and to indicate the presence of and distance to the additional emergency areas.

The base operational situation for this type of smart motorway is that of a standard dual three lane motorway (D3M) equipped with AMI and MS4 for queue protection and congestion management. The difference in this case is that there is an AMI over the emergency lane as well as over each full-time running lane, as shown in Figure 8.8. Under free flow traffic conditions all the AMI and MS4 will be blank although the latter may be displaying a more general driver safety or strategic information legend. Should a queue form then the signal system operates as described in the section on queue warning for controlled motorways. However, in addition to the speed limits over the running lanes, the AMI over the emergency lane will display a 'broken' Red X with no additional flashing signals. This indicates that the emergency lane is closed to traffic and removes the potential for a driver to interpret a blank AMI as just being defective. This reflects an operational policy that if one AMI is illuminated then all AMI on the same gantry must also be displaying an appropriate aspect so that there are no blanks (except for genuinely faulty devices).

Drivers are informed that the emergency lane can be used as a running lane by the presence of a speed limit on the AMI over that lane supported by a legend such as 'CONGESTION USE HARD SHOULDER' on the MS4. In theory drivers should understand that they must not drive in the emergency lane when the AMIs and MS4 are blank or there is a broken Red X displayed over the emergency lane. Whilst the majority do seem to follow this behaviour there is a minority who will drive in the emergency lane irrespective of the signs. What is more, when one

Figure 8.8 Dynamic hard shoulder smart motorway section on M42, Birmingham, UK

driver does this then a significant number of others will follow their lead. The time and locations of these incursions are impossible to predict but clearly represent a potential danger to those using the emergency lane for its intended purpose and for maintenance crews who use it for access. To address this issue, it has been found that it is better to display the legend 'HARD SHOULDER FOR EMERGENCY USE ONLY' on MS4 at frequent intervals, although not necessarily at every one, see Figure 8.9. It has also been found beneficial to have a variable display at the downstream end of any on-slip road merge that clearly shows whether or not the emergency lane is open using large clear arrows, an example of which can be seen in Figure 8.10.

When opening the emergency lane to traffic it is important that it is confirmed as free from obstructions and stationary vehicles. A dedicated CCTV system has been installed that provides 100% coverage of the emergency lane and is used by the control room operators in conjunction with the signal control system.

The sequence for opening the emergency lane between junction A (upstream) and junction B (downstream) is:

1. Set a 60 mph speed limit over all running lanes, and broken Red X over the emergency lane on all signalling gantries between junctions A and B. This pre-warns driver that the lane is about to be opened and 'conditions' traffic by reducing the speed differential between lanes. On the gantry within Junction B, or the next downstream gantry, the AMI should display the national speed limit symbol over the permanent running lanes and a broken Red X over the emergency lane in order to signal the end of the section under control.
2. Check that the emergency lane is free from stationary vehicles or other obstructions between the end of the junction B off-slip road and the first gantry upstream using

Figure 8.9 *MS4 displaying message confirming status of emergency lane on M42, Birmingham UK*

Figure 8.10 *MS4 displaying which lanes are open at end of on-ramp, M42, Birmingham UK*

the CCTV system. If it is clear then immediately set the AMI over the emergency lane on that gantry to 60 mph. It is important to do this straight away to prevent anyone legitimately stopping in the emergency lane in the time interval between the CCTV check and the AMI changing from a broken Red X to a speed limit.

3. Repeat the above step from the first gantry upstream of junction B to the second gantry upstream of junction B.
4. Repeat again for each gantry until junction A is reached.

If at any stage an obstruction is viewed on the CCTV then the opening sequence can safely be frozen at that time until the blockage is removed as the AMI upstream will still display broken Red X.

This sequence is repeated for each junction pair, always working from downstream to upstream. Once the emergency lane is open the queue protection and congestion management algorithms will continue to work across all lanes, including the newly opened one, using the same signalling patterns as described earlier for controlled motorways.

Typically, it will take an operator in the control room up to a minute to check and open a section of emergency lane between two successive gantries if no obstruction is found. Gantries are typically 800–1,000 m apart so opening the emergency lane on one direction along a 10 km length of motorway can take more than 10 minutes.

In order to obtain the most benefit from providing additional capacity the extra lane needs to be open before the demand on the full-time lanes has reached a level where the probability of congestion induced by flow breakdown is high. Given that is takes minutes to open a section, the control room needs to plan to carry out the action in advance. Timing is important in maintaining the credibility of the signals in the eyes of drivers. Open too late and there maybe congestion and the flow may have broken down to the extent that even adding the additional lane is not enough to recover the situation. Open to early and the flow may not seem high enough to justify the lowered speed limits to drivers passing through the section. During the period approaching peak demand flow levels can change very rapidly so it can be difficult to find either a predictable time to open or a reliable flow threshold to use as a trigger.

The traffic pattern and road alignment at junctions may mean that the emergency lane is not suitable to be used as a running lane through the junction. In such cases, the emergency lane is used as running lane lead into the off-slip road and it will only be for traffic leaving at that junction. The MS4 upstream of the junction are used to indicate the alignment ahead, e.g. 'HARD SHOULDER FOR JNC 6 ONLY'. The next on-slip ramp will be aligned to run directly into the emergency lane in the next section with a MS4 near the end of the ramp that shows if the emergency lane is open or not.

Within the junction itself, over the main carriageway, there will be a signal gantry whose AMIs display the speed limit being applied over the permanent running lanes and a broken Red X over the emergency lane.

Once the peak demand period is over, flow levels have lowered to the point where the risk of flow breakdown is reduced, and traffic can safely travel at the national speed limit then the emergency lane can be closed to traffic and returned to its normal function. Closing the lane is more straightforward as there is no need for manual supervision because we do not need to check that the lane is clear. The sequence starts from the downstream junction and proceeds in the following steps:

1. Set lane divert arrows over the emergency lane on the two gantries immediately upstream of the junction to indicate to drivers that they need to move into the next full-time running lane.
2. After a short time set the signal over the emergency lane at the farthest downstream gantry to broken Red X and set the signal over the emergency lane on the third gantry upstream of the junction to lane divert.
3. Set any VMS on, or downstream of, the gantry now showing the broken Red X to a legend that reinforces the closure of the emergency lane, e.g. HARD SHOULDER FOR EMERGENCY USE ONLY.
4. Repeat the previous two stages, stepwise, upstream until all the signals above the emergency lanes show a broken Red X and the VMS display a legend showing the lane is now only for use in emergencies.
5. Once the emergency lane has been closed the signals can be cleared down by progressively. From the downstream end, on each gantry change the speed limit signs to the national speed limit symbol for a few tens of seconds followed by setting all the AMI blank.
6. Once all AMI are off, i.e. set to be blank, then clear the legend from the VMS leaving sufficient of them to deter inappropriate use of the emergency lane.

This sequence can be automated in the signal control system so that all an operator needs to do is to initiate the operation and acknowledge when it has been completed. The time steps need to be sufficient that a driver obeying the speed limit and just passing a gantry would not see a change in signal on the next downstream gantry. This is about 40 s for a 60 mph speed limit and gantries spaced at 1,000 m. Again, it will take a few minutes for a section between two junctions to have the emergency lane closed and the speed limits removed for all lanes. However, there is little downside effect over and above the loss of credibility from delaying starting the closing sequence for too long after the traffic demand warrants it.

The VMS and lane signals provide a key tool in managing incident situations. In addition to the techniques outlined in the section on controlled motorways, they may be called upon to support further actions by the highway operator. The first is to provide some degree of protection to any vehicle that breaks down and is unable to reach an emergency area when the full emergency lane is open to traffic. The AMI can be set to show the lane as closed using a full Red X with flashing red lights immediately upstream of the stopped vehicle, with lane divert signals on the two gantries further upstream. The speed limit can be lowered to 40 in the remaining lanes and the VMS can show a legend describing the problem (e.g. ‘STRANDED VEHICLE’, ‘OBSTRUCTION AHEAD’).

A similar technique may be used if it is necessary to remove a broken-down vehicle from an emergency area when the emergency lane is open traffic. Although waiting until traffic flow has reduced and the emergency lane has been closed so that it can be used as an acceleration lane looks to be a safe option, this will mean that the emergency area is occupied and any further vehicle having problems has no safe refuge available. Each situation has to be risk assessed individually. If it is decided the best course of action is to move the first stranded vehicle then the signal sequence of full Red X with flashing signals and lane divert arrows can be set upstream of the emergency lay-by in question. As before, speed limits can be lowered in the immediate area and context information can be provided via the VMS.

A different scenario is where an incident occurs when all the lanes are open and the emergency service require access. In this case the signals can be used to close the emergency lane from the scene back to the next upstream junction to provide a clear path for rescue and recovery vehicles. The signal sequence for closing the lane is the same as above but with the VMS providing supporting information as to why the lane is being closed. The lane will be closed for the whole section between junctions to facilitate egress of ambulances and, eventually, to provide a route for removal of debris.

8.4.3 All-lane running

Dynamic hard shoulder schemes provide a method to introduce additional capacity at times of peak demand by temporarily converting the emergency lane into a full running lane. However, it is not without its drawbacks. Some drivers, albeit a minority, do not seem to be able to understand whether the emergency lane is open to traffic or not. Their ambiguity creates a potential hazard that could lead to collisions between vehicles driven by those who fail to correctly interpret, or who wilfully ignore, the signs and signals and vehicles that have stopped in the emergency lane. Furthermore, the act of opening and closing the lane, particularly opening it, requires significant specific technology and human operator time. A dedicated CCTV surveillance system is needed, and AMI are needed over each lane in order to provide comprehensive and coherent signal sequences to drivers.

This raised the question of whether it was feasible to simplify operations by inverting the logic of dynamic hard shoulder smart motorways. Instead of providing additional capacity at times of peak demand by adding a lane it could be more straightforward to provide extra capacity all the time and remove a lane only when required to deal with a breakdown, incident or to provide maintenance access. There would be no emergency lane, although emergency areas would be in the form of clearly marked lay-bys, at regular intervals. This design is termed 'ALR' and Figure 8.11 shows one example.

Clearly, there are safety implications because of the loss of the emergency lane. Emergency lanes are themselves a site of accidents with many recorded instances of vehicles wandering or veering from the nearside running lane across into the emergency lane and colliding with stationary vehicles. It has been observed that during the periods when all lanes are open to traffic in dynamic hard shoulder schemes that the number of drivers using the emergency areas or stopping in live lanes was very much less than when an emergency lane is available.

This still leaves the question of vehicles that come to a halt in a live lane. In dense traffic conditions these will impede the vehicles immediately behind them but, as they do not come to a halt instantly, the risk of a collision is low. This situation will cause a queue to form and the queue protection system will come into action to mitigate the risk of a tail-end collision. However, when traffic conditions are less dense, with low flow and

Figure 8.11 ALR smart motorway, M1 Northamptonshire, UK

high speeds, this scenario is clearly more hazardous than before. Some vehicles might not have made it to the emergency lane under any operational regime but there will be some who would have made use of the emergency lane. Whilst they may not have been completely safe there, the risk to them would clearly be lower than if they were in a live lane.

Initial analysis suggests that the reduction in hazard level from removal of the emergency lane outweighs the higher risk for the very small number of drivers who are forced to stop in a live lane during free-flowing traffic. The signs and signals can be used to warn approaching drivers and protect the scene as this is an incident. The key factor in maximising the mitigation provided by the driver information systems is the time to alert the control centre. Highways England are rolling out a system for stopped vehicle detection using scanning radar to improve response times. However, mobile phone calls from the driver involved or one passing the scene are often the alerting mechanism.

ALR does not have the same need for signalling over each lane as is required for dynamic hard shoulder operations. Advances in the capability of variable message panels that permit colour graphics and multiple display areas on one sign, e.g. MS4, mean that many gantries can be replaced by cantilever-mounted display panels. This reduces the capital costs of the scheme. Furthermore, the distance between signs can be increased to around 1,500 m as longitudinal control does not need to be as rigorous. Care should be taken not to increase the distance too far as secondary signals for queue protection could end up being a long way from the tail of the queue. In these circumstances, drivers will feel they are being slowed unnecessarily and the credibility of the information will be undermined.

The arrangement of signals for an ALR scheme is different from those in the previous sections. Gantries with AMI and associated MS4 are installed at the upstream and downstream ends to give more precise control at the junctions. Between them, cantilever-mounted MS4 are installed at nominal 1,500m spacing. The MS4 are capable of displaying multiple information simultaneously. They can display a variable speed limit that applies to all lanes alongside a graphic that provides some context, e.g. an electronic version of a standard road sign such as a red triangle containing the graphic for queuing traffic. Short textual messages can also be displayed beneath the graphic such as 'STRANDED VEHICLE'. The display can also show if any lanes are closed in the graphic panel using a 'wicket' style diagram.

Maximising the use of graphics over text ensure information can be understood by those who are not familiar with the language. This is beneficial because the limited number of characters available means that information is compressed into a form that may be ambiguous or unclear to a non-native speaker., A well known example is the legend 'DEAD SLOW ANIMALS'. This is meant to advise drivers to travel very slowly because there are live animals in the carriageway but could equally be interpreted as describing the state of the animals not how to drive. This is one of the reasons why operators cannot enter text freely but instead must select from a pre-defined list of approved text messages. It is also why graphics are often preferable to text.

All the information displayed is connected to the same traffic condition or incident to avoid any confusion. It is not good practise to mix information about a queue ahead in the same motorway link with, say strategic information regarding delays on a different link further away, despite this being technically possible. The control system for the VMS has an internal prioritisation mechanism that places tactical warning messages ahead of strategic information, and the latter ahead of general information such as generic road safety warnings, e.g. DONT DRINK AND DRIVE.

The queue protection and congestion management algorithms operate in exactly the same way as they do for controlled motorways and dynamic hard shoulder schemes. The display panels are regarded as gantries although they will display a single speed limit sign that applies to all lanes. The rules regarding primary and secondary signal settings also apply. The graphics display capability means that context information will usually be displayed by means of an image of a road sign instead of text.

8.5 Maintenance

8.5.1 Fault prioritisation

In smart motorway schemes of any type, the AMI and MS4 play an important role in making drivers aware of conditions ahead, controlling the movement of traffic and mitigating potential hazards by, for example, closing a traffic lane to protect a stranded vehicle. The loss of signing capability because of a fault with the technology can compromise the ability to control traffic adequately or mitigate hazards effectively. Hence, the ability to detect, diagnose and repair faults and restore the overall system to service as quickly as possible, subject to safe working constraints, is very important. If faults go undetected and unresolved then the scheme benefits will not be realised, the credibility of the operation could be undermined and mitigation of hazards will not be as high as it intended. For example, if a signal over the emergency lane in a dynamic hard shoulder smart motorway cannot display a broken red-X to indicate the lane is closed then a driver might misinterpret the blank signal next to three signals displaying a speed limit as merely being faulty and drive into the emergency lane believing it to be open. To avoid such situations the operator may follow a policy of not opening the emergency lane to traffic if any AMI mounted above that lane is faulty. This would lead to congestion during peak hours starting earlier and lasting longer thereby removing the benefits related to improvements in traffic flow and journey time stability.

Signs and signal assets are subject to a range of faults. A fault will cause the asset to exhibit some sort of symptom and some loss of function. Certain losses of function will mean that there is a reduction in service level from the technology for customers and the highways operator. This may mean that certain hazards can no longer be fully mitigated against and the operator has to put in place some form of 'work around' that may be both costlier and less effective. If the symptom of the fault is visible to the public this may also have a negative effect on the credibility and reduce compliance with traffic signals.

Such faults are termed 'Service Affecting Faults' and are to be given priority attention by the technology maintainer. They are defined through a combination of the type of asset, the function affected and the operating regime of the section of road where the asset is located. For driver information asset to have a Service Affecting Fault there must be a loss of function (e.g. cannot display a mandatory speed) that causes a reduction in the effectiveness of an operational function (e.g. congestion management) within an operational regime (e.g. Smart Motorway – ALR).

Therefore, we can identify those signs and signals that are significant because they perform particular functions, within operational functions, as part of an operational regime, e.g. MS4 at marker post X on M4 Westbound delivers speed and lane restrictions as part of incident management, and advisory speed as part of queue protection, on an ALR scheme. The maintainer needs to identify which assets are operationally significant and needs to classify (on the Fault Management System) the reported fault as either a Service Affecting Fault or Other Fault as soon as the fault symptom is reported.

8.5.2 Time to repair

Clearly a key operational parameter is the time is takes to return the device to full working order or at least correct any Service Affecting Faults. The time is usually measured from the time the fault is reported until the device is confirmed as free from the specified fault. Note that this does not mean the device is fault-free as there may be faults present that do not directly affect the service it is delivering at that time. Also, it is worth noting that what is initially detected and reported is one or more 'symptoms' within the system as a whole. The precise nature of the underlying fault has to be investigated and it may be that the symptoms are cleared, and the device starts to work again without the real fault having been addressed.

The elapsed time from report to return to service will have several sub-periods; logging of the fault, transfer of fault report to maintenance organisation, scheduling of resources to investigate the nature of the fault and the repair needs, scheduling and mobilisation of the repair resources including arranging safe site access, suitable plant and spares as well as the technicians. Finally, the repair is undertaken followed by any testing to confirm the fault has been cleared before departing the site and removing any temporary traffic management that had been required to ensure safe working. Total repair times can vary hugely from a few hours to several weeks or months. The latter are clearly undesirable.

We have seen in the previous section how the presence of a fault can reduce the benefits that are being realised. Another important factor is the potential increase in risk of a hazard occurring when a device is faulty and, as a consequence, the full hazard mitigation is not possible. For example, if a MS4 on an ALR smart motorway is unable to display lane specific signals then it cannot display a full Red X over a lane that has an obstruction. This will mean that part of the early mitigation to warn approaching drivers will not be available at that location. The longer that a device is faulty, the greater the chance of a problem arising where it is needed as part of the mitigation.

Targets for time to repair are therefore a balance between the increase in costs of shorter restore times and the loss of benefits plus the increased risks of not being able to mitigate fully against a hazard should it occur.

8.5.3 Access to technology at the roadside

One of the challenges for maintaining the technology used in smart motorways is getting access to the equipment for repair and regular maintenance. Devices mounted on gantries or cantilevers are over live lanes so either the lanes will need to be temporarily closed during maintenance or the gantry or cantilever must be designed to allow engineers to access the device without any risk of dropping tools onto live traffic. The latter may increase the cost of the civil engineering because they need extra steel to support the weight of maintenance personnel and this will also increase the size of the foundations. However, installing and removing temporary traffic management on motorways costs several thousand pounds each time, can only be done at night without causing excessive congestion and has workforce safety implications. Mobile, hydraulic access platforms will also be required to reach the required height. Scheme designers need to look at the whole-life costs and compare savings in capital costs from mountings that do not support maintenance access against the operational costs of temporary traffic management and mobile plant.

Ground-level access is required to service electronics, communications and power supply equipment which is usually installed at that level. Where an emergency lane is never used by moving traffic short-term access can be gained from that lane without needing temporary traffic management. Alternatively, maintenance access bays behind the emergency lane can be installed. Where live traffic may use the emergency lane as indicated by the electronic signalling, or where there is no emergency lane i.e. ALR, then access without temporary traffic management is not safe. The only exception is where access can be provided from off the network by a footpath from an adjacent non-motorway road. This can be very useful for fault investigations and repair or maintenance activities that do not involve heavy tools or spares. However, there is a capital cost in installing steps and footways, they must be secure so as to prevent use by the general public and they must also be maintained.

8.6 Awareness, compliance and enforcement

Motorways in the United Kingdom are special roads that have their own regulations attached, typically banning pedestrians and restricting the types of vehicles that may use them as set out in the Road Traffic Regulation Act 1984. The ability of vary mandatory speed limits and use of the emergency lane as a live traffic lane requires additional regulations that are specific to those sections of motorway [13]. The Highway Code [14] makes reference to obeying lower limits on motorways where the signals exist. The support of regulation is vital for mandatory signals as without it there would be no ability to enforce sanctions against those who failed to comply.

Driver education and awareness are essential and key parts of implementing smart motorway schemes. At a general level, drivers need to be aware such schemes exist, know how to recognise them and then to drive in response to the patterns of signals that they see. They also need to be aware specifically when they are entering and leaving an individual smart motorway section. The former requires public information combined with publicity whilst the latter requires local advertising campaigns when the scheme opens, and fixed signing at the roadside at the entry and exit points, including all slip roads.

There are a number of factors to take into account when considering the degree to which the public are complying with the motorway management regime that is in place on a particular section of motorway. The proportion of traffic that complies with a lowered mandatory speed limit is clearly of key interest as controlling speeds is the primary intervention mechanism. Consideration needs to be given to how this varies over time and also along the section. As discussed earlier, compliance may be poor if signalling takes place too far upstream or continues too long after the reason that caused the limit has dissipated. Where the emergency lane is dynamically opened and closed the volume of vehicles that drive down that lane when it is closed is an indication of a lack of understanding of the signalling and the schemes intention.

The number of vehicles that pass under a full Red X with flashing red lights is an indicator that drivers do not fully understand the signalling. In the United Kingdom, the Highway Code [14] is explicit that this is an offence under the Road Traffic Act 1988 [15]. Failure to comply could have serious consequences so tracking driver compliance with this signal is important.

Another aspect that is worth tracking is the extent to which additional emergency areas, that is to say laybys installed for emergency use when there is no emergency lane or it is in use as a live traffic lane, are used inappropriately. One of the issues for highway operators is that if there is a visible place to stop then drivers will use it for any reason, very few of them being a real ‘emergency’. Excessive inappropriate use is problem for two reasons. First is that, because of the restricted geometry of the entry and exit points, drivers using the bay will be travelling much more slowly than the rest of the traffic in a live lane as they enter or leave. This increases the risk of a collision. Secondly, the size of the layby is limited and if it is already occupied a driver with a genuine emergency may not be able to enter safely; the worst case being that they still stop but with their vehicle wholly or partially in the live lane. Associated with this is the time it takes to recover a broken-down vehicle from an emergency refuge. This needs to be treated carefully as it may be better to wait until traffic volumes are lower before allowing the recovery and recovered vehicle to re-join the carriageway.

Although compliance can be encouraged through driver education and publicity campaigns, it is also necessary to be able to identify and punish those who do not comply willingly. It is unsafe and infeasible to stop motorists within these motorway sections so automated and remote capture of evidence is necessary. Speed enforcement using radar detectors and digital cameras is well-developed and provides a method for identifying and prosecuting those who fail to comply with mandatory speed limits. The technology is now readily available, and evidence is accepted in courts with little question. The main risk to prosecutors is that the speed limit visible to drivers and the threshold used to trigger a prosecution do not match.

Automated evidence collection for other issues, such as passing under a Red X with flashing red lights or driving along the emergency lane when it is closed to traffic are more complex. The evidence requires a whole sequence of images or a continuous video sequence that can demonstrate that the driver has no reasonable excuse for not having complied with the signs and signals. The technology is available to collect the data and it is now a matter of determining what is an acceptable data set to support prosecution. Unlike speed enforcement, which can be effective even if only a small number of locations are equipped, the majority of locations would have to be kitted out if Red X and emergency lane violations are to be captured in sufficient quantity to act as a deterrent. In the case of speed compliance this could be improved through the use of average speed enforcement instead of spot speed enforcement. However, this would also require many more cameras and is technically difficult if speed limits vary from gantry to gantry.

8.7 Outlook

Smart Motorways, mainly using the ALR design, are widespread on the motorway network in England and by the mid-2020s around 800 miles of the most heavily used sections will have been installed. There are also plans to deploy similar technologies along sections of the all-purpose trunk roads, i.e. main roads that are not motorway standard, to be called 'expressways' [16].

Drivers are increasing able to access a variety of sources of information. Alongside the roadside signs and signals are sources inside the vehicle such as radio, navigation system and traffic information apps accessed via smartphones linked to the in-vehicle media systems. At present there is virtually no linkage between the on-board and roadside systems. Navigation and traffic information apps are now capable of warning drivers of queuing traffic with good spatial and temporal accuracy. Conversely, drivers are potentially sceptical of roadside signs warning of congestion if their on-board app is not showing any and will have their confidence in the roadside systems undermined if the app proves to be correct. Unfortunately, this lack of credibility then extends to other warnings, such as for an obstruction in the carriageway. System operators need to be aware of the other sources and the impact that poor signing practise can have on the credibility of their information systems. Prompt removal of expired signal patterns and avoiding signalling too far upstream are two areas where credibility can easily be lost.

A significant amount of research and development has taken place into the concept of 'connected vehicles and highways' whereby vehicles' on-board electronics communicate directly with other vehicles and roadside systems. Communications can be via local short-range methods such as DSRC or ITS-G5 or via cellular communications using 5G [17]. This is being combined with developments in automated control of vehicles to become 'connected and autonomous vehicles' (CAV); the focus of much research and testing.

Transmitting the information, both advisory and mandatory, directly into the vehicle and using vehicle to vehicle communication pathways to pass messages along the traffic stream has some major attractions. The need for expensive roadside infrastructure can be eliminated, information can be sent in respect of any location not just where the sign is located, the available bandwidth for information is not restricted to simple graphics or short text strings and information can be passed along the stream of traffic about hazards ahead. If the information is capable of being directly interpreted by an on-board system such as a speed limiter, advanced cruise control or a fully automated driving system then compliance with mandatory instructions such as speed limits can be assured.

Communicating roadside sign and hazard warning information into vehicles is not technically difficult and many demonstrations have been shown, for example, the Road Traffic Advisor project described in [18]. However, the regulations regarding mandatory signs do not envisage the display of the signs inside a vehicle nor the direct use by a driver assistance application or fully automated driver. Variable, mandatory speed limits apply at a specific location and time and any sign within a vehicle would have to have the same temporal and spatial information. With existing signs, the location, format and timing are all in the control of the highway operator so they can be confident that all drivers are receiving the same information at the same time. This is much harder to guarantee when they are only in control of the transmission of a wireless data message. Even if the receiving vehicle has to acknowledge receipt of the message it is difficult to ascertain that all vehicles did receive actually receive it. Once inside the vehicle the way the information is used is outside the highway operators control unless there are regulations in place that determine the acceptable and required means of processing the data transmitted. Regulation in this area, which can be seen as similar to the requirement to have a working speedometer, can be difficult to enforce.

Wireless communication of information directly to vehicles is likely to be implemented to support the deployment of driver assistance systems and fully automated driving applications including automated platooning. However, until all vehicles are equipped to receive and process the information via this route direct communication will supplement rather than supplant physical infrastructure. This will place an additional burden on the roadside systems operator as they will need to ensure that both methods are always synchronised.

8.8 Conclusion

Smart motorways manage traffic by influencing driver behaviour through the use of a mix of advisory and mandatory signs and signals. The pattern of signal setting in space and time creates a consistent and coherent set of information about the road immediately ahead. The main challenge is educating drivers about how to respond to the signs and maintaining credibility in order to maximise compliance. Wireless infrastructure to vehicle communications will supplement physical signs in order to support driver assistance and automated driver systems. The increased bandwidth creates not only an opportunity for improved information but also challenges for the system operators in ensuring that information is coherent across multiple channels.

References

[1] Jiang, R, Chung, E, and Lee, J (2011) 'Variable Speed Limits: conceptual design for Queensland practice'. *Proceedings of the Australasian Transport Research Forum 2011*, PATREC, Adelaide Hilton Hotel; Adelaide, 28–30 September, 2011. pp. 1–11.

[2] Department for Transport. *Road Lengths in Great Britain 2018* [online]. Statistical Release 14th May 2019. Available from https://assets.publishing.service.gov.uk/government/uploads/system/uploads/attachment_data/file/801357/road-lengths-in-great-britain-2018.pdf [accessed 6 Sep 2019].

[3] Highways England. *Customer Service Strategy* [online]. 2016. Available from https://assets.publishing.service.gov.uk/government/uploads/system/uploads/attachment_data/file/490538/S150470_Customer_Service_Strategy.pdf [accessed 6 Sep 2019].

[4] Transport Focus. *Road users' priorities for the Road Investment Strategy, 2020-25*[online]. 28th June 2017. Available from https://www.transportfocus.org.uk/research-publications/publications/road-users-priorities-for-the-road-investment-strategy-2020-25/ [accessed 6 Sep 2019].

[5] Harbord B. 'M25 controlled motorway – results of the first two years'. *Proceedings of the 9th International Conference on Road Transport Information and Control (No. 454)*; London, UK, 21-23 April 1998. IET.

[6] Sultan, B., Meekums, R., and Brown, M., 'The impact of active traffic management on motorway operation'. *Road Transport Information and Control -RTIC 2008 and ITS United Kingdom Members' Conference*; London, UK, 20–22 May 2008: IET, 2008. pp.1–8

[7] Remeijn, H., 'The Dutch Motorway Control System, 13 Years of Evolution'. *7thIFAC/IFORS Symposium on Transportation Systems*; Tianjin, China, August 1994: IFAC, 1994.

[8] Kwon, J., Petty, K., and Varaiya, P., 'Probe vehicle runs or loop detectors? Effect of detector spacing and sample size on the accuracy of freeway congestion monitoring'. *Proceedings of 86th Transportation Research Board Annual Meeting*, Washington, DC, January 21–25, 2007: TRB.

[9] Highways England. *INTERIM ADVICE NOTE 161/15 Smart Motorways (SM)* [online]. Available from http://www.standardsforhighways.co.uk/ha/standards/ians/pdfs/IAN161_15.pdf [accessed 9 Sep 2019].

[10] Papageorgiou, M., Kosmatopoulos, E., and Papamichail, I., 'Effects of variable speed limits on motorway traffic flow'. *Transportation Research Record* 2008 (2047), pp. 37–48.

[11] Highways England. MCE 2103, NMCS Signal Sequencing Rule Specification, Issue P, December 2010.

[12] Highways England. MCH 1744, NMCS2 MIDAS Signal and Sign Setting Strategy, Issue H, July 2009.

[13] Department for Transport, *UK Statutory Instruments with a subject starting with SPECIAL ROADS* [online]. Available from https://www.legislation.gov.uk/uksi/special%20roads [accessed 6 Sep 2019].

[14] Department for Transport, *The Highway Code, 2019 edition*[online]. Available from https://www.highwaycodeuk.co.uk/ [accessed 6 Sep 2019].

[15] Department for Transport. *Road Traffic Act 1988, Section 36* [online]. Available from http://www.legislation.gov.uk/ukpga/1988/52/section/36 [accessed 6 Sep 2019].

[16] Department for Transport. *Shaping the Future of England's Strategic Roads, Consultation on Highways England's Initial Report Moving Britain Ahead,* [online] December 2017. Available from https://assets.publishing.service.gov.uk/government/uploads/system/uploads/attachment_data/file/666965/shaping-the-future-of-englands-strategic-roads.pdf [accessed 6 Sep 2019].

[17] Transport Technology Forum. *Local and Cellular Communications: A guide to help policy development* [online]. March 2017. Available from https://its-uk.org.uk/wp-content/uploads/2017/04/A-comparison-of-ITSG5-with-cellular-comms-policy-implications.pdf [accessed 6 Sep 2019].

[18] Ellis, S., Lansdown, T., and Richardson, J. 'Road traffic advisor project'. *Tenth International Conference on Road Transport Information and Control*; London, UK, April:2000: IET.

Chapter 9

Contribution of cooperative ITS to road traffic and travel information

Alan Stevens[1]

9.1 Introduction to chapter

9.1.1 Intention and content

In this chapter, the overall intention is to describe the state-of-the-art in Cooperative Intelligent Transport Systems (C-ITS) relevant to traffic and travel information (TTI). Information services for drivers of passenger cars are particularly highlighted although several other C-ITS services are dedicated to freight and logistics. Geographically, the focus is on the United Kingdom and relevant European developments.

The chapter begins by taking an engineering approach to explore C-ITS through its hardware and communication requirements including short range and long range (cellular) options. European coordination and financial support have been instrumental in recent C-ITS developments and so several European-level structures and activities are then described. There follows an introduction to the range of C-ITS services with more detailed description of five of them (floating vehicle data (FVD), automatic crash notification, road works warning (RWW), incident ahead warning and green light-optimised speed advice (GLOSA)). These five have been chosen to illustrate specific characteristics and issues with C-ITS deployment. The chapter includes descriptions and discussion of several key technical and non-technical issues which are fundamental to implementation of services. These include how data and communication are protected through public key infrastructure (PKI) security and how legal issues, such as liability, data protection and privacy, are being addressed.

The chapter continues with a section describing how the impacts of C-ITS for TTI can be evaluated and discusses the benefits and limitations of field trials. There are many current C-ITS trials and pilots and the chapter takes two examples; one from the United Kingdom and one European. Other examples of how C-ITS supports automated driving can also be found [1].

At the time of writing there is only very limited take-up of C-ITS services so it is important to look at business cases for deployment and to consider the roles of different stakeholders. In a final outlook section, the prospects for sustainable

[1]Transportation Research Group, University of Southampton, UK

deployment of information services are discussed as well as the challenges in developing new services in support of vehicle automation.

9.1.2 Introduction to cooperative ITS

The objective of C-ITS is '... to give advice or facilitate actions to improving safety, sustainability, efficiency and comfort beyond the scope of stand-alone systems' [2]. C-ITS involves communicating and sharing information between four (so-called) ITS-stations; these are personal devices, vehicles, the roadside infrastructure and the back-office.

C-ITS are also often referred to as 'connected vehicles' and also by the relatively new term of connected autonomous vehicles (CAV). However, as CAVs and, to a lesser extent, connected vehicles tend to focus on *automation* of vehicles, this chapter will use the more specific term C-ITS to concentrate on information exchange and the services built thereon. Such systems are now being trialled in cities and along transport corridors with first applications being provision of real-time information and entertainment to vehicle occupants and collection of position data (FVD) to assist traffic management.

From a technology perspective, a lot of efforts in developing C-ITS have been devoted to wireless communications between vehicles and the roadside. Typically, this communication is achieved through roadside beacons providing short-range connectivity to moving vehicles or using smart-phone cellular networks. Information exchange enabled through connectivity is then used as the basis for services to drivers/passengers and to road network operators. Of course, provision of traffic and travel services through C-ITS requires a substantial value chain and involves many stakeholders including vehicle manufacturers, in-vehicle product developers, electronics companies, app developers and communication network providers as well as road operators.

This plethora of communication technologies and services leads to various implementation options with different benefits and drawbacks for the various stakeholders involved. Thus, evaluating C-ITS and developing sustainable business models for implementation presents a current and on-going challenge with (as yet) few success stories.

9.2 Communications for C-ITS

Communication between personal devices, vehicles, the roadside infrastructure and the back-office (known as the C-ITS 'stations') is vital for C-ITS. More specifically, communication for exchange of information between vehicles is called vehicle-to-vehicle (V2V) and that of vehicle-to-infrastructure (V2I), leading also to the more generic term V2X. In terms of media for exchange, two key technologies are short range communications, referred to as 'ITS-G5' and cellular communications. Cellular networks operate through existing commercial cellular radio infrastructure whereas ITS-G5 requires the deployment of dedicated roadside beacons. Having both pathways available (or, indeed, others such as Bluetooth and ZigBee) is referred to as 'hybrid communication'.

In Europe, the ITS-G5 short-range V2X technology is based on the IEEE 802.11 standard (often referred to by its brand name WiFi, and, more specifically, IEEE802.11p) and standardised in Europe as ETSI EN 302 663. In the United States, the ITS-G5 technology is also referred to as WAVE (wireless access in vehicular environments) or DSRC (dedicated short-range communication). EU regulation provides spectrum in the 5.9 GHz band for traffic safety and management related services on a non-exclusive basis. This is where IEEE802.11p/ETSI ITS-G5 will operate but as spectrum is also required for other applications there are potential competition pressures which could interfere with robust and safe operation of C-ITS services. For Transport and Traffic Telematics (TTT), frequencies in the 5.8 GHz band are used, for example for road charging applications.

Cellular terminology can be confusing. The 'G' in 3G or 4G stands for 'generation'. 4G is the fourth generation of mobile data technology, as defined by the radio sector of the International Telecommunication Union (ITU-R). LTE stands for 'Long-term Evolution' and applies more generally to the idea of improving wireless broadband speeds to meet increasing demand. 4G LTE provides major improvements over 3G speeds, but, technically, it is not 4G (although most cellular carriers advertise their networks as 4G LTE, because it sounds the same as 4G or even better).

The cellular standardisation organisation 3GPP [3] is developing cellular short-range communication technologies. One of them, LTE-V2X, is based on 4G cellular standards. LTE-V2X is sometimes also referred to as 3GPP. LTE-V2X proof of concept testing has started recently but few independent results are publicly available. Cross-border (e.g. international) tests or cross-brand tests are unknown so far. Another development from 3GPP is C-V2X which effectively bundles a LTE wireless broadband service with a short-range point-to-point service. The 5G Automotive Association [4] has been set up specifically to promote C-V2X.

For many years there has been a search for a convincing business model for ITS-G5. It was thought that as standards emerged governments (at least in the United States and Europe) would mandate that technology. However, in November 2017, the US government effectively side-lined a mandate, and in Europe the C-ITS delegated regulation [5] did not mandate ITS-G5 but only specified the technology to be adopted if fitted. A mandate would have created a strong business case for ITS-G5 so early adopters would not be penalised with high development costs for a potentially unused technology. Without such a regulatory requirement the business case for implementation is much less clear.

With cellular communication, car makers can anticipate millions of cars with cellular connections but not ubiquitous coverage of cellular service and there are questions about latency and security. There is also a cost for the service to be paid to the cellular providers. Nevertheless, for non-safety critical services such as weather and traffic information (see below) current 3G/4G/4G-LTE cellular technology work well. For time-critical safety services it is possible that future cellular technology will have sufficient performance but that is currently unproven [6]. The two communication options also have different approaches to security and it is currently unclear if one communication pathway for TTI to vehicles will dominate.

9.3 European coordination activities

9.3.1 C-ITS platform

The European Commission has a strong belief that the information and driver alerts generated through C-ITS have great potential to improve road safety and efficiency of road transport. Because of these expected benefits and relatively moderate deployment costs, there is substantial motivation to enable the pace and scale that will translate into European market production and early deployment.

The Platform for the Deployment of C-ITS in the European Union (C-ITS Platform) was launched by the European Commission in July 2014, and met for the first time in November 2014 [7]. It involved key stakeholders of C-ITS including the Commission, public stakeholders from Member States and local or regional authorities, road operators, vehicle manufacturers and suppliers, service providers and telecommunication companies.

The Platform was created to develop a common vision for C-ITS, exchange technical knowledge and to facilitate cooperation on technical, legal, organisational, administrative and governance aspects. One specific objective was to identify and agree on how to ensure interoperability of C-ITS across borders and along the whole value chain, as well as to identify the most likely and suitable deployment scenario(s). These included the first V2V and V2I services to be deployed across the EU and their most beneficial geographical environments (such as inter-urban corridors, urban roads, tunnels etc.).

The C-ITS platform addressed important technical, legal and implementation issues involved in deployment. The technical issues covered communication frequencies, hybrid communications, cyber-security and access to in-vehicle data and resources. The legal issues included liability, data protection and privacy. The other implementation topics included standardisation, benefit-cost analysis, business models, public acceptance, road safety and international cooperation.

On most issues a shared vision and a route to solutions emerged; for some others possible solutions were mapped out but further work on consensus building was needed. This led to a second phase of work which reported in September 2017 [8].

9.3.2 C-Roads platform

The C-Roads platform [9] is a joint initiative of European Member States and road operators for testing and implementing C-ITS services particularly focussing on cross-border harmonisation and interoperability. C-Roads oversees significant investments at national and EU level and works closely with industry including the CAR 2 CAR Communication Consortium [10].

9.3.3 Delegated act

The European Commission has drafted a legal framework on C-ITS [5]. The 'delegated act' is subordinated to the ITS Directive 2010/40/EU. Its focus is on C-ITS services to be deployed in the short term that will provide driver information that contributes particularly to road safety and traffic efficiency.

It addresses direct short-range communication between C-ITS stations, be they installed in vehicles, at the roadside or on infrastructure, such as traffic lights. The objective is to guarantee that communication is interoperable, reliable, secure and backward-compatible – all key principles which are defined in the ITS Directive.

An Annex to the delegated act defines the C-ITS priority services by describing the service profiles i.e. the specific configuration of standards, defining the implementation of various options of standards (see Section 9.4).

The delegated act requires that C-ITS stations use the EU C-ITS security credential management system. This comprises a certificate policy, which sets out the requirements for the management of public key certificates for C-ITS services (see section on security below), and a security policy, which sets out the requirements for the management of information security in C-ITS.

9.3.4 Display and human machine interaction

With the proliferation of information services to the driver made possible through C-ITS there is a need to carefully consider how information is presented to drivers. In fact, this has been an issue since vehicles were first designed and where style and manufacturer's brand ethos have sometimes taken precedence over practicality and usability. With the advent of traffic information and route guidance systems, stand-alone (nomadic) or built-in displays became common-place in vehicles. This led to the development of a range of human machine interaction (HMI) standards for vehicle information [11] covering minimum requirements for presentation of information (including, for example, text size and symbology and auditory sounds).

As well as the visual (and audible and tactile) representation of information, it is clearly important to design *how* drivers interact with information system; this is called 'dialogue management' and where the following key principles have been identified:

- Appropriate for use while driving
  - compatibility with driving
  - simplicity
  - timing/priorities
- Appropriate for the information or control task
  - consistency
  - controllability
- Appropriate for the driver
  - self-descriptiveness
  - conformity with driver expectations
  - error tolerance.

ISO standard 15005 [12] provides an explanation of these principles and related requirements, and recommendations are given, together with examples demonstrating how the principles can be applied. Wherever there are requirements, compliance verification conditions are also given. The standard is applicable to information systems comprising either single or multiple devices, which can be either independent or interconnected.

As C-ITS applications are deployed, the number and frequency of messages presented to drivers can be expected to increase. So as not to overload the driver with information, and present that which is most important at the time that it is needed, some form of message prioritisation is required. If these messages are not managed properly, drivers could fail to obtain critical information, which may degrade safety. An ISO Technical Specification [13] offers two methods to prioritise system-initiated or driver-requested information presented to drivers while driving. The primary method takes criticality and urgency ratings of messages into consideration when calculating a priority index. An alternative method involves paired comparisons of all possible messages to form a priority matrix. The ISO-TS provides a consistent basis for the management of messages competing for the driver's limited information processing capability.

9.4 Services

This section describes a number of C-ITS services which provide TTI to drivers or provide information to road operators from which TTI can be derived. As space is limited, and many services are similar, this section identifies the expected 'first wave' C-ITS services and then describes a small set of them in more detail.

The descriptions provided here are derived from actual pilot implementations which may differ in detail from the formal service definitions in Annex I of the European C-ITS Delegated Act [5]. The Act provides 'service profiles' which specifies the standards and options required to formally define the services. The Act also classifies services into different categories and, of course, not all services are applicable on all roads; for example, a service providing information about traffic signals might be part of a bundle of urban services but not applicable on inter-urban highways.

9.4.1 'Day 1' and 'Day 1.5' services

The C-ITS Platform Phase 1 final report [7] identified a number of (so-called) 'Day 1' services i.e. a range of services which are most likely to be implemented in a first wave of deployment.

The C-ITS Platform also agreed on a list of 'Day 1.5 services', considered as mature and highly desired by the market, though, for which specifications or standards might not be completely ready.

Table 9.1 Day 1 C-ITS services

#	**Day 1 service**
1	Emergency electronic brake light
2	Emergency vehicle approaching
3	Slow or stationary vehicles
4	Traffic jam ahead warning
5	Hazardous location notification
6	RWW
7	Weather conditions
8	In-vehicle signage (IVS)
9	In-vehicle speed limits
10	FVD
11	Shockwave damping
12	GLOSA
13	Signal violation/intersection safety
14	Traffic signal priority request by designated vehicles

Table 9.2 Day 1.5 C-ITS services

#	**Day 1.5 service**
1	Information on fuelling and charging stations for alternative fuel vehicles
2	Vulnerable road user protection
3	On street parking management and information
4	Off street parking information
5	Park and ride information
6	Connected and Cooperative navigation into and out of the city (first and last mile, parking, route advice, coordinated traffic lights)
7	Traffic information and smart routing

9.4.2 Description of five key services

This sub-section describes five key C-ITS services which illustrate different aspects of TTI using C-ITS and also illustrate the range of challenges with this relatively new information channel.

- floating vehicle data (FVD)
- Automatic crash notification (eCall)
- road works warning (RWW)
- In-vehicle signage (IVS)
- green lightoptimised speed advice (GLOSA).

FVD and eCall are services providing information *from* equipped vehicles *to* the roadside. FVD is used as an information source for TTI; it introduces two important message formats. eCall is the first actually deployed C-ITS service and illustrates how providers of vehicles, communication services and infrastructure managers can work together. The other three C-ITS services provide information *to* drivers. RWW illustrates the importance of coordination between roadworks providers to generate quality information; IVS duplicates both mandatory and other external signage within vehicles but raises legal and operational issues; GLOSA informs drivers about signals at intersections but may present trade-offs between individual information and network-wide flow optimisation.

9.4.2.1 Floating vehicle data

As a result of the need to evaluate the impacts of traffic control schemes in the 1960s, the technique of collecting journey times using vehicles 'floating' in the traffic stream was developed. Drivers of these vehicles (usually cars) measured journey times between specific origins and destinations whilst trying as far as possible to adopt the speed of surrounding traffic and by overtaking as many vehicles as overtake them. Now, with the advent of mass surveillance, the positions in time (and hence movement) of individual vehicles can be measured (e.g. by phone Bluetooth signals or by video-matched number plates) and 'real-time' or average journey times derived. This approach, by analogy with earlier data collection, is also called FVD and provides a substantial source of modern traffic information.

FVD can now also be obtained from cooperative vehicles using V2I communications; specific standards and protocols have been developed to convey not only positional information but a range of other traffic data. Such cooperative vehicles 'know' not only their position, speed and direction but typically have embedded sensors and systems capable of describing their driving experience (e.g. windscreen wiper status, temperature, ABS brake performance, electronic stability control feedback, etc.). Also, drivers can interact with on-board interfaces to indicate other situations such as animals in the roadway.

Vehicles fitted with C-ITS systems can transmit multiple message types, two of which are CAM (cooperative awareness message) and DENM (decentralised environmental notification message). CAM messages relate to the status and position of the vehicle and are transmitted continuously. DENM provides information about the environment, as observed by the vehicle or by the driver, and are only transmitted as required. Both message types can be considered as FVD and can contribute to traffic information.

A vehicle will generate a CAM [14] when the driving direction changes more than 4°, when the distance between current and past position has changed more than 4 m or the speed has changed more than 0.5 m/s compared to the last time a CAM is sent but at least once a second and, at the most, once every 0.1 second under normal conditions. The important information elements are shown in Table 9.3.

The process of FVD collection from CAM messages is relatively straightforward: the vehicle indicates its speed, position and direction up to 10 times per second. These messages are received; the data is collected, anonymised, aggregated and consolidated within road side units or back-offices according to a parameter defined by the road manager. This process is totally invisible to the road user; there are no alerts or information displayed on the vehicle's information screens. The data can be used by road operators to derive more comprehensive knowledge about their networks which

Table 9.3 Cooperative awareness message information elements

CAM information element	Meaning
Date and time	Date and time at which the data was generated
Vehicle position	Position of the vehicle
Vehicle speed	Speed of the vehicle
Vehicle direction	Direction of the vehicle (0°–360°)
Driving direction of the vehicle	Forward/backward
Vehicle lights status	Status of the vehicle's lights (fog, headlights, etc.)
Vehicle length	Length of the vehicle
De-/acceleration	De- or acceleration of the vehicle

compliments data gathered by embedded road loops, cameras and mobile patrols. Thus C-ITS FVD provides a valuable source of data for TTI.

A DENM [15], by contrast with a CAM, contains information related to an event that has potential impact on road safety or traffic condition. An event is characterised by an event type, an event position (not the transmitting vehicle position), a detection time and a time duration and is sent in addition to the CAM. These attributes may change over space and over time. DENMs are only triggered in relation to an event but can then be sent up to 20 times per second and the sending period can last for several seconds. The messages may initiate, update or terminate an event. DENM, in contrast to CAMs, are 'multi-hop' messages; that is, they could be sent from an 'ITS Station' (vehicle, roadside, mobile or back-office) to a certain area and there be 'hopping' to other ITS station.

Although DENMs are being used to provide information to vehicles (see RWW and IVS services below), the use of DENMs to provide FVD is not widely established and, meanwhile, other message sets such as IVI (in-vehicle information) have been drawn up. The IVI standard [16] provides better and more extensive options for transmitting data from vehicles to enhance FVD.

The primary expected benefit of FVD is better information about traffic conditions which contributes directly to better traffic information for drivers. Data provides road managers with real-time average speeds at each point in the network; they can thus identify congestion in a timely and more precise way and can characterise the impact of events on traffic (such as development of congestion or end of queue evolution). FVD can also be used to assist in the development and evaluation of traffic management strategies and measures. As C-ITS becomes more widespread, road managers may be able to economise on other data gathering systems such as road loops. No specific behaviour is expected from road users, to whom the operation of the service is totally invisible. Thus, no behavioural drawbacks can be anticipated.

9.4.2.2 eCall

eCall is service which initiates a mobile phone call in the event of a significant vehicle crash, establishes a voice link to the emergency services and transmits data that specifies the vehicle's exact location, the time of incident and the direction of travel (most important on motorways), even if the driver is unconscious or unable to make a phone call [17]. An eCall can also be triggered manually by pushing a button in the car, for example by a witness of a serious accident.

eCall is not included in the lists of Day 1 and Day 1.5 services described above because it is considered as an already implemented service. It is one of the first C-ITS services actually deployed and can be seen as a special case of FVD which provides information on the most severe road crashes.

The primary expected benefit of eCall is saving lives. eCall should improve the timing of emergency responses to vehicle crashes as a result of both the reduction in time between an accident occurring and a call to the emergency services being made, and the improved accuracy of location information contained within the eCall message (compared with the location accuracy provided by a caller and cell ID location from a mobile phone). Safety benefits then arise from the more rapid, and in some cases more targeted, response by the emergency services which helps to preserve life and reduce the impact of injuries. The European Commission estimates that, once the system is fully deployed, eCall could save hundreds of lives every year and help injured people quicker. There are also expected to be benefits in terms of traffic management as a result of more rapid deployment of road assistance and, indirectly, improved driver information.

Deployment of eCall has not been straightforward and illustrates how multiple stakeholders are involved around the, so called, 'chicken and egg' issue: vehicle manufacturers do not want to install transmitting equipment until infrastructure is in place to respond, and infrastructure owners do not want to implement back-office services until there is substantial demand from vehicles. In the case of eCall, the communication providers also needed to be convinced – they are required to provide emergency message calls without cost and so had little incentive to upgrade hardware to carry new digital flags and messages from eCall equipment. For more than ten years implementation of eCall was encouraged on a voluntary basis and although there was good progress on technical standards, some private services offered by high-end vehicle manufacturers, a 'Memorandum of Understanding' between European governments and sponsored research projects, commitment was insufficient for a sustainable service. In part this was because a business case was lacking but also because the economic case for the service was not convincing [18].

To avoid market failure, the European Commission enacted Regulations, which finally came in to force in mid-2018, to ensure eCall deployment by requiring:

- mandatory fitting of eCall devices in new vehicle types (personal cars and commercial light vehicles) according to required specifications
- cellular operators to transmit and handle the eCall flags and data free of charge for the users
- Member States' emergency services to deal with receipt and handling of eCall, in accordance with the required specifications.

To help assuage concerns over privacy, eCall transmits data only in the case of a severe crash (e.g. triggering the vehicle's airbag). Information is not stored any longer than necessary and vehicles are not tracked. It is now expected that eCall will be progressively deployed as new vehicle types come to market and after-market devices become more popular.

9.4.2.3 Road works warning

RWW has been identified as providing important traffic safety information because there are frequently cases where drivers come into conflict with traffic management equipment, such as traffic cones, barriers and cushions which can lead to crashes and injury. An alert sufficiently in advance of the road works would help prevent this type of situation by informing drivers about changes to road conditions and suitable driving behaviour to adopt. There is a particular risk of rear shunt type collisions at mobile road works with impact protection vehicles and the danger is greater at work site that are 'lighter' in terms of protection and signalling. RWW covers only the situation of restricted flow or very temporary road closure so does not include information on alternative routes.

Roadworks may be planned many months in advance or may need to take place at very short notice and coordination between road operators and contractors will be vital if timely and accurate RWW messages are to be delivered to drivers. Messages will typically originate centrally from a Traffic Operation Centre but may also be constructed and transmitted from a stand-alone roadside trailer or a mobile unit (e.g. a salt spreader). RWW are encoded using the DENM message format [15]. The message contains a description of the work site (start/end position, expected duration) and additional information can be added including the speed limit of each affected section. The DENM messages may initiate, update or terminate a RWW.

A number of 'use cases' of RWW have been defined and are being evaluated in field trials:

- Lane closure or other restrictions – The RWW provides information on the closure of part of a lane, a whole lane or several lanes (including hard shoulder), but without full road closure.
- Temporary road closure – The RWW provides information on a road closure due to a set of static roadworks.
- Mobile road works (e.g. grass cutting, salt spreading, line painting) – The RWW provides information on a zone within which, at some point, a lane is narrowed or closed, but without full road closure, due to planned mobile work.

The primary expected benefit of RWW is increased safety as a result of more attentive and pre-warned drivers. Drivers should be able to anticipate reductions in lane availability or other restrictions (such as temporary speed limits or width restrictions) and to adapt their speed and position on the road. This should mitigate sudden braking or steering/swerving manoeuvres. RWW is also expected to reduce the number of collisions with traffic management measures associated with road work and increase the safety of road workers and emergency service personnel. Indirectly, it is also expected to improve traffic flow.

However, reliably delivering the benefits of RWW will be a challenge for a number of non-technical reasons. Firstly, the data provided needs to be timely (near real-time), accurate (corresponding to the physical location) and complete, which requires close coordination between road operators, contractors, emergency services and other stakeholders. When RWW messages are generated from multiple locations this also requires mechanisms to ensure consistency if driver confusion or poor service reputation are to be avoided. Secondly, the information has to be presented to drivers in a form which is non-distracting and easily understood; information which has changed e.g. about long-term roadworks may need to be particularly highlighted. Thirdly, the reaction of non-informed drivers to the behaviour of informed drivers (e.g. early lane changes or speed adjustments) may need to be considered and appropriate roadside signage provided. Nevertheless, despite these challenges, the benefits of RWW are expected to be worthwhile both for drivers and road operators.

9.4.2.4 In-vehicle signage

IVS is an information service to inform drivers about actual, static or dynamic road signs via in-car systems. The external mandatory or advisory road signage to be duplicated within vehicles may include:

- Static speed signs and other signage, e.g. bans on overtaking or lane advice
- Lane open and closed signs, variable direction signs and dynamic speed limits
- Variable message signs (VMS) displaying operational, tactical or strategic information
- Other text panels.

Messages will typically originate centrally from a Traffic Operation Centre and may contain text (in a variety of languages) and images. IVS is intended to display the same information as the external signage, rather than conveying additional information. IVI messages [16] have been used in field trials to implement this service.

A number of 'use cases' of IVS have been defined and are being evaluated in field trials:

- Static and dynamic or temporary speed limit information – the speed limit can be specific to a vehicle type (e.g. heavy goods vehicles)
- Embedded VMS 'free text' and images – which may be displayed in different languages
- Dynamic lane status information – to inform drivers of a dynamic lane opening, and notify them whether their vehicle is allowed to use it or not.

The primary expected benefit of IVS is increased safety through increased driver awareness of road conditions and driving requirements. As the information is displayed within the vehicle, drivers and passengers have plenty of time to view the information if external signage is obscured or only momentarily visible, thus increasing comfort. With contextualised information directed to specific vehicle types (restrictions for trucks) or to specific application areas, better regulation adherence and traffic management might also be expected. In the future, the availability of information within the vehicle might allow external signage to be simplified or removed thus reducing road management costs.

As well as the importance of delivering timely, accurate and complete information (discussed above for RWW) provision of IVS presents additional challenges concerning the presentation of information. This is not specified or standardised but it is left to the provider of the in-vehicle information system as to how information is presented. Information presented within the vehicle is not legally binding but, clearly, should be consistent with actual regulatory signs even if presented in different languages. Service providers (e.g. road authorities, car manufacturers) might have limited control over the display options on an HMI as they do not know the content of the message. This can lead to driver distraction and overload if messages are not well managed. To address these HMI challenges, drivers can be informed at system start-up about the pre-eminence and legally binding status of external signage and some form of prioritisation of in-vehicle message could be developed and implemented [19] : 'higher priority messages' should be displayed to warn the driver and prevent accidents; 'lower priority messages' should be displayed to inform the driver without causing distraction.

9.4.2.5 Green light-optimised speed advice

As initially envisaged, GLOSA is a service that informs drivers approaching an intersection about the status of the traffic signals, and also advises them of an optimum approach speed that should minimise delay and ensure a smooth transition through the conflict area of an intersection. However, implementation of GLOSA is complex and it may be adapted for a number of technical and legal reasons. Clearly, any speed advised should not be greater than the speed limit and in-vehicle displays should not conflict with the roadside signals. There is also a question as to the legality of advising a speed in excess of a vehicle's current speed or advising a stopped vehicle when the lights will become green (the legal situation may be different in different countries).

In order to implement GLOSA, the road operator or highway authority need to have a digital representation of the intersection layout (known as a MAP file) which might include road topography and road attributes, such as number of lanes on each approach to the intersection and the manoeuvres possible between them. This digital representation provides the framework upon which signal phasing and timing (SPAT) information is overlaid. Both the signal timing and intersection layout information need to be conveyed to vehicles either locally from the traffic signal controller, or centrally from an operator's traffic signal control system. SPAT/MAP infrastructure-to-vehicle communication messages are specified by the SAE-J2735 standard. Thus, the SPAT/MAP information that is broadcast, together with other information, reflects the real-time (current and/or future) signal phase and timing status for each lane of the intersection.

When approaching a signalised intersection, the received SPAT/MAP messages, combined with information on the position and direction of the vehicle, its speed and the distance to the traffic lights, enables an algorithm in the vehicle to calculate an optimal speed of approach. According to the delegated Act [5]: 'Traffic conditions, such as queues or traffic jams, affect the validity of speed advice and shall therefore be taken into account'; although, in practice, it is difficult to see how this can be accomplished with current technology.

The presentation of information on an in-vehicle display is not specified or standardised and it is left to the provider of the in-vehicle information system as to how information is presented to the driver about traversing an intersection. This may include lane advice, the current and anticipated future state of traffic signals or advice on speed. It may also extend advice to cover a sequence of intersections and suggest an optimal 'green wave' trajectory. Of course, what a driver does with the information provided is for them to decide within their traffic context, but a number of potential scenarios and driver responses can be anticipated:

- When the approaching traffic signal is green, and the driver can pass it without exceeding the speed limit, the driver receives speed advice and arrives at the intersection during a green signal phase. (The advised speed may be the current speed; it may also be a higher speed if legally permitted; it may or may not be achievable depending on traffic conditions).
- When the approaching traffic signal is red but will shortly change, the driver receives speed advice and maintains or decreases speed to arrive at the intersection at the start of a green signal phase.
- When the approaching traffic signal is green but the driver cannot pass it without exceeding the speed limit, the driver is informed, no speed advice is given and the driver then prepares to stop.
- When stationary at a red traffic signal (and where legally permitted) drivers may be provided with 'countdown to green' advice in order to minimise their start delay.

The primary expected benefit of GLOSA is smoother vehicle flow through signalised intersections. Other benefits may include:

- Improving compliance with traffic signals (reduced desire to chase green signals or the need to rapidly decelerate)
- Reducing driver anxiety
- Decreasing start delay based on 'Countdown to green' information
- Optimising vehicle engine stop based on likely duration of the red signal.

So, GLOSA should result in reduced energy consumption, reduced vehicle emissions (including CO_2), reduced vehicle wear and tear, increased safety, improved driver and passenger comfort, reduced stress and better average speeds, particularly in urban environments where there is generally a higher density of traffic signals. However, the extent to which the expected benefits are delivered are yet to be fully explored in field trials and a number of potential dangers and challenges to be overcome can also be envisaged:

- GLOSA depends on accurate (correct, precise and current) SPAT/MAP data provided by the traffic light controller or from central traffic control sources. Not only has the source data to be correct but it has to be provided by infrastructure-to-vehicle communications in a timely manner; this may be a challenge in some urban conditions depending on the communication channels.

- Locational accuracy (usually provided by GPS) could be insufficient at complex intersections that employ multiple traffic phases on approach (e.g. left turn and right turn traffic phases in addition to a straight-ahead movement). Vehicles will need to be able to identify their position and intended direction, or route, with sufficient accuracy in order to identify the relevant traffic phase.
- The highly dynamic nature of some traffic control systems, which are able to make late signal phasing decisions, combined with other adaptive systems (e.g. selective vehicle detection bus priority), could limit the viability of GLOSA in some applications or require efficiency compromises.
- Similarly, area-wide dynamic traffic control systems may need to optimise on a more strategic level and require specific intersections to skip stages or constrain turning movements in a dynamic way. The practical compatibility of GLOSA with dynamic traffic control is an issue for future research and field study.
- A more frequent current limitation is that of other traffic behaviour, particularly in congested urban conditions. If there is a traffic queue in front of a signal, GLOSA's speed advice is of little use. However, perhaps future cooperative systems may be able to adapt advice based on dynamic traffic conditions.
- As noted above, driver behaviour is an important factor in understanding the impacts of GLOSA and this may change and develop as the proportion of equipped vehicles increases. Whilst it can be expected that most drivers will readily and safely adapt to GLOSA the behaviours of a small minority may become of concern. One example could be 'gaming' based on the information e.g. increasing speed to beat the lights or anticipating a 'countdown to green' with a drag-racing mentality. Information on the length of a red phase could also prompt undesirable and distracting in-vehicle behaviours such as checking a smartphone. Identifying negative behaviours might prompt road authorities to introduce additional safety margins thus countering at least part of the traffic flow benefits.
- Also, of safety concern is the adverse reaction of other drivers, e.g. unaware that the driver ahead is knowingly slowing down (in anticipation of not be included in the current green phase). Such 'trickling along to a red stop' behaviour could provoke frustration or aggressive overtaking in others. Perhaps vehicles will need to signal to other road users that they are responding to anticipated future signal behaviour.

9.5 Security and legal issues

9.5.1 Security and public key infrastructure

Security is considered paramount in deployment of C-ITS to prevent the spread of erroneous information that could compromise safety and traffic management. This requires a suitable and effective 'Trust Model'; i.e. a set of processes among the main entities of C-ITS which aims to ensure that information can be exchanged safely, securely and privately. At the heart of most modern trust models are cryptographic techniques involving digital signatures. A digital signature is a mathematical scheme for verifying the authenticity of digital messages; these give the message recipient very strong reasons to believe that the message was created by a known sender (authentication), and that the message was not altered in transit (integrity).

For several reasons including communications and computational efficiency, most work on C-ITS trust models have focussed on cryptographic techniques involving PKI. A PKI provides a set of roles, policies, and procedures to create, manage, distribute, use, store and revoke digital certificates and manage public-key encryption. Digital certificates are electronic documents containing the public key for a digital signature and specifying the identity associated with the key, such as the name of an organisation; hence, digital certificates are used to prove the ownership of a public key. In essence a PKI creates digital certificates which map public keys to entities, securely stores these certificates in a central repository and revokes them if needed.

For Europe, the European Commission has developed a common standardised C-ITS trust model (i.e. one single cryptographic algorithm and certificate format) and certificate policy [20] all over the EU, based on a PKI and defined in an appropriate regulatory framework [21]. Meanwhile, the C-ITS platform [8] identified an urgent need for international cooperation to discuss how interoperability of multiple trust domains can be realised. The security of C-ITS and implementation of PKI within C-ITS are still under discussion at the time of writing.

PKI has been incorporated within the ITS-G5 communication standards to provide security at the message level. CAM and DENM messages (described above) include cryptographically signed certificates. In fact, the certificates make use of pseudonyms which preserve security without revealing the identity of the vehicle or its driver and, as the pseudonyms are changed regularly, no long-term tracking of individual vehicles is possible. By contrast, security for cellular communication used by C-ITS services relies on channel security (i.e. protecting the communications infrastructure) rather than PKI.

9.5.2 Liability

Working Group 3 of the C-ITS Platform discussed legal issues including liability in relation to traffic information provided through C-ITS [7]. Their conclusion was that Day 1 (and Day 1.5) services provide only information and so control of the vehicle remains the driver's responsibility. Therefore, the Working Group concluded that there were no changes concerning liability compared with the current situation.

Nevertheless, a number of issues were raised concerning the extent to which drivers might begin to trust and rely on the technology and information, particularly that provided by public authorities, beyond the use for which it was designed. They could even become dependent on it – as, might be argued, some SatNav users are today. A partial solution to this 'over-trust' would be a general disclaimer that the information provided inside the vehicle is for guidance purposes only. This would need to be implemented through the vehicle's HMI in some way and whether this is sufficient is a matter of judgement. Additional measures could involve public awareness and driver training or periodic more specific warnings to the driver from the in-vehicle system.

Future C-ITS services are likely to include more safety critical messages for the driver. Also, future developments could see some traffic signage migrating from the physical infrastructure to only being provided digitally inside the vehicle. In these cases, quality of information and potential liability issues will become even more critical. It will then be important to fully analyse the data flow/value chain (from information providers, such as traffic management centres, to road side units and end users) to better appreciate how issues of responsibility and liability will be distributed and resolved.

9.5.3 Data protection and privacy

C-ITS make use of information and communication technologies that enable different parts of the road transport network to share information. C-ITS vehicles are capable of broadcasting and receiving data that allow them to communicate with each other and/or with the infrastructure. In addition to what drivers can immediately see around them, and what vehicle sensors can detect, all parts of the transport system will increasingly be able to share information to improve driver decision-making and optimise transport operations and safety.

Most drivers would agree to provide their data in anonymous form for traffic related issues (e.g. road safety or congestion management). However, many would raise concerns regarding commercial use of driver data and not all would be prepared to provide information in order to have access to commercial services. Having robust mechanisms to guarantee that data is secure and that existing Data Protection legislation can efficiently be enforced is therefore vital. Privacy performance depends critically on the security performance of C-ITS. For example, the use of PKI which pseudonymises the certificates associated to the vehicles protects vehicles from direct identification and so supports increased privacy.

Working Group 4 of the C-ITS Platform [7] has looked specifically at Cooperative Awareness Messages (CAM) and the DENM and concluded that those messages can be considered as 'personal data' because of the potential of indirect identification of users. The European legislation on Data Protection 95/46/EC is therefore applicable and so finding workable solutions and building public acceptance around data protection is thus of key importance.

The applicable legal framework concerning privacy and data protection for all individuals of the European Union (EU) and the European Economic Area (EEA) is the General Data Protection Regulation (GDPR) (EU) 2016/679 [22]. GDPR offers a robust legal framework applicable to C-ITS; it sets out the principles relating to processing of personal data and different grounds for lawful processing of personal data.

One legal basis to process personal data would be if informed consent is given by the data subject. However, the practicality of this given the multitude of services, choices and potential processing purposes is questionable. A second legal basis could be a contractual relationship between the data subject and the data controller (e.g. a road authority). A third could be processing in the public interest or in the exercise of official authority (in this case the purpose would be road safety and traffic efficiency); however, this might only apply for public authorities or if there were national laws. A further legal possibility would be on the grounds of legitimate interest where the data controller would need to balance the benefits with the impact on the data subject. As well as being lawful, data collection and processing needs to respect the GDPR principles including fairness and transparency, and be for an explicit and specified purpose.

This is clearly a complex area in which further work and case law will be developed as C-ITS is deployed. More details can be found in a working paper from the C-ITS platform [23].

9.6 Assessment and evaluation of C-ITS

9.6.1 Purposes and methodologies

The term 'assessment' tends to be used when forward looking (also called ex-ante). 'Evaluation' is more used when referring to the effects based on actual data rather than forecasts (also called ex-post). Impact is also a general term more used in an ex-post sense. As C-ITS services are relatively novel, with results from pilot trials only now emerging, there is a dearth of real-world data on which to base a comprehensive evaluation of their impacts.

A number of distinct purposes for evaluation of information services through C-ITS can be distinguished:

- Technical evaluation is concerned with issues such as efficiency and effectiveness of message delivery, verification of standards.
- Business evaluation by service providers tends to focus on customer needs and wants (see Acceptance above); finally, if customers like it and buy it, it is a success. Innovations in one form of information provision may displace other forms of TTI but the development is evaluated only in terms of the new provision.
- Evaluation by road operators will focus on operational costs and potential impact on their key performance indicators (which, for example may include road safety, congestion management and providing TTI).
- Socio-economic evaluation estimates the gains or losses for society resulting from deployment compared with a reference case or other options; these typically investigate the expected impacts of a system for a future time horizon.
- Strategic/Strategy evaluation provides a long-term political analysis of entire technologies or applications.

To address these different evaluation purposes a number of methodologies and tools have been developed and improved through a succession of international projects. For example, the European Commission's Impact Assessment is one approach to Strategic/Strategy evaluation [24] and the COBRA tool has been developed for Road Authorities [25].

To support the design, conduct and analysis of field operational tests (FOT), the EC sponsored development of a FOT methodology called FESTA (field operational test support action). The original FESTA project, completed in August 2008, developed a handbook of best practice for the design and implementation of FOTs in Europe. This has been subsequently updated as a result of project experience [26]. The handbook describes the key issues to be addressed including: (1) planning and running a FOT; (2) legal and ethical issues; selection of functions to be tested, and definition of use cases, research questions and hypotheses relating to those functions; (3) performance indicators; (4) experimental procedures (including participant selection, study design, study environment and pilot testing); (5) data acquisition; (6) databases and analysis tools; (7) data analysis and modelling; and (8) socio-economic impact.

FESTA describes some of the specific issues involved in conducting field trials with C-ITS [27] including: integration of the C-ITS into the vehicles; the need to train drivers in the C-ITS usage; the need to equip vehicles and infrastructure and suitable logging equipment; and ensuring a sufficiently high penetration rate for interaction effects to be observable. It also provides general advice (not specific to C-ITS) concerning data acquisition, databases and analysis tools, data analysis and modelling.

In the area of socio-economic evaluation, the main methodology is Benefit Cost analysis (BCA) [28]. This can be applied in a forward-looking way to support investment decision-making, and was used as a key input to build the Day 1 and Day 1.5 services lists. Such prospective studies typically make use of literature reviews, simulation work and expert estimation and, although often comprehensive in scope, have little access to data from real-life conditions. BCA can also be used in planning and retrospective analysis of field trials and pilot studies [29]. Here, data comes from real-world measurements taken in specific situations from the deployed technology and participants involved. Some of the specific challenges with C-ITS field data are described below.

9.6.2 Specific challenges with C-ITS evaluation

Evaluating C-ITS services presents some unique challenges as a result of their relative novelty and the diversity of hardware components, system architectures and stakeholders that need to be involved. The paragraphs below concentrate on issues that are specific or significant for C-ITS field trials beyond those that are challenges in evaluating technology innovations more generally, such as monetary valuation of outcomes and distributional effects of benefits between stakeholders in society.

Size of the effect: Estimates of the potential of C-ITS have been based on different definitions of services/bundles of services and different methods of evaluation. These are typically based on sparse data, estimations from relevant sources and expert opinions and then scaled up to estimate national or regional impacts. From these estimates the size of effect (e.g. travel time or vehicle crashes) is expected to be less than 10% and often less than 5% even if all vehicles are benefitting from the cooperative service.

In pilots and field trials it is typical to equip a few hundred or even a few thousand vehicles in a city or corridor. The total number of vehicles provides a 'headline' but for evaluation purposes a more important metric is vehicle-km in the area of interest within a defined time period. Any measurable effects will depend on the pilot fleet's usage of the network, and more specifically its density at any one time and place. So, an important question is what density of equipped vehicles is required to 'see' the effect of a C-ITS service, bearing in mind that the period over which a C-ITS service is relevant may be limited (e.g. during congestion/incidents) or relevant only at specific points in the network.

For services providing information to drivers, a change of behaviour would be required in order to measure effects and it therefore seems very unlikely that flow data typically obtainable from traffic counters (e.g. providing 2-minute averaged traffic counts) would 'see' any change at a bulk traffic level. A similar argument could be applied to environmental data and to safety data such as vehicle crashes.

Therefore, in order to measure an effect, it will be important to monitor the behaviour of individual vehicles/drivers and correlate this with the information provided. This can be achieved through interrogation of the driver (e.g. questionnaires, logs) or interrogation of the vehicle (e.g. by data logging). If, and when, effects are measured, there needs to be some assumptions and modelling about how the change in behaviour will translate into impacts such as traffic flow and safety. Thus, there is a long and, arguably fragile, chain of assumptions between the effects measured and the impacts anticipated.

Bundling the effects of individual services: Impacts may initially be considered (or measured in trials) for individual C-ITS services. In practice, deployment would likely involve a 'bundle' of services implemented together, rather than in isolation.

A simple addition of impacts would only be reasonable if the individual functions' impacts did not overlap. For most of the functions within a bundle, this will not be the case, for example hazardous location warning and traffic jam ahead warning do overlap when a traffic jam originates from a hazardous location. Therefore, in general, it will be necessary to identify overlaps and the combined effects of multiple services during evaluation.

Overlap with existing infrastructure: C-ITS can offer some services that are entirely new; however, C-ITS can also replace or supplement non-cooperative services so the comparison is more difficult (indeed, C-ITS can completely change some business models so may have wider consequences).

Where C-ITS services overlap with existing technologies, the impacts of these existing technologies also need to be taken into account. For example, FVD overlaps with data from inductive loops, Bluetooth and numberplate matching flow data; RWW and IVS overlap with static and dynamic signage and radio reports; Automatic Crash Notification (eCall) overlaps with video surveillance and static vehicle sensors.

'Hotspots' and transferability: Some C-ITS service will only be relevant at specific points in a network. Similarly, some services may be relevant only to certain types of roads, vehicles, drivers or journey purpose. Therefore, care has to be taken when scaling-up impacts measured in one region or corridor. The scaling-up will involve obtaining data on benefits achieved, with informed assumptions to make estimates where that data is not available, increasing the fragility of the estimates of effects.

Proportion of equipped vehicles and road network: Deployment variables include the number of equipped vehicles and, for services communicating with road infrastructure, the proportion of road network that is equipped with communications (e.g. cellular coverage or wireless beacons). For most C-ITS services which receive information from the roadside, it may be assumed that the relationship between deployment level and overall impacts is linear, i.e. each additional equipped vehicle benefits equally. This may not actually be the case as two examples illustrate:

- RWW might cause a large number of vehicles to seek an alternate route whereas, in practice, the better option may be for most to stay on the original road. So, as an information service becomes widespread it needs to take account of the number of vehicles likely to act on the advice.
- Information services that cause vehicles to slow down (such as RWW or GLOSA) might have the effect of also slowing other traffic, particularly if overtaking opportunities are limited.

For C-ITS services that involve information exchange between vehicles, the relationship between deployment and benefit will certainly not be linear; it is likely that for very low penetration rates there are no impacts at all and it is only when the number of equipped vehicles reaches a certain threshold that any impacts are achieved. This threshold depends on the C-ITS service, but also other factors. Similar to the 'hotspots' assumption for existing infrastructure, the number of equipped vehicles within range would be greater in busier traffic conditions.

Driver acceptance and compliance: A vehicle may be equipped with an in-vehicle unit and be travelling along an equipped section of road, but the user may choose to turn off the service. In other cases, the user may choose to turn on the service (or it may be permanently on), but then choose to ignore any information from it. This would clearly affect the impact of C-ITS services.

Business models: Uncertainties abound concerning how new infrastructure, including communications technology and equipment on-board vehicles, will be deployed and funded. So it is important to explore how capital expenditure and operational expenditure are borne by different stakeholders, which in turn depends on the business models for deployment of the C-ITS services.

Time horizon for evaluation: One of the specific challenges in evaluating C-ITS is deciding on the appropriate time horizon. This is always an issue in evaluation but C-ITS involves three components (of devices, vehicles, and infrastructure) with very different time horizons. Road infrastructure has an expected life of perhaps 30 years whereas vehicles are typically replaced in 5-10 years and IT devices, such as mobile phones, have a typical replacement frequency of 1-2 years. So deciding an appropriate period for a C-ITS service is problematic and depends on the perspective of the stakeholders.

9.7 Field trials and C-ITS pilots

Owing to the relative novelty of C-ITS there are numerous on-going and planned trials, often financially supported by national governments or international bodies. The term 'Field trial' or 'Field Operational Test' has come to mean a distinct time-limited monitored deployment, whereas the term 'Pilot trial' has implications of an indeterminate end-point and (hopefully) transition into a fully operational and sustainable service.

This section does not attempt to review all on-going activities but after considering why trials are important, selects one wholly UK example and one European example which includes a UK connected corridor. Finally, a brief review of some international initiatives and further references are given.

9.7.1 Learning from trials

There are many reasons for conducting pilot trials and much information to be gained from them in different areas including:

Technical

- Test equipment in real-life conditions
- Implement an evaluation methodology (adoption of a common methodology facilitates consolidation and comparison of results from different pilots)
- Undertake data collection and management (learning how to deal with scalability from small scale to large scale)
- Improve standards ('learning by doing').

Behavioural

- Appreciate real needs and wants of service users (drivers)
- Assess design of driver interfaces (human machine interface)
- Identify (negative) behaviours and attitudes
- Measure real performance of C-ITS services.

Organisational

- Develop long-term partnership between public and private stakeholders
- Identify governance for successful C-ITS service operations
- Address legal aspects for C-ITS deployment on public roads.

Political

- Obtain facts about the impact and evidence of benefits to shape C-ITS transport policy and to take fully informed decision on large-scale deployment
- Raise awareness of policy-makers.

Commercial

- Raise awareness of decision-makers, investors and end-users
- Undertake pre-commercial procurement of C-ITS equipment
- Develop business models with all stakeholders along C-ITS service chain
- Identify business opportunities where demand meets with supply.

9.7.2 UK CITE

CITE (Connected Intelligent Transport Environment) was a 30-months £7.1m project which completed in early 2019 [30]. With investment from the UK government's

Intelligent Mobility Fund and Highways England (HE), UK CITE created a vehicle-to-infrastructure (V2X) test environment that included the installation of a network of ITS G5 (wireless) roadside units in Coventry and on the M40, M42, A45 and the A46 road network. The roadside units provided the C-ITS service 'In Vehicle Information' (IVI) which enabled messages displayed on smart motorway overhead gantry signs to be replicated in vehicles using in-vehicle information (IVI) messages. The messages transmitted also include RWW, traffic condition warning, emergency electronic braking, and emergency vehicle warnings.

The UK CITE consortium was jointly led by Visteon Engineering Services Limited and Jaguar Land Rover (JLR) and included Coventry City Council, Coventry University, Highways England, Horiba Mira, Huawei Technologies (UK), Siemens, Transport for the West Midlands (TfWM), Vodafone Group Services, and WMG at University of Warwick. The project team equipped more than 40 miles (64 km) of urban roads, dual carriageways and motorways with combinations of three connected vehicle technologies, and tested a fourth LTE-V system.

In addition to on-road trialling of the V2X communications using a fleet of JLR connected vehicles, simulation and modelling enabled the consortium to study the impact of connected technologies in different road, traffic and weather conditions, comparing them to non-connected vehicle scenarios. The project also included cybersecurity testing alongside the large-scale trials on a road network.

The project demonstrated that C-ITS technology and services can improve journeys, reduce traffic congestion, and provide entertainment and safety services through better connectivity. Whilst not leading directly to a sustainable business model, there are plans to build on the installations as part of a future testbed for connected infrastructure and ecosystems.

9.7.3 InterCor

InterCor [31] stands for Interoperable Corridors, and linked the C-ITS corridor initiatives of the Netherlands C-ITS Corridor NL-DE-AT, the French SCOOP@F, the UK's A2/M2 connected corridor and the Belgian C-ITS initiatives. The €30m project achieved a network of approximately 1,500 km of corridors providing continuity of C-ITS services in four Member States.

Data communications were implemented through cellular, ITS G5 and a combination of cellular and ITS-G5 (hybrid) networks and the project focussed on deployment of 'Day-1' services including RWW, GLOSA, IVS and FVD. InterCor undertook testing of interoperability, security and certification, and seamless continuity of service, operation and maintenance. This was achieved at national sites and through a series of open 'Testfests' also involving external commercial and research organisations and other C-Roads Member States.

A major contribution, led by the United Kingdom, was development of a common evaluation framework which demonstrated the cumulative, real-life benefits of C-ITS applications of increased safety, efficiency, flexibility, user acceptance, and sustainability of road transport. The ultimate aim was to support public and private stakeholders to invest in C-ITS, and beyond the project's completion in 2019 InterCor aims to offer a testbed for beyond Day-1 C-ITS service development.

The UK element of InterCor was provided by the Department for Transport, Highways England, Transport for London and Kent County Council and their contractors. It involved the A2 and A102 in London and the M2 and A229 in Kent, a total of 110 km. It became known as the 'A2/M2 connected vehicle corridor'. The UK implemented and evaluated the four core services, led evaluation throughout the InterCor project and was responsible for organising and delivering the successful 'Testfest' of hybrid communications which took place in Kent in 2018.

9.7.4 Other international developments

Although there are more historic examples of pioneer projects which would now be called C-ITS, a key milestone project in Europe was CVIS [32] which ran from 2006 to 2010. Information is available [33] about current (up to June 2019) and recent C-ITS projects supported by the EU under the Connecting Europe Facility (CEF) for Transport and those now overseen by C-ROADS (see above). These projects will install C-ITS equipment in various locations throughout the EU, with pilot schemes designed to test the solutions in different operating environments (for instance in highways and urban areas). There are also numerous national C-ITS pilots; for instance, in France, Germany, Austria, The Netherlands, Finland, Sweden, Denmark, Norway, the Czech Republic, Spain and the United Kingdom.

The United States foresees using short range communications similar to those within Europe and has a number of deployment pilots [34]. Until recently it was widely believed that the United States would mandate the introduction of C-ITS Stations (referred to as V2V in the US policy context) into vehicles for the US market, although progress on this seems to have stalled.

In 2013, Australia started participation in international harmonisation efforts on C-ITS security, together with the United States and the EU. The first C-ITS demonstrations took place in Australia in 2016 and there are currently several C-ITS pilots [35], the largest of which is in Queensland. Many other countries are also undertaking or launching C-ITS pilots including as Canada, China, Singapore, Japan, and South Korea.

9.8 Business case and deployment of C-ITS

9.8.1 Market for C-ITS traffic and travel information

Predicting the market for C-ITS TTI is difficult for a number of general and specific reasons. General issues in prediction are uncertainties about the future and the impact of global trends and novel technologies. More specifically, market size depends on the definition of the services and technologies involved and the geographic area being considered (these factors are not always specified in enough detail in published studies). Also, there is often conflation of, or confusion between, markets (e.g. different traffic and travel modes, C-ITS, Connected and Autonomous Vehicles etc.). Finally, it has to be recognised that 'hype' is a factor as some stakeholders 'talk up' the benefits or are very optimistic over timescales and take-up in order to encourage government and private investment. With these caveats, three UK studies can be cited:

- In 2015, work by KPMG and the SMMT [36] predicted that the annual economic benefit from connected and autonomous vehicles (CAVs) in the United Kingdom would be £51bn per year by 2030.
- Research into the CAV market and associated technologies has been commissioned by the Centre for Connected and Autonomous Vehicles and carried out by the Transport Systems Catapult, Element Energy and Cambridge Econometrics. The report [37] published in 2017 suggested a UK market worth at least £28bn in 2035, capturing 3% of the £907bn global market.
- The Transport Technology Forum have calculated [38] that the value to the United Kingdom from applying technology to roads is already about £7.3bn per year and that exploiting connected vehicles and other technologies gives an additional potential annual £6.5bn value to the United Kingdom in the near term.

Even with the caveats expressed above it can be appreciated that TTI is now seen as part of a substantially larger market involving connected and, in the future, increasingly automated vehicles. This has recently become particularly prominent with the launch of the UK Industrial Strategy [39] and its 'Grand Challenges' involving future mobility.

9.8.2 Driver acceptance and compliance

Driver acceptance is important if the full benefits of TTI provided through C-ITS are to be achieved. This is an area in which there is relatively little research specific to C-ITS services although there is general consensus on some important issues [40]:

- Acceptance is a complex construct which has many facets and dependencies.
- Acceptance is based on individual judgements, so a driver-centric view is required to measure or predict acceptance at an individual level (assessing societal acceptance requires an additional broader perspective).
- Two key determinants of acceptance of new technology are usefulness and ease of use.
- Acceptance depends on the individual, so issues such as gender, age, culture and personality are likely to be important.
- The context of use is also important, including the supporting infrastructure (in its widest sense), whether use of the technology is voluntary and also broader social/cultural influences.
- Drivers do not have to actually like a technology/system to be accepting of it (but liking it may increase use of the technology).
- Acceptance should be regarded as a continuous variable, not a binary concept.
- Acceptance is not invariant; it may change (even for one individual) depending on the specific time/context in which the new technology is used and as experience with the technology develops.

Whether drivers want a service (and perhaps whether they will pay for it) is important in developing business cases for investment and deployment. If specific acceptance issues are identified, such as privacy and security, these can be responded to during system design; similarly, the driver interface can be designed to maximise acceptance. Acceptance is also important to increase use of a service where taking notice of the information supplied may change driver behaviour and so improve traffic flow and promote safer operation of the road network.

9.8.3 Business models

TTI delivered through C-ITS involves many stages and participants and these working together constitute a so called, 'value chain'. In many countries the value chain involves both the public and private sector because the deployment of C-ITS cannot rely on public funding alone. For the value chain to be initiated, and hopefully become sustainable, the participants will need to make initial investments (including in technology, people, changes to working practices) and such investment decisions need to be based on sound business cases for all the participants along the value chain.

Large-scale implementation of new C-ITS systems, which in many cases are administrated and financed by the public sector, requires substantial public funding, and hence acceptance from a policy making perspective. However, policy makers can sometimes (and often rightly) be sceptical about not fully proven technologies and so reluctant to make such investments. As both scale and interoperability across borders are required to support business cases, there can be a tendency to wait for other partners to be first movers rather than having fragmented and non-interoperable system a few years after investment.

One way to help develop business cases is to promote a shared and common understanding of how a service can be delivered and the role of each participant; development of standards is an area where good progress has been made.

Another is to ensure that all pilots fully analyse and report on the benefits and costs. However, even where the potential benefits of C-ITS can be demonstrated to outweigh the costs, it is important to understand how the benefits and costs are distributed: typically a substantial part (increased safety, less time spent in traffic, lower fuel consumption) go directly to the users and local communities, while the costs of investment and operation need to be borne by road operators and vehicle manufacturers. Some commentators have suggested [8] that a 'governance' layer needs to be established to help redistribute costs and benefits such that the needed contribution of every participant in the value chain is sustainable.

The current situation revealed by recent field trials and pilots is that the C-ITS market is not yet sufficiently mature and independent to operate without governmental support and some form of transitional assistance to large-scale commercial deployment seems to be needed. Therefore, further sponsorship and close monitoring of pilots and early deployment initiatives around the world is required.

9.9 Outlook

The C-ITS concept offers a new technology and architecture upon which many services can operate. Some, but not all, of these services are in the area of TTI. C-ITS is a relatively new concept which seems likely to develop in several ways that are discussed in the following paragraphs.

Firstly, there will be a refinement of existing information services through further trials and pilots and possibly an accommodation between the different communication options which will allow the development of commercial services. More precise and detailed information from vehicles (e.g. through the IVI standard) and more developed driver interfaces will improve the quality of information throughout the system and help promote sustainable business cases. Most vehicle manufacturers are convinced that future cars will be 'connected' although the multiple options available for communications cause uncertainty and robust business cases for deployment are yet to be established.

Secondly, the development of newer information services can be expected for not only passenger cars but also public transport and fleet and logistics services. Examples include information on parking and loading, hazardous loads and routes, truck routing within areas such as ports, tunnel management and passenger information. These will be a focus for future field trials and demonstrations before pilot services can begin to operate.

Thirdly, TTI will increasingly be used by vehicles themselves rather than by their drivers as C-ITS and automation converge to become cooperative, connected and automated mobility [41]. For example, advanced driver-assisted systems could directly make use of IVS to set vehicle speeds and could use GLOSA to optimise vehicle engine stop based on likely duration of the red signal. Many further automated vehicle safety and enhanced traffic flow services have been proposed on the assumption that C-ITS communications can be made sufficiently secure, reliable and rapid and that legal issues such as liability can be solved.

References

[1] Lu, M. *Cooperative Intelligent Transport Systems: Towards high-level automated driving*. s.l.: IET, 2019. ISBN: 9781839530128.

[2] ISO. *ISO/TR, 17465-1 Intelligent transport systems – Terms, definitions and guidelines for Cooperative ITS standards documents – Part 1: Terms, definitions and outline guidance for standards documents'*. s.l.: BSi. Standard.

[3] 3GPP. 3GPP A Global Initiative. [Online] https://www.3gpp.org/.

[4] 5GAA. 5GAA. [Online] 5gaa.org.

[5] European Commission. Commission Delegated Regulation. [Online] https://ec.europa.eu/transport/sites/transport/files/legislation/c20191789.pdf.

[6] Vermaat, P. Gs and 5s updated. *Smart Highways*. 2019, Vol. 7, 2.

[7] MOVE, EC – DG. C-ITSplatform final report January 2016. [Online] 2016. https://ec.europa.eu/transport/sites/transport/files/themes/its/doc/c-its-platform-final-report-january-2016.pdf.
[8] MOVE, EC. C-ITSplatform final report September 2017. [Online] 2017. https://ec.europa.eu/transport/sites/transport/files/2017-09-c-its-platform-final-report.pdf.
[9] C-Roads platform. [Online] https://www.c-roads.eu/platform.html.
[10] Car-2-Car Communications Consortium. [Online] https://www.car-2-car.org/.
[11] ISO. ISO/TS22/SC39Standards Committees. [Online] https://www.iso.org/committee/5391225.html.
[12] ISO. Road vehicles. Ergonomic aspects of transportation and control systems. Dialogue management principles and compliance procedures. BS EN ISO 15005:2017.
[13] ISO. Road vehicles – Ergonomic aspects of transport information and control systems (TICS) — Procedures for determining priority of on-board messages presented to drivers. ISO/TS 16951.
[14] ETSI. Intelligent Transport Systems (ITS); Vehicular Communications; Basic Set of Applications; Part 2: Specification of Cooperative Awareness Basic Service. EN 302 637-2.
[15] ETSI. Intelligent Transport Systems (ITS); Vehicular Communications; Basic Set of Applications; Part 3: Specifications of Decentralized Environmental Notification Basic Service. EN 302 637-3.
[16] ISO. Intelligent transport systems – Cooperative ITS – Dictionary of in-vehicle information (IVI) data. 2015. ISO/TS 19321.
[17] European Commission. European eCall. [Online] [Cited: 10 June 2019.] https://ec.europa.eu/transport/themes/its/road/action_plan/ecall_en.
[18] Stevens, A. and Hopkin, J. *UK eCall impact assessment*. TRL. s.l. : TRL, 2010. Published Project Report (PPR). ISSN 0968-4093.
[19] ISO. Road vehicles — Ergonomic aspects of transport information and control systems (TICS) — Procedures for determining priority of on-board messages presented to drivers. 2004. ISO/TS 16951:2004(E).
[20] EC. Certificate Policy for deployment and operation of European Cooperative IntelligentTransport Systems (C-ITS). [Online] June 2017. https://ec.europa.eu/transport/sites/transport/files/c-its_certificate_policy_release_1.pdf.
[21] EC. Security policy and governance framework for deployment and operation of cooperative intelligent transport systems C(ITS) Release 1. [Online] December 2017. https://ec.europa.eu/transport/sites/transport/files/c-its_security_policy_release_1.pdf.
[22] EU. GDPR. https://eur-lex.europa.eu/eli/reg/2016/679/oj. [Online]
[23] Working Group of the European C-ITS Platform. *Processing Personal Data in the Context of C-ITS*. 2017.
[24] European Commission. Better Regulation Guidelines. [Online] https://ec.europa.eu/info/law/law-making-process/planning-and-proposing-law/better-regulation-why-and-how/better-regulation-guidelines-and-toolbox_en.

[25] Ball, S.D, van Noort, M. and Nitsche, P. *Costs and benefits of cooperative ITS for road authorities: the COBRA decision-support tool*. Dublin: ERTICO, 2013. Vols. 9th ITS European Congress 4-7 June Paper ID: TP0081.

[26] FESTA Handbook. [Online] http://wiki.fot-net.eu/index.php/FESTA_Handbook.

[27] Regan, M.A. and Richardson, J.H. *Planning and implementing field operational tests of intelligent transport systems: a checklist derived from the EC FESTA project*. 2, s.l. : IET, Vol. June 2009, pp. 168 – 184. PrintISSN 1751-956X, Online ISSN 1751-9578..

[28] Stevens, A., Hopkin, J. and Ball, S. Evaluating Cooperative ITS. [book auth.] Meng Lu. *Evaluation of Intelligent Road Transportation Systems*. s.l.: IET, 2016, 10.

[29] Stevens, A., Hopkin, J. and Ball, S. *Impact Assessment Methodology for a Connected ITS Corridor*. Melbourne: ITS Workd Congress Association, 2016. 23rd ITS World Congress. paper ITS-EU-TP0239.

[30] UK CITE project. [Online] [Cited: 10 06 2019.] https://ukcite.co.uk/.

[31] InterCor. InterCor. [Online] https://intercor-project.eu/.

[32] EC. [Online] https://cordis.europa.eu/project/rcn/79316/factsheet/en.

[33] EC. [Online] https://ec.europa.eu/inea/en/news-events/newsroom/eu-reinforces-commitment-to-connected-and-cooperative-road-mobility.

[34] US Department of Transportation. [Online] https://www.its.dot.gov/pilots/index.htm.

[35] ITS Australia. Trials, pilots and project dashboard. [Online] https://www.its-australia.com.au/reports/trial-and-pilots/.

[36] KPMG. Connected and Autonomous Vehicles. [Online] https://home.kpmg/content/dam/kpmg/pdf/2015/04/connected-and-autonomous-vehicles.pdf.

[37] Transport Systems Catapult. [Online] https://www.gov.uk/government/publications/connected-and-autonomous-vehicles-market-forcecast.

[38] Transport Technology Forum. [Online] https://ttf.uk.net/wp-content/uploads/2018/03/Connected-VP.pdf.

[39] UK Government. [Online] https://www.gov.uk/government/topical-events/the-uks-industrial-strategy.

[40] Regan, M., Horeberry, T. and Stevens, A. Driver Acceptance of New Technology: theory, *Measurement and Optimisation*. Farnham: Ashgate, 2014. ISBN: 978-1-4094-3984-4.

[41] Zenzic. [Online] https://zenzic.io/what-we-do/uk-connected-and-automated-mobility-roadmap/.

Chapter 10

Multimodal traffic and travel information

Nick Illsley[1]

There has long been a need for travel information and this has progressed from the early days of timetables and route descriptions by railway and coaching company through to advanced systems that link information in real time from a multitude of sources in order to offer the 21st Century traveller the easiest and most efficient route from place to place.

Nick Illsley, as the Chief Executive at Transport Direct, was responsible for specifying and then operating what was probably at the time the world's most sophisticated travel information system. A Wikipedia entry on Transport Direct exists for those who wish to read it and look at referenced sources but here Nick offers his personal recollections of the development of multi modal travel information in the UK.

Editors

10.1 Early travel information

Time has always mattered to people and influenced their lives, as has the locality in which they chose to live. Originally issues such as the sunrise and sunset, the seasons and the weather, and the security and usefulness of the land were key considerations. Archaeology and anthropology show that people have moved around to escape from adversaries or to seek out more fertile land or to exploit natural resources, but they did not move around as part of their day-to-day life. There is also evidence of trade across countries, or even continents, but generally most trade was localised and most people lived on resources that were grown or sourced in their immediate locality. This was about to change and to change the world and people's lives for ever!

10.1.1 Travel on rail

10.1.1.1 Early days

Primitive wagonways developed, often to transport coal from mines to the nearest rivers. By the late 1700s, these had cast-iron plateways, many of which were feeders to

[1]Consultant then (ex-DfT), Wetherby, West Yorkshire, UK

Britain's expanding network of over 4,000 miles of waterways. This network offered huge capabilities for moving raw materials from one end of the country to the other but was limited by the pace of its horse-drawn power, a 29 mile journey on the Worcester & Birmingham Canal taking up to a couple of days.

The invention of the steam engine heralded the railway and transport revolution, changing the face of the world forever. The world's first passenger carrying public railway was the Oystermouth Railway (later Swansea & Mumbles Railway) opened in 1807, but still horse drawn. The Stockton & Darlington Railway, built to link collieries near Shildon to the River Tees at Stockton via Darlington, was the world's first public railway to use steam locomotives. It was authorised to use 'moveable engines' and for the 'carriage of passengers'. On 27 September 1825, 'Locomotion' hauled a passenger-carrying steam train on a public railway for the first time in the world.

10.1.1.2 The need for uniformity

The core building blocks for the future digitisation of information about rail travel were geographical referencing, a structured description of the movement of trains referring to this referencing and the means of purchasing travel, ideally across ownership and geographical boundaries. From the start railways were referenced, using a similar methodology to the mileposts and milestones associated with roads going back to Roman times. For the internal purposes of the sector this was perfectly acceptable, however it was to cause issues when it became desirable to associate the railway and notably the station, to the towns and villages that they served.

The world's first Inter-City railway was the Liverpool & Manchester Railway opened on 15th September 1830 and was an instant success, particularly with the travelling public. Soon the Grand Junction Railway, from Birmingham to a junction with the Liverpool & Manchester at Newton-le-Willows and the London & Birmingham Railway opened in 1837 and 1838, respectively, setting the basis for a national network to be developed. Each line operated to a timetable, based on the referencing system set out above and soon control regimes involving early signalling systems came in to regulate the movement of trains and locomotives, primarily to stop them colliding. The concept of a departure and an arrival time was now set in stone and in the public mind, detailed to the exact minute at each station, a practice that railwaymen in the future were probably going to come to regret!

Human experience was no longer circumscribed by physical distance. It had to obey not the sunrise but the national clock. Time had to be standardised as 'railway time', based on Greenwich. An early GWR timetable warned 'London time is kept at all stations, which is 4 minutes earlier than Reading time, 7 minutes before Cirencester time and 11 minutes before Bath and Bristol time.' Universities and the Church generally opposed the coming of the railways as immoral, so much so that Oxford's Christ Church cathedral to this day refuses to acknowledge railway time, still starting services five minutes 'late'.

10.1.1.3 Impacts on society

The speed of rail transport fractured old patterns of trade. New markets opened to agriculture and industry alike. Track-side electric telegraphs were soon able to take the communication of information from the age of the horse to that of the internet. The new union of England, Scotland and Ireland was joined as never before, in bonds of steel. The centralisation of British government on the nation's capital received an almighty boost.

The railway tore open contemporary society. It reduced geographical identity and rewrote the conversation between London and the nation, town and country, rich and poor. It also introduced a new realm of classlessness. The Liverpool & Manchester's segregation of passengers in different buildings did not last. Two (initially three) classes of waiting-rooms and carriages survived, but the concourse, ticket office and platform became, like the church nave and the market place, a common space.

Travel beyond the bounds of one village or town had been confined to a few, mostly prosperous, people. Most humans lived, worked, played and died within single communities. Travel now became an obsession, abetted by Thomas Cook's innovation of rail excursions in the 1850s. A train journey merged a private voyage with a public one, a coming together of people in the act of moving apart. Bradshaw's railway timetable, first published in 1839, was the bible of the new mobility. It enjoyed pride of place and was hallowed by Sherlock Holmes. Britain had acquired a national glue, under the aegis of the great god, time.

10.1.1.4 Timetables and traveller information

George Bradshaw was born in Pendleton, Lancashire in July 1801 and began his career as an engraver in Manchester. He became an accomplished cartographer and in 1830 published a series of maps of the canals of Yorkshire and Lancashire.

In 1839, George Bradshaw published the world's first railway timetable in partnership with William Adams. Sold in a cloth cover, Bradshaw's Railway Time Tables and Assistant to Railway Travelling was initially a fairly small publication of eight pages and cost 6d (2.5p), but as new railways quickly opened it grew in size until it had reached a staggering 1,000 pages by the end of the century. The title had shrunk to Bradshaw's Railway Companion in 1840 and from 1841 it was published monthly as Bradshaw's Monthly Railway Guide and included information on places to visit, hotels and shipping services.

From information supplied by the plethora of railway companies, Bradshaw compiled his updated timetable guide every month along with a map of Britain's rail system. Each line on the map was cross-referenced with the number of the page on which the timetable appeared. It was not an easy guide to use – there were so many companies that it was difficult to standardise the information and the multiplicity of footnotes and miniscule typography accompanying each timetable added more confusion. However, despite being the butt of Victorian cartoonists and music hall jokes, it became a national institution and was an invaluable tool for the travelling public for over a hundred years.

Bradshaw did not face any real competition until the 1930s when the 'Big Four' railway companies started issuing their own timetables. Despite this and the publication of regional timetables by British Railways after 1948, Bradshaw remained the only all-line timetable until its final publication in 1961. The first British Rail all-lines timetable was not issued until 1974, continuing, in book form, until 2007 (published by Network Rail).

10.1.1.5 Purchasing Travel and Sharing Receipts

The final piece in the digital jigsaw was also soon put in place, although from commercial necessity and with a more painful gestation period. Through carriages from one company to another were not allowed in many cases and some companies refused to carry third class passengers from other lines. The through booking of a journey that involved using several railways was not possible either, there being no standardisation of tickets. Similar issues existed with goods traffic and vehicles and amazingly even the gauge of the tracks differed with most companies opting for the 'standard' gauge of 4 ft 8.5 in, with the notable exception of the Great Western and its acolytes that chose Brunel's 7 ft 0.25 in 'broad' gauge. Even station clocks had no common synchronised time.

This all caused enormous confusion and was certainly a stumbling block to the continuing successful development of the railways. In 1842, painfully aware that some form of organisation and standardisation needed to be introduced, the Chairman of the London & Birmingham Railway invited representatives of eight other railway companies (many of them owned by the 'Railway King', George Hudson) to a meeting at Euston. Thus was the Railway Clearing House formed, although membership was voluntary and some companies, such as the Great Western, waited another twenty years before joining.

The agreed RCH objectives were to organise the through booking of passengers, carriages and horses between different companies and to divide receipts based on mileage. The RCH also organised a classification for goods traffic and tried to encourage a rate for these on a per-mile basis. The RCH had standardised railway timekeeping using GMT in 1847 and before long had published distance tables and encouraged standardised ticketing, promoting the Edmondson ticket-printing and date-stamping machines. The RCH was given legal status in the Railway Clearing Act of 1850 and remained in operation until 1963, with Edmondson card tickets finally being phased out by British Rail in the 1990s but living on at numerous heritage railways.

10.1.1.6 From Analogue to Digital

The referencing, timetable and fare structure served the railway in its analogue form with ever increasing numbers of printed media being accessed by passengers and the railway timetable, normally in form of local line schedules being a fixture in briefcases and handbags. Towards the end of the twentieth century, two seismic changes in relation to train travel occurred, the commencement of the digitisation of the three information building blocks and the privatisation of the industry. The two changes happened pretty much simultaneously; the underlying industry systems and its data and information structure were therefore digitised and repurposed to support the new industry structure with new Train Operating Companies running train services and selling tickets to customers.

At the core of the industry two new organisations were created to facilitate the retailing activity and the information needs of passengers. These were the Rail Settlement Plan (RSP), a modern-day Clearing House and National Rail Enquiries (NRES), the digital successor to Bradshaw. RSP set out the process and systems to create fares and retail products, to sell the tickets, including for other parties and other services and how to allocate and settle the resulting revenues between the new industry players. New centralised systems were put in place and contractual arrangements were implemented to regulate the new structure.

In the British Rail era, the growth of demand for information, increasingly by telephone, ranked very lowly in terms of management attention and allocation of financial resources and as a result the service offered was shambolic with very high levels of unanswered calls and a poor quality of response when callers actually got through. In the immediate aftermath of privatisation, as the new industry struggled to get to grips with the new Licence condition, mandating the provision of information to new, demanding targets, the situation continued, with the operation of NRES being seen as the Achilles heel of the industry and the Rail Regulator taking draconian enforcement action against the industry.

The threat of financial penalties and the outsourcing of the call answering function, primarily to SERCO and BT, focused minds so that in approximately 18 months, NRES went from being the black sheep of the industry to Rail Business of the Year. In addition to properly resourcing the call centres with call handlers, the companies also invested in improved systems support in terms of timetables, fares and supporting information such as station facilities and nascent real-time information provision.

The scale of the challenge was perhaps shown by the fact that BT considered that in 2000/2001, at a level of over 60 million calls per annum, NRES was the most phoned single phone number in Europe. Fortunately, research commissioned by the industry showed that the provision of good information services by telephone had a positive financial benefit for all train operators, ranging from very positive for long distance operators such as Virgin West Coast to much more marginal, but still positive, for local operators such as Merseyrail. This period of telephone being the pre-eminent channel for information provision was also very brief as the emerging dominance of the Internet was just dawning.

10.1.1.7 Privatisation

At privatisation, the industry was divided between infrastructure and train operation with Railtrack becoming infrastructure owner and operator and Train and Freight Operating companies (TOCs and FOCs) running the trains over the Railtrack infrastructure. Thus Railtrack became the overall owner of the consolidated timetable, although the services were operated by the TOCs and FOCs and was responsible for the allocation of train paths between operators and hence the overall timetable and its associated systems. One of the earliest major Internet sites, with a rapidly growing user base was the Railtrack national timetable site. As discussed previously the referencing and timetable building blocks of the rail industry were ideally suited to digitisation and this made navigation and the user interface easy to adapt to the new Internet age and up to the demise of Railtrack in 2002, the site was a real success in the new technological revolution. Following this the timetable baton passed on to the National Rail Enquiries site.

The responsibility for retailing and ticketing lay squarely with the TOCs and one of these, Virgin Trains, seized the opportunities of the new digital age and produced 'The Trainline', one of the first mass usage retail websites. Once again the work done by the industry in the fares structure, the legacy of the Rail Clearing House and the adaptation of the systems to meet the needs of privatisation, meant that the transition to digitisation and the navigation and user interface were enabled. The major success of the 'Trainline' was, however, as much to do with the marketing and branding expertise of the Virgin group as it was with the systems and technology. Virgin, operators of the West Coast and Cross Country franchises, apparently spent significantly more on creating and advertising the brand as on the development of the technology and 'Trainline' was established as the de facto rail industry retailing site, a position that it holds to this day.

10.1.2 Travelling by road

10.1.2.1 Early days

The age of the stagecoach arrived in the mid-1600s. The first public coach service had begun in 1657, running the 190 miles between London and Chester in four days at a cost of 35 shillings. Weekly runs to York and Exeter were also launched from London, each carrying six paying passengers. In 1700, York was a week's journey from London with coaches simultaneously leaving the Black Swan public houses in Holborn and Coney Street, every Thursday at 5 am. Due to the state of the roads, the services only ran in summer and even then passengers had to walk stretches that were impassable or make unscheduled overnight stays whilst the road dried out!

Although all routes suffered, nevertheless stagecoach routes proliferated – to Dover in three days and Tunbridge Wells, Salisbury and Oxford in two. In addition to the state of the roads, passengers had to contend with the threat of highwaymen and also arrange overnight stops, although these were frequently facilitated by the coachman (probably with a vested interest). Efforts to speed up the journeys were constant, with new coaches and even the beginnings of marketing of services, often with the promise of impossible timescales. The state of the roads precluded major improvements to the schedules until the coming of the turnpikes, commencing with the Old North Road in 1663. By 1770, the turnpike mileage had reached 15,000 and travelling times dropped sharply. In the early years of the turnpike, a journey from London to York by stagecoach took four days and cost 25 shillings inside and 18 shillings outside. The travel time to Cambridge reduced from an initial two days in 1750 to just seven hours in 1820.

As railway mania spread to every corner of Britain, the coaching industry withered and collapsed, and with it the impetus for either new roads or the upkeep of the old ones. There was a new transport revolution around the corner, however, and it arrived from France at the very end of the 1860s – the mass produced bicycle. All developments in transport had thus far been in ways that were communal and dependent on prescribed routes; cycling opened up all the bits in between. More to the point, it opened up the road network to the ordinary man, who was determined to make the most of his holidays.

10.1.2.2 The rise of road transport

In the 1890s, there were half a million bicycles on the roads of Britain. There was an explosion of cheap road maps, packaged towards cyclists, though many of these were printed from plates that were nearly a century old. Colouring of roads to indicate their quality and the addition of symbols to indicate steep hills and other features of interest to cyclists were added.

The best maps were produced by John Bartholomew and Ordnance Survey. These were pricy but reliable and generally accurate. Scale was an issue; OS produced one-inch to a mile (too large for cycling) and also quarter-inch (too small perhaps) and by the time they produced their half-inch series in the early twentieth century, the market had been cornered by Bartholomew's half-inch series, including the very popular use of colour shading to indicate the height above sea-level.

The advent of the motor car at the end of the nineteenth century and taking hold at the start of the twentieth century changed transport for ever and with it the need for information, notably around route finding. However cycling initially remained a much larger potential market for map makers and publications for cyclists had mushroomed at the end of the nineteenth century. By the second decade of the twentieth century cars had, literally and metaphorically, pushed bikes to the side of the road and map makers turned their attention to the growing and affluent driver market.

10.1.2.3 Classification and mapping

One of the first tasks of the new Ministry of Transport, launched in 1919, was to classify the road system, based on the French equivalent, starting with the A1 running from London to Edinburgh, and first published in map form by Michelin in 1922. The road atlas market peaked in the years following the turn of the millennium, with many variations on the theme to tempt motorists. This also included publications for cyclists, harking back to the origins of the road atlas a century earlier. Perhaps the most unusual was the publication of the 'Upside Down Atlas of Great Britain (2000)' produced by Ashley Smith, a young innovator inspired by his father John, a trucker.

10.1.2.4 Digitisation of the Roads

All of these products were however about to be engulfed in a cartographic revolution. Digital technology was sweeping all before it, and in the world of mapping its effects were revolutionary. For many of those steering their way around Britain's increasingly crowded road network, the satellite navigation device, or SatNav, was a godsend. Until the popularity of the smartphone in the 2010s, the satnav was the fastest-adopted piece of new gadgetry in British history; from initial stand-alone device, purchased at Halfords or Tesco, to being subsequently built in as standard to most new cars within a relatively short period of time. Paper maps and road atlases were jettisoned with gusto as drivers fell in love with their new gadgets.

Since the earliest days of the car, inventors have dreamt of something like the SatNav, one of the earliest analogue prototypes, the Jones Live Map, was patented in the United States in 1909 and was formed of prescribed paper routes mounted on a turntable powered by the car's odometer. The SatNav and its associated devices developed at a staggering pace, from the early error-prone devices, loved by their adopters, that resulted in HGVs being stuck in medieval gatehouses, cars plunged into rivers or driven along railway lines, with drivers believing or misinterpreting the instructions. The explosion in locational awareness made possible by the SatNav and the applications embedded into smartphones means that being able to locate destinations and routes to reach them, by vehicle or on foot, has made navigation of unfamiliar places much easier that was previously possible.

As the digital revolution has progressed, the options for digital mapping and route finding have become almost limitless. Long established brands such as Ordnance Survey and Michelin have been joined by new entrants such as Bing (2009), City Mapper (2011), Google and Open Street Map (2004). Open Street Map could be termed as being the Wikipedia of cartography and is a vast and growing database whose aim is to map the entire world, for free and for infinite re-use. Using freely available aerial photography, mapping and, most importantly, data supplied by its volunteers, it is editable by any of its millions of registered contributors. Google has inevitably created many of the headlines in the mapping world, although its service in Britain did not exist until 2006. From a fairly rudimentary beginning, it has added superb satellite imagery and also services such as Street View, using imagery captured by 360 degree cameras mounted on cars.

10.1.3 Towards multi-modal journey planning

'Traveline' was the first geographically large-scale programme that aimed to provide multi-modal public transport journey planning options. Previously, authorities in the larger urban areas, notably Transport for London, had done good work to provide city wide multi-modal options but these had the benefit of being done by a single body and in London by the provider of the vast majority of services. 'Traveline' was very much a product of the John Prescott objective of providing joined-up public transport and attempting to reverse the long-time growth in car ownership and in driving, largely at the expense of public transport use. From the mega-agency of the Department for the Environment, Transport and the Regions (DETR) emerged a number of initiatives including the Commission for Integrated Transport.

The Deputy Prime Minister made it quite clear in the late 1990s that he expected the transport operators (largely the big bus companies) and the local authorities across Great Britain to work together to produce user-friendly public transport information. The shape of 'Traveline' was probably determined by Bill Tyson, the deputy head of Greater Manchester Passenger Transport Executive (PTE), who sensibly suggested that the only way to meet Prescott's demanding aspirations within any acceptable timescale was to build on what already existed on a regional basis.

Although successful from a timescale and political achievement point of view, the resulting service had huge variation in terms of governance, funding and quality of service. There were in fact 11 'Traveline' regions, with tentative links to 'Translink' in Northern Ireland. Scotland and Wales established their own national services with the strong and enthusiastic involvement and support from the emerging Devolved Governments. In London, Transport for London continued pretty much as before, not really bothered about 'Traveline' but doing sufficient to tick the box of being a Great Britain-wide service.

The remainder of England adopted different approaches, ranging from the truly collaborative between bus operators and Local Government in the South West, to operating separate local services in the North West where Manchester, Liverpool, Lancashire, Cumbria and even Halton in Cheshire continued to operate their local call centres and journey planning sites under a thin regional veneer. In Yorkshire and the North East, the existing PTEs took on the service with local funding arrangements with operators and other local authorities. The East Midlands, East Anglia and the South East had no existing PTEs so had to cobble together arrangements, often struggling for funding with operators sometimes reluctant to participate and finally in the West Midlands the PTE, other local authorities and operators appeared to be in a state of almost constant disagreement.

Nonetheless, despite some real teething issues and with some very challenging quality of service issues, the 'Traveline' service was launched in mid-2000 with an event at the Millennium Dome at North Greenwich and had achieved Great Britain wide coverage in a matter of a few months from conception to delivery. As set out earlier, the quality of service varied massively and the approach to journey planning, funding and service provision was also far from a common standard. However, the development of standards such as NaPTAN and TransXchange and the digitisation of stops and services was a major advancement, even if it took almost another decade before the service could truly be described as a national service that was fit for purpose.

10.2 Transport Direct

10.2.1 Historical context

When the Transport Direct website was closing down in 2014 a fair amount of comment was made on the internet from supporters and opponents but one comment struck me and rather took me by surprise. A blogger not personally known to me posted 'to put Transport Direct in historical context, the UK Government created this several years before Google even tried'.

Wow, that sort of put the challenge and achievement of the initial site into some sort of context for me!

But back to the beginning; the first mention of Transport Direct in the public domain came in the Labour manifesto for the 2001 election where the only transport commitment was the establishment of this as a national, multi-modal travel planning site. On starting at the Department for Transport (DfT) in August 2002, I was told about the legendary letter from Number 10 stating the desire for such a service to be developed (and we did find the physical copy of the letter for the close down party).

In truth, Transport Direct came from two general directions of Government policy, the first from John Prescott's Integrated Transport Strategy and secondly from the desire to develop a suite of electronic public services, each of which included Direct (such as NHS Direct). But with only a skeleton of an idea the first key task was to set out what Transport Direct would be when it was delivered to the public.

10.2.2 Connecting People to Places

As the specifiers, we eventually settled on the user only having to be able to identify a start point and an end point and have some idea of when they wanted to travel. This firmly established the principle that it was about people and places and not the transport modes that were offered. With the subsequent rapid development of mobile telephony and e-services, this linkage to geographical location turned out to be most fortuitous.

In defining place we wanted the scope to be as wide as possible so any address, postcode, place (town, suburb etc.), transport node (bus stop, railway station) or attraction could be identified as an origin or destination. Although this seemed eminently user-friendly, the complexity of having around 35 million origins (O) and destinations (D) gave both definition and mathematical dilemmas that needed to be resolved. This is more fully discussed later.

In addition, the databases of these origin and destination points did not exist in 2002, at least not to a satisfactory level of accuracy and completeness. Therefore a core initial task was to complete the National Public Transport Gazetteer (NPTG) and the National Public Transport Access Nodes (NaPTAN) databases to give the basic materials to enable the vision of Transport Direct to be delivered. This was a largely manual task with each local authority dispatching staff to visit bus stops and click their GPS device to give an accurate location.

This painstaking task was fundamental and it is great to reflect that today NaPTAN goes from strength to strength, is a core database not just for journey planning but also for asset management and to measure things like accessibility to public transport services. It was a proud day when the then Prime Minister Gordon Brown, as part of a major announcement on Transparency and Open Data, headlined his speech with 'today we will release the database of the location of all bus stops and railway stations in Great Britain' and it remains one of the most downloaded databases on data.gov.uk.

10.2.3 Designing, building and operating Transport Direct

The contract to Design, Build and Operate the Transport Direct portal was awarded to Schlumberger Sema (later to become Atos) at the very end of 2002. Although it is unfashionable to say, this was undoubtedly the key moment in the genesis of the service. From day one, Rob Price (now UK COO of Atos) and his team were totally committed to delivery and the team work between the consortium and the DfT was the basis of success.

The key technical challenges were to make a theoretical concept into a user-friendly reality in a relatively short period of time and in a very public environment. There was no existing service to plagiarise or even learn lessons from and many experts were of the opinion that the Transport Direct concept was mathematically impossible. The other technical challenge was to set the system default between comprehensive coverage, ease of use and IT processing time. Wherever this was set, all three would be sub-optimal so it needed to be comprehensive enough, easy enough to use and to process quickly enough and to the consortium's credit, they pretty much got this spot on (as judged by user satisfaction measurement).

10.2.3.1 Data Acquisition and Stakeholder Management

For those of us in the Government at DfT we had two major issues to resolve, the supply of data and the management of stakeholders. The data supply may seem to be a strange task for the client to be tackling, why wasn't this a supplier responsibility? In 2002, the only organisation that the myriad of operators, local government and other key data suppliers (Ordnance Survey etc.) would happily and reasonably supply data to was the Government. We had also realised that unless the coverage of modes and locations was comprehensive, then Transport Direct would lose much of its service value. Hundreds of data agreements were put in place ranging from the whole geography of Great Britain (OS), through bus and train timetables, to individual toll bridges in deepest Berkshire.

Recruiting and retaining the stakeholders was probably the task that exercised me the most, under the inspirational leadership of Sir Malcolm Field, the former CEO of W H Smith and Chair of the Civil Aviation Authority. Malcolm, who was appointed as the Chair of Transport Direct was no expert in electronic services, or in transport operations, but he knew about business, governance and people. With stakeholders as varied as the newly devolved Governments in Scotland and Wales, through local government and Transport for London to the transport operators and the IT industry, Malcolm led a masterclass in engagement and collaborative working. I consider this to be one of Transport Direct's key achievements and the fact that no stakeholder withdrew support throughout the life of the service may be unique in the delivery of a public service.

10.2.3.2 Core IT Systems

There were many high-risk aspects to the development of Transport Direct, one of which was the number of practically new IT systems that sat at its heart, in addition to the untested standards and protocols. Prime amongst these were the operating system, Microsoft.net, the GIS package, ESRI Arc GSI and Ordnance Survey

Integrated Transport Network Layer, at the heart of road referencing and planning. These products were at the start of their subsequently successful life and as such Transport Direct was something of a testbed for their application and development. Both Microsoft and ESRI in the United Kingdom are akin to franchises with the products being created in California and on more than one occasion system issues seemed impenetrable here, only for a solution to appear overnight from the United States that resolved the issue totally. Dealing with huge multinational organisations and also being UK Government undoubtedly helped in de-risking this aspect of scheme development.

10.2.3.3 Data and Geographical Referencing

It was also necessary to complete and to significantly enhance the geographical referencing work already referred to around NaPTAN and NPTG and to develop and test the Journey Web interface standard that underpinned the multi-modal journey planning capability. In deciding how we would know that we had 'done Transport Direct' we concluded that only if the user could request a journey knowing only three facts could we have been deemed to have succeeded. These three facts were:

- Where do I want to start from?
- Where do I want to go to?
- Approximately when do I want to travel?

In order to meet this challenge the largest possible definition of origin and destination needed to be provided, together with both depart at and arrive by time choices. The gazetteers for origin and destination therefore included:

- Address (around 30 million)
- Postcode (around 3 million)
- Place (town, village, suburb, etc.)
- Public transport stop (bus stop, railway station, tram stop, ferry terminal, airport terminal, potentially taxi rank and coach stop – around 350,000)
- Attraction, point of interest, amenity such as a hospital or council office, etc.

Meeting this objective meant a mixture of existing gazetteers including NaPTAN and NPTG but also databases via Ordnance Survey including Address Point and Code Point and the Point X points of interest gazetteer. New data bases were also created around car parks, airports and ferry terminals, including interchange models of how they worked and were navigated taking into account check-in times and walking allowances.

They also created mathematical and accuracy challenges, if one just considers public transport access points, there are 350,000 origins and destinations that create 100 billion potential journey pairs. Even before variations around times of the day, days of the week, modes, timetables and interchanges if we aimed at an optimistic 99% accuracy, there would be 1 billion incorrect pairs in the basic Transport Direct system that would inevitably produce a wrong or sub-optimal answer to any journey request.

10.2.3.4 Journey Planning

In order to create the capability to produce the basic origin to destination multi-modal public transport journey plan, the Journey Web standard had been developed to enable the existing Traveline journey planners to be utilised as the source of public transport options. This had the benefit of not replicating the existing Traveline systems but needed to be very carefully developed and implemented as each Traveline region had performed its own procurement and as a result there were a number of suppliers that developed different solutions and worked in subtly different manners.

It was also necessary to build in future proofing so that development in journey planning and also in procurement rules and competitive benefits could be accrued by the regions, so that support for the Journey Web standard could be incorporated and would not be an overly onerous requirement. Journey Web was developed as a question and answer regime that enabled the Co-ordinated Journey Planner (CJP) at the heart of the Transport Direct logic to break down user journey requests, ask remote systems for specific itineraries and other supporting information and then to join together the various segments of the response to create a coherent journey plan.

Having decided to tackle the most difficult task in the first delivery, the public transport planner, upon receiving a request for a journey plan, the Transport Direct portal sent a request to the originating and destination Traveline planners (there were 11 Traveline regions in all). This Journey Web request was seeded so that normally the originating planner was asked for the journey from the origin to the Trunk Exchange Point nearest to the destination (where the train or coach journey ended) and the destination planner from that point to the destination. Following the responses, the CJP logic spliced their responses together to create multi-modal travel plans. At the same time the portal asked National Express for coach journey options and also any fares information that was available to cover journey legs. A

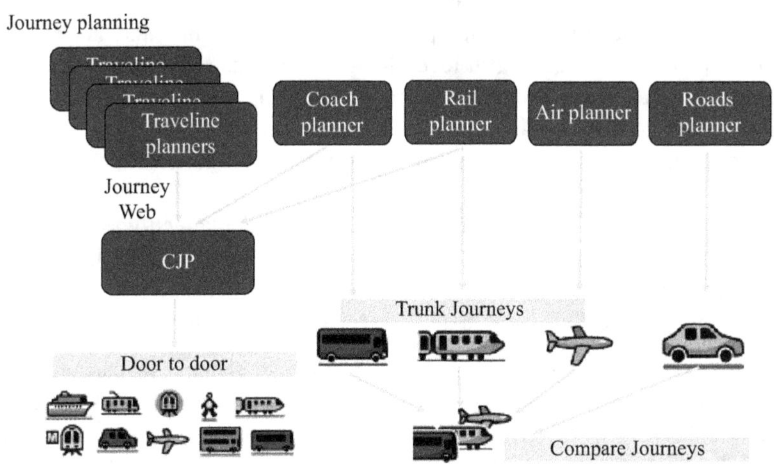

Figure 10.1 Journey planning components within Transport Direct

check was also run against the rail industry planner as the Traveline regions took weekly, not daily, updates and should the plans not tally then the rail industry response replaced the Traveline response. The resulting journey plans, up to 5 in total, were presented to the user with the one that closest matched the request (from A to B at such a time) being the top option.

At the same time the road journey planning function created a car journey option that included many new features compared to the existing car planners that were around at the time. The journey was time specific and calculated as accurately as possible, the journey time including known congestion and other relevant information. Most planners were not time specific and used the speed profile of the road to calculate time. This resulted in grossly optimistic journey times and also poor selection of roads such as across Central London or on farm tracks. Additionally a new hierarchy of roads was developed to keep the core part of the journey on suitable and appropriate roads and only allow minor roads to be included close to the start and finish of the journey.

Throughout 2003, the consortium and the departmental team delivered increasing amounts of data and developed software and capability so that by the end of the year the core technical principles of the service had been developed and proved to be feasible; it did work! Work during 2004 broadened and deepened the coverage with a soft launch (no publicity) in the summer and a formal launch by the Secretary of State at the end of the year.

10.2.4 Launching Transport Direct

At the time of the launch in 2004 Transport Direct was the best journey planner in the world. It had to be as it was the only national, multi-modal journey planner that existed. But that did not mean that it was without glitches and imperfections. Two key factors need to be considered in this context:

- The simple scale of the data and the journey planning parameters meant that practically assessing data and journey planning quality and accuracy was impossible to any meaningful level. Most of our data providers had no serious quality and accuracy measure and most were not charging for the provision. Instead we adopted a wide industry spread of knowledgeable testers to assess the outputs of the service and no reported faults were tolerated, each remained open until it was resolved.
- Had a totally risk free approach been adopted where launch was delayed until errors were eliminated, then the service would probably never have seen the light of day. In any event the propensity of data providers and software engineers to fix errors is much greater in a live service than one under test.

So the inevitable happened; there was a smallish element of bad press in the early part of 2005, mostly aimed at John Prescott and the weakness of Government IT projects. Having worked in the rail industry for almost 30 years I had expected this and rather than being front page news it was on page 16 or 20. But the Civil Service seemed to go into meltdown at this turn of events with calls to cease the service and to take the contract off the suppliers. Many of the articles picked out journeys that made great headline pictures but which a user would never seriously consider, such as Grimsby to Scunthorpe by coach via Leicester, rather than the really worrying faults which would appear correct but in reality leave someone stranded say overnight at Severn Tunnel Junction.

After an uncomfortable few days the Secretary of State and the Permanent Secretary were reassured that the service should be continued and the consortium and departmental team went into overdrive to tackle all known issues. Although the press coverage dwindled away to nothing and was never seriously re-ignited, some of the reputation of the service was damaged and some opinion formers were still referring the episode when the service finally closed down.

10.2.4.1 Indications of Success of the Service

To conclude the development and launch part of the Transport Direct story, at a DfT review with the then Head of the Civil Service, Andrew Turnbull, who clearly stated that the biggest single achievement of the department in 2005/6 was the launch of Transport Direct, a view repeated by David Rowlands the Permanent Secretary. This world first service had been delivered on schedule, with full functionality and at a cost around 30% lower than the original estimates.

Most importantly, the end-user could specify any of 35 million origins and destinations and get public transport and driving journey plans to compare and contrast. In customer research evaluation the service had very high levels of user satisfaction including the intention to use the service again in the future. And there was clear evidence of behaviour change with around one third of users who were asking about journeys they had made before stating that they would do things differently in future. There was some modal change from car to public transport but a greater proportion stated they would change the time and route of their journey as a better option had been suggested in the journey plans that were presented.

10.2.4.2 Initial Service Content

The initial Transport Direct service was heavily based around the Great Britain multi-modal public transport and car journey planner with the other notable part of the service being the Live Travel News. This latter service re-used some of the existing industry services such as that delivered by National Rail Enquiries for the rail industry and the Highways Agency service. Importantly however it also included an all modes travel news desk service, provided by TrafficLink (later to become part of the INRIX empire) that gathered and edited traffic and travel news across all of Great Britain. Much of this effort re-used the work done to create traffic and travel bulletins for the BBC and other local media outlets and importantly provided journalistic input to transform the language of the industry into user friendly output and added prioritisation of news stories on the Home Page.

10.2.5 Operating and developing Transport Direct

There were many subsequent upgrades and developments to the overall Transport Direct service over the next nine years, normally at the rate of around two per year, of which most included some new functionality, although some were more IT and systems based, such as refreshes to the Transport Direct environment to virtualise the servers and upgrades to mapping and the Microsoft.net operating system.

10.2.5.1 Find A Journey Planning

The first, and perhaps most successful of these was the 'Find A' series of services:

- Train – the most used, even though National Rail Enquiries was, with Transport for London, the pre-eminent transport site, later including fares and station facilities
- Car route – a simplified version of the car planner within the main Transport Direct functionality, later with traffic levels and representations of congestion levels
- Flight – including the air interchange model and allowances for check-in, luggage, customs and walking time within the airport
- Ferry – similar to the above albeit a more simple interchange model
- Coach – including National Express, Scottish CityLink, MegaBus and a selection of long distance scheduled bus services such as the X5 between Oxford and Cambridge
- Bus – simplified version of the Traveline service

Later additions included:

- Car park – new national database showing local authority and private car parks, also stations and other transport facilities
- Cycle ride – based on a new standard, CycleNetExchange, and featuring an emerging national dataset of cycle routes and roads suitable for cycling including the National Cycle Network and canal towpaths
- Place – surprisingly successful and enabling journey planning following the discovery of the desired location, apparently heavily used by SME (small and medium-sized enterprises) delivery companies that then planned a car route to the delivery
- Nearest – any key location of function based on the NaPTAN and Points of Interest gazetteers

Figure 10.2 shows the Transport Direct Homepage in 2014, illustrating many of its features.

10.2.5.2 Fares and Retailing

In the original specification for the Transport Direct service the question of retailing (ticket selling) and facilitating an end-to-end service for users had been the subject of much discussion and one of the few areas of contention. Although actual retailing within the site had never been seriously proposed, it was also clear at the initial Advisory Board meeting with the key commercial stakeholders that any

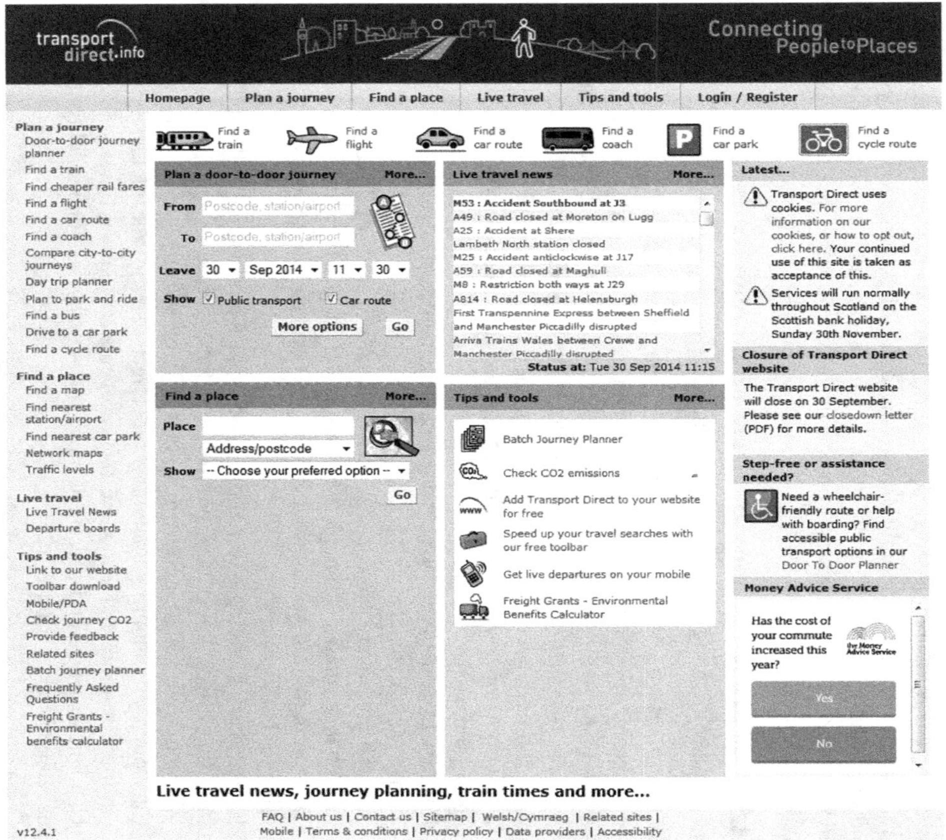

Figure 10.2 The Transport Direct home page

suggestion of a potential foray into their commercial activity would result in rapid and irreversible withdrawal of interest and involvement.

However, facilitating seamless navigation to purchasing remained an aspiration, and a fairly elegant (for the time) hand-off mechanism was devised and implemented for rail, National Express and Scottish CityLink. This took the relevant parts of the users request and pre-populated the retail sites in question at the start of the potential purchasing transaction. Two ground rules were agreed with parties, there could be no attempt to replace part of the itinerary provided by another party with the retailers own service, and where there were multiple retailers (in rail) the site would randomly change the order of presentation to the user to avoid bias. Sadly, very few users chose to use this facility, it would appear that window shopping for options and purchasing products might be largely separate activities in the mind of the user.

10.2.5.3 Journey Planning and Data Enhancements

Other journey planning developments included:

- City-to-City – a development that we thought reflected how many people planned, giving the trunk options between main towns and cities (rail, coach, air and car) with the ability to then extend these by adding local connecting journeys to actual origins and destinations. Little used, which either means that users did not understand the concept, or they really don't plan like this
- Day Trip Planner – ability to add legs together and create a multi-stop itinerary, again little used
- Mixed Car and Public Transport – the mathematics of driving part of the journey and then accessing public transport get too complex to calculate, and research previously showed that users generally would not drive part way and then change modes, although they may travel locally to access a better route. The option offered was either drive to a Park and Ride or choose a car park and drive to that location
- Fares Based Planning – for train, coach and car including tolls and ferries, including lowest/lower fares compared to faster routes
- Walking – with Walkit.com where they offered step-by-step walking routes
- Car Journey Planning – including the effect of real-time incidents with the ability to re-route and also the effect of events on journey times
- National step-free and accessible planning – using the functionality and data developed for London 2012

There were a number of background information improvements:

- Train Taxi – a proper labour of love from a small business based in Peterborough that created a dataset for every station in Great Britain showing if there was a taxi rank and also providing contact details for companies that would collect users from stations
- Departure Boards – for each station and links to real-time bus information and London Underground real-time information

- Carbon Calculator – to compare journeys by public transport mode and by car between origins and destinations. In the development of this functionality it became clear that users did not understand the different types of pollutant, generally they referred to carbon and absolute levels did not matter, rather they wanted measures of big and little and good and bad

10.2.5.4 Other Services and White Labelling

In addition, a number of IT and functionality-related developments were implemented:

- Mobile and Digital TV Versions – these were obvious technical developments. The mobile version proved reasonably popular, but the Digital TV version, aimed at those that had no mobile or internet, was almost impossible to navigate due to the very limited bandwidth; each user session lasting around 10 minutes for a simple enquiry
- White Labelling, APIs and Web Services – at the onset of Transport Direct it was thought that there would be a demand for 'White Labelled' sites. Such sites sell products directly but under a different domain, brand or company name. Four of these were produced for BBC, DirectGov Business Link and Visit Britain. Although moderately successful they were soon displaced by the use of less complex and more flexible linking mechanisms that did not require multiple stylesheets and which also avoided interesting licensing issues with some of the key system suppliers. Numerous Application Programming Interface (API) users and web service contacts were developed including Google, which in advance of Google Transit downloaded the entire NaPTAN dataset with page landing capability from each stop that pre-populated a Transport Direct search as either an origin or destination
- Batch Journey Planner – perhaps the most successful of the third party services, developed for Medical Appointment Planning, where all attendees were presented with a Journey Plan to their appointment to reduce the no-show level. This was pretty successful and was later adopted by a number of agencies covering topics such as major office relocation and academic research

10.2.6 Political Viewpoints

As previously mentioned, the fact that Transport Direct was a Government initiative (technically three Governments, as it closely followed devolution and was supported by the DfT, Scottish Executive and Welsh Assembly) was very important in terms of funding, stakeholder involvement and key supplier delivery. It did however have a major downside, once delivered and having met the political promises made in manifestos and government programmes, the interest in running it as a service and providing helpful user guidance diminished very quickly, with a few honourable exceptions. From 2007 onwards, each new budgetary review and spending round increasingly questioned 'Why are we doing this?'.

The emergence of new digital bodies, initially Central Office of Information, with DirectGov and Business Link, also questioned what was this behemoth doing in the dynamic information market place. It was reluctantly agreed that Transport Direct could continue alongside the new all singing, all dancing Government websites as a jointly-owned asset with the industry, local government and the devolved assemblies. Some respite was offered by the emergence and growing importance of Open Data with the industry and Transport Direct datasets forming the core of early releases in the transport sector. As previously noted, the then Prime Minister, Gordon Brown, in a speech about Open Data, hailed the release of Great Britain-wide bus stop data (NaPTAN) as the key achievement in 2010.

With the advent of the Coalition Government in 2010 and the emergence of Francis Maude into the Cabinet Office with the new Government Digital Service (GDS), the future looked grim for Transport Direct. It was produced by one of the large IT companies (Atos Origin) despised by GDS, it did not meet 'Agile principles' and worst of all it was the child of John Prescott and the last regime. It would surely join the bonfire of the quangos and linked services. But in fact, perhaps the greatest hour for the Transport Direct technology and data was just about to dawn.

10.3 London 2012, an Olympian task

10.3.1 The requirement

When London was bidding for the 2012 Olympic and Paralympic Games there was a general consensus that transport represented a major Achilles Heel and that mitigation was a key objective in securing a winning bid. The memory of travel and transport chaos from previous Games (e.g. Atlanta) and the very clear fact of an underground system running close to and at capacity during peak hours in London had to be addressed.

The success pulled off by the bid team in 2005, including the dramatic journey from St Pancras to Stratford in Land Rovers through the nascent High Speed 1 tunnels, was to be a foretaste of the transport successes to come, but prior to this much needed to be put in place to guarantee the promises made by the team.

The scale of the challenge can be simply expressed as London finding a way of moving millions of Games ticket holders through a complex and extremely busy public transport network. In addition, the London 2012 bid promised to be the first truly public transport Games.

To achieve this ambitious objective, the Olympic Delivery Authority (ODA), in conjunction with Transport for London (TfL) and the DfT established the Travel Demand Management programme (TDM). The programme consisted of three key streams:

- Advice to businesses
- Marketing and communications
- Travel information services.

Although closely linked to the first two streams, the Spectator Journey Planner (SJP) was a key part of the third Travel Information Services theme. The SJP was commissioned by the ODA to support the key objective of informing spectators and at the same time supporting the operation of the London travel network, including the Games Network of Accessible Travel (GNAT).

The creation of the SJP, as with that of the overall TDM programme, was a cross organisation initiative that required the agreement, participation and commitment from a huge number of organisations including transport operators, planners, local and national Governments and industry experts. It was Travel Demand Management on a truly industrial scale.

The TDM programme concentrated on short-term changes to travel behaviour during the summer of 2012 in the vicinity of Games venues, notably in East London but also in other parts of the capital, around Windsor and Eton and at Weymouth, and in the case of football and cycling, across large swathes of the South East and the rest of the country.

The SJP was a core deliverable of the Travel Information Services work stream and the strategy of this work stream was to:

- Provide a pre-planning and real-time tool, primarily for spectators, as part of the end-to-end Games experience
- Provide traveller information (through all appropriate channels) to assist in the management of the transport network.

The initial intention had been to procure a new travel planning package from the rapidly developing market but the increasing pressure on funding (the global financial issues that beset the end of the first decade of the century had impacted on the overall Games funding) and the need to de-risk delivery led to a new approach. ODA, TfL and DfT agreed that a pragmatic and sensible approach to both of these issues would be to form a partnership between London 2012 and Transport Direct. This was also facilitated by ATOS holding the principle IT contract for the Games and hence excellent working relationships with the key decision makers.

If this partnership de-risked delivery in terms of operations, timescales and finances, the key transport risk of increased levels of demand resulting in significant delays to spectators (and regular travellers) and the possibility of missing events and poor publicity for the Games remained. Therefore at the core of the SJP were Travel Guidance and TDM Routeing Strategies.

Over the period between the summer of 2009 and that of 2012, spectator routeing strategies for the Games were designed, developed, managed, updated and refined in partnership with a range of stakeholders including LOCOG (Organising Committee), train operating companies, TfL and other transport operators. These were essentially the recommended routes for the mass movement of spectators in London to Games venues. The routes reflected the preferred operational choice of routes between distinct geographical areas, and sought to better match supply and demand over the network, notably at pinchpoints such as London Bridge on the Jubilee line.

It was the first time that this type of approach had been undertaken, and on a huge scale, given the wide range of Games venues and the number of options for public transport access, including from a large variety of National Rail terminals and stations. These routeing strategies underpinned a range of new products and capabilities for the Games and were subsequently encoded as the TDM logic that sat within the SJP and produced journey options to the Games.

10.3.2 Games time delivery

The SJP family included the following core products:

- SJP main portal – a multimodal planner that was integrated into the information architecture of the London 2012 website and allowed planning by public transport (including tube, rail, bus and river) together with a road and cycling transport planner. The road planner focussed on Blue Badge and Park and Ride sites, as the venues were car-free and the Games were set up as the first accessed by public transport. Included in the main SJP were extensive accessible journey planning capabilities
- SJP mobile – the mobile version of the planner was integrated into the London 2012 mobile platform and included features such as dynamic cycle mapping and locational awareness so that users could plan from their current location using the GPS capability or locational functionality on their device
- Travel news – the creation of venue specific travel news with advisory information supported by the journey planner which could be dynamically updated to re-route spectators around transport network closures or major disruption to services
- London 2012 Join In App – the SJP was integrated with the official London 2012 Join In App to allow easy 'plan me to this venue' functionality as well as real time travel alerts
- The TfL journey planner was updated to integrate seamlessly with the SJP so customer enquiries were redirected to the SJP, if the user chose to plan to a Games venue
- TfL's telephone-based journey planning service was updated so that customers were able to receive results produced by the SJP, should they be planning to the Games venues
- The Games Transport Control Centre (TCC), based in Southwark, used embedded SJP and Travel News capability to inform and monitor the information being provided to millions of spectators that were on the transport network, en-route to venues
- Travel Ambassadors had access to mobile versions of the SJP to assist with queries that they received on the ground from Games spectators

The SJP was launched 12 months prior to the commencement of the Games, a huge logistical and technical achievement, and was continually refined right up to the actual start of the Games.

In addition to the Travel Demand Routeing strategies that have already been discussed, a number of other challenges required innovative solutions:

- Accessibility planning – to enable the SJP to present accessible routes to Games venues, an approach was adopted that combined the existing detailed accessible routes through the transport network in London with a new data set that defined, for stops and services outside London, whether they met the accessibility standards as defined by the ODA, This data set, the Games Network of Accessible Travel (GNAT) enabled the delivery of accessible journey plans across Great Britain and was the first UK planner that delivered this capability on a nationwide basis.

- Timetable planning horizons – most timetables are available no more than 3 months prior to the date of travel, sometimes significantly less, and are the subject to later alteration due to engineering work or other external factors. This had to be addressed in the SJP which was to be launched 12 months prior to the Games, to enable ticket sales and travel arrangements to be finalised. In agreement with Traveline Board, the SJP was built and launched in July 2011 with a mixture of actual 2012 schedules for trunk operators including the rail industry and National Express, and summer 2011 schedules rolled forward for local transport operators. These were updated and rebuilt as more operators published Games times schedules (although changes were kept to a minimum) and so ticket purchase could be supported a year ahead of the Games.
- Scaling and performance – the SJP was a world first, so there was no experience on which to model the patterns and volume of usage. The delivery team developed a capacity model for the 15 month life of the applications based on spectator numbers, projected ticket sales and predictions on the regional split of spectator origins. It was clear that usage would vary greatly and needed to be highly scalable and achieve very high levels of availability. To do this the team needed to adapt the mature Transport Direct ASP.Net application. Therefore the team ported the application so that it ran as a set of identical work units that could be deployed on virtual machines as capacity required. These were deployed in the cloud across multiple data centres. To manage session state across the work units the application was modified from a central session database to a distributed memory based service that still enabled session state to be accessed from each work unit and to maintain response times and availability. Finally, the application had to meet the very stringent security requirements of London 2012 even though it was deployed in a cloud environment. Despite all these innovations and requirements, the SJP worked and performed very impressively
- Data to Support Travel Demand Management – a London 2012 specific user interface was developed for the SJP to support spectator journey planning to, from and between venues. To ensure that spectators received journeys that gave them realistic journey times to venues, a venue data set was created that allowed for potential delays due to security checks and also to allow for navigation around the venues themselves, including any crowding that may be experienced. This data was created to the NeTEX standard and was loaded into the journey planning algorithm such that the journeys planned reflected the agreed routes across London to venues and avoided places where congestion was expected to be high.
- Adapting Existing Data Structures – the SJP was built largely re-using the logic and the technology and standards that supported the existing Transport Direct portal. This gave the application a head start and also de-risked much of

the development, providing capability such as Great Britain wide multi-modal journey planning. The Transport Direct approach of using the Journey Web standards for distributed journey planning was continued into the SJP, which enabled the key Traveline data sets to be utilised for SJP journey planning requests. The existing NaPTAN data sets were used to describe stops and to link service data to stops, but with important accessibility characteristics added to the existing parameters. The existing Journey Web interface was updated to version 2.4 to include features new to the SJP such as TDM routeing, accessible journey planning and venue characteristics so that these could be utilised and influence the journey plans returned.
- Real-time information – during the Games a real time layer was added to the journey planner. This was created using some of the technology from the TfL Real Time team and also the Transport Direct Travel News feed, supplied by INRIX. For the Games this was extended through the creation of a SIRI-SX feed so that regional Travelines were able to integrate a consistent set of news to journeys planned. For the SJP specifically, the news feeds were linked to venues so that spectators travelling to and from events could see specific travel advice generated from the Traffic Control Centre in Southwark.
- Car journey planning – although the Games were to be accessed by public transport and the car journey options were limited to Blue Badge and Park and Ride access, a couple of fairly meaty road related issues needed to be addressed. These were the effect of the Olympic and Paralympic Route Networks on traffic in London and the major closures of roads across swathes of the South East to enable the cycle road races to take place. Transport for London does not offer a road planning service and therefore drivers and residents that wished to discover the likely effects of the networks and the planned closures were directed to the main Transport Direct site. The geometry of the networks and the closures were uploaded to the road planning data set, including issues such as banned turns as well as road closures, and the time penalty of either diversions or merely having to cross the networks was factored into the journey times presented. For a few poor souls in Surrey and Hampshire they discovered that they would be unable to exit their homes for a period until the road races passed and the roads were re-opened.

10.3.3 Measurement of success

As ever with the delivery of a service, the real measure of success is not the actual existence of the service but whether it met its goals and achieved the desired outcomes. The SJP, as part of the wider TDM effort and indeed TDM as part of the mammoth transport provision for the Games was generally considered to be an ‘Olympian Success’. Specifically:

- There were over 2.6 million unique visitors, which equates to around 75% of all ticket holders
- Less than 0.1% of failed requests or ‘sorry’ pages
- There were nearly 0.8 million mobile sessions

- Thirty thousand accessible routes were requested
- Two hundred thousand travel news updates were accessed
- There were significant numbers of cycling and river service requests
- The busiest Olympic hour was 21:00 on 30th July and the busiest Paralympic hour was exactly one month later at 21:00 on 30th August
- Virtually no customer complaints were received regarding journey planning by the SJP, ODA or TfL during the Games

In the wider transport and Travel Demand Management context, the results were at least equally impressive:

- Evidence indicated that Londoners and spectators listened to the TDM advice and modified their journeys
- An average of 35% of Londoners changed their travel behaviour every day during the weekdays of the Olympics
- Predicted levels of crowding were averted even though the transport system carried record numbers of passengers with nearly 50 million people carried between 27th July and 10th August
- Nearly 4.5 million passengers were carried on the London Underground alone on Tuesday 7th August, an all-time record

Numerous compliments were received in the press and from individuals, including athletes, about the operation of the transport network and the absence of any of the predicted travel problems. This not only mitigated the reputational risk to London and the risk that spectators may miss their allotted events due to transport issues but probably enhanced the reputation of London as a city that can be navigated by public transport with ease.

Specifically regarding the range and value of the travel advice received by passengers:

- Nearly 75% of respondents to survey research indicated that the messages delivered via the TDM programme had been relevant to them.
- Nearly 80% found the advice about busy stations to be useful with a similar figure valuing the advice about delays to the road network and their car journey.

At the July 2011 launch of the Games travel pages, Lord Coe said 'Getting spectators to events on time and back home again is going to make a huge difference to how people remember the Games in years to come'. Transport Minister Norman Baker added 'The last thing any spectator wants is a sprint finish to get the venue in time. This tool will give people the best route and how long their journey will take' and Paralympian Baroness Tanni Grey-Thompson concluded 'The fact that the website also gives information about which stations and services are accessible means that everyone will be able to take advantage.'

Following the Games in September 2012, Hugh Sumner, the Head of ODA Transport, stated 'The development of the SJP ensured that Games spectators had the correct information to get to their events on time, whenever and wherever they were travelling from. The innovative approaches to Travel Demand Management, accessible travel information and the overall system capabilities resulted in a first-of-its kind customer information tool that helped make transport during the Games a major success'.

10.3.4 London 2012 legacy and the demise of Transport Direct

At the conclusion of the Paralympic Games the Spectator Journey Planner had no operational future, however it was thought that there might be significant legacy benefit from what was generally considered to be the best journey planner in the world. Specifically, six key developments were identified as having real potential for being ported into the mainstream of journey planning:

- Contingency IT capacity via cloud computing – the Spectator Journey Planner showed the flexibility and potential for rapid reaction to emerging situations, such as bad weather or major disruption, through replicable application design and limitless potential capacity. It was also noted that this was complex systems integration and also required, and achieved, very high levels of security and resilience to malicious attack.
- Planning to venues with complex access arrangements – it was possible to model complex destinations and to factor these into the journey plans provided to customers. This allows routeing to be adapted to the precise location in the venue and also to ensure time is allowed for access to venues. Potential applications could be widespread event venues such as the Notting Hill Carnival.
- Matching transport supply and demand – probably the biggest Games challenge and also the biggest potential for significantly influencing travel behaviour and mitigating locations with excess demand with Travel Demand Management rules and outcomes identifying hotspots and overcrowded routes. This was seen as a major day-to-day goal by Transport for London.
- Accessible journey planning – initially simple yes/no logic to meeting accessibility requirements within journey planning algorithms, with extension into full digitisation of geometry and parameters in future. This could include detailed location specific accessible information and even extend beyond the transport network into destinations and origins.
- Real-time information – extension of public transport real-time information from London and the rail industry to other modes and the rest of Great Britain which would require some co-ordination of control as typified by the TCC at Southwark.
- Road-based changes to the network – the mapping of the Olympic Route Network (ORN) (see Figure 10.3), Paralympic Route Network and cycle road races showed the potential for coding and assessing the potential effects on the network and journey times from issues such as major road works on motorways and special events such as races, carnivals and festivals.

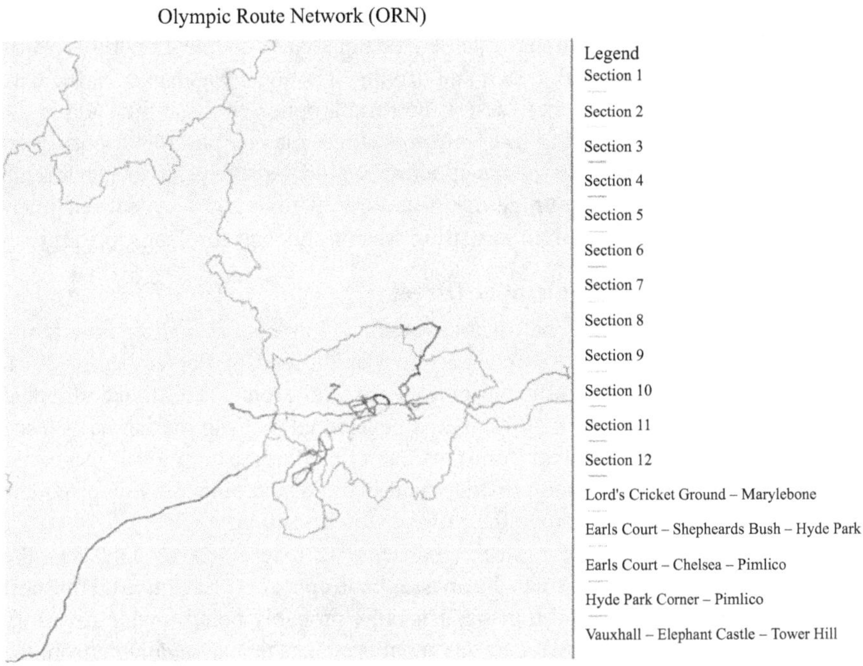

Figure 10.3 The Olympic Route Network

10.3.4.1 Legacy Achievements

A number of post-games events were held considering the legacy of the Games and Peter Hendy, the Commissioner of TfL established a sizeable team within the company to further the development of TDM principles within the day-to-day operation of transport within London. Despite this high level support, there was little, if any, enthusiasm within DfT and especially within the Cabinet Office, which had taken over complete responsibility for digital services within central Government, to continue with the provision of the Transport Direct service, let alone to build on the legacy of the SJP.

Of the six legacy themes only one has become pretty much mainstream in transport information provision, the hosting of services in a cloud environment, which was likely to have happened anyway. A second theme was partially achieved, the provision of accessible information within the Traveline service; this was funded by DfT prior to the closedown of the Transport Direct service. Sadly, the implementation of some of the TDM routeing logic into day-to-day planning has not progressed, nor have the other three potential legacy themes, which seems very disappointing given the clear results in terms of behaviour change and matching supply and demand across the network that were so clearly evident in 2012.

One final potential development to the Transport Direct service was the production of an initial international planner, demonstrated on the London to Paris corridor and covering air, rail, coach and driving. The logic was based on the City to City planner and asked users to assess their trunk option first and then add in the British domestic journey, using the Transport Direct planner and the airport interchange model. The final part of the itinerary would be a request to the foreign planner for their domestic journey using the arrival time and an idealised interchange model. But realistically, the writing was on the wall for Transport Direct.

10.3.4.2 The End of Transport Direct

After a number of reasonably half-hearted attempts to identify an alternative future for the service, the service was closed down at the end of September in 2014, having serviced over 200 million user sessions and around 2 billion individual pages following the launch in 2004. It was considered that the market had closed the gap in service provision that Transport Direct had been set up to fill, primarily Google and the on-line service providers and many of the other service providers sensed an opportunity to improve their own market position.

A surprising number of critical feedbacks were received by DfT and the Cabinet Office, many from small businesses that appear to have used Transport Direct as their navigation and logistics package, probably being unable to justify one of the excellent but expensive professional packages that are available from the market. Transport Direct closed at lunchtime on 30th September and in the view of many in the business, has never been properly replaced.

10.4 Conclusion

The future is always impossible to predict; however, the inexorable direction of travel is to increasingly bespoke and personalised services, that are location and context specific and are also cognisant of the facilities and services available nearby and which meet the preferences of the individual. Advances of technology like location services in mobile phones greatly simplifies the processes of traffic and travel information provision. Already multi-modal journey planning has supplanted single mode planning, leading to potentially significant efficiencies in travel time and mileage. Integrating all modes through a common fare and payment system will also open the door to travellers being able to use the most efficient (and environmentally least damaging) journey from point to point.

Chapter 11

Social media and traffic and travel information

Susan Grant-Muller[1], *Frances Hodgson*[1] *and Phil Cross*[2]

11.1 Introduction

This chapter considers the use of publicly available social media data as a potential additional source of traffic information. Social media data with geographical information may be useful for estimating the speed of traffic. Information on traffic flow, delays, infrastructure and environment-related traffic issues may be obtained from studying the textual content of the messages. This chapter is concerned with assessing the relevance of these social media data to the needs of road administrations, particularly in the context of traffic management. We aim to focus on the potential of one commonly available type of social media data, Twitter, as a new source of travel time information. We consider the efficacy of the data, its availability and different business models for accessing and processing the data. A case study is used to provide detailed illustration of some of the issues with the functional contribution of Twitter data and the surrounding eco-system.

11.2 Background and context

One of the main sources of information regarding the flow and speed of traffic on the road network is inductive loop traffic detectors built into the road surface. Each inductive loop is a closed electrical circuit generating a magnetic field, the inductance of which changes as a vehicle passes over the detector. Traffic speed may be determined by placing two loops a known distance (approximately 4–5 m) apart, enabling a time comparison to be made when a vehicle arrives at each detector, leading to an inferred speed. However, inductive loops are expensive to install and maintain, with access requiring the road surface to be cut and trenched; consequentially, only major roads are monitored in this way, with many sections of the road network not being monitored at all. It can also lead to inequitable levels of service for the travelling public, with some geographic areas able to provide up-to-date dynamic travel information while other areas provide a low level of service.

[1]Institute for Transport Studies, University of Leeds, Leeds, UK
[2]School of Geography, University of Leeds, Leeds, UK

To date, the use of social media as a potential source of data for traffic information has been recognised as offering a huge opportunity [1,2]. On-going research [3] has highlighted how technology and social media have given rise to new and emerging forms of data for use in Traffic and Travel Information.

There are a number of different types of social media data, for example, Facebook, Flickr, Instagram, Twitter, YouTube, LinkedIn, WhatsApp and Pinterest. All offer a mix of textual and pictorial content as well as spatio-temporal metadata such as the location through geo-tagging services. Some of the forms of data also offer additional networking data allowing the connectivity of individuals and messages to form part of the value chain in application in traffic and travel information. Twitter generates data for traffic and travel information through the content of the message that is posted by the individual, the data attached to the post including the geo-location and the date and time and through the networking of different posts. However, some of these features are not present in all messages. The 'geo-tagging' of a tweet is an optional feature and many users choose not to geo-tag their posts [4]. In addition, there is variation in the volume of messages tweeted by individuals.

Whilst we are primarily concerned with the contribution of Twitter to Travel Time Information and traffic management, research has highlighted the potential to use Twitter data for a broader range of transport contexts: commuting patterns estimation [2], transport system management for managed events [5], transport system incident detection [6,7]; tourist destination choice pattern observation [8,9], emergency or disaster management [10,11], adaptive filter techniques for disaster and crisis situational awareness [12]; inferring individual activity patterns [9], and combining with other 'Check-in' data from other social media platforms [13]; inferring residence location and mobility patterns [4]; examining the link between urban activity and land use [14] and that between employment and land use; [15]; and investigating real-time sensing of traffic information [16]. In the United Kingdom, the Office of National Statistics completed four studies into the potential to use 'big data' in the delivery of official statistics. This included an examination of the potential for use of Twitter to infer residency and mobility patterns. The study concluded that with large enough data sets Twitter was sufficient to be able to identify residential location. They were also able to identify large-scale patterns of movement between local authorities showing the mobility flows based around the academic year.

The overview of recent research shows the wealth of concepts, techniques and proposals for exploiting Twitter data in the transport sector. The advantages of using Twitter data that have been highlighted include the volume. Researchers often mention that the data is collected continuously over long periods of time rather than being a cross-sectional snapshot in time. In addition, others mention that compared to orthodox methods it is collected passively, that is whilst the individual traveller is engaged in other activities. A further advantage mentioned is that the costs can be relatively low.

However, some disadvantages are also mentioned including the amount of 'noise' [17], that is, because Twitter data is not bespoke and is collected whilst individual travellers are engaged in other activities there can be a lot of data that is not of value to the transport researcher and on-going research, for example that of reference [12], which aims to find techniques to identify the data from the 'noise'. Other disadvantages mentioned are that details on the 'content' [3,13], that is the 'who' and the 'what are they doing' and 'why' are scarce and have to be inferred from patterns in the data or from combination with other datasets. This lack of 'content' limits the development of causal explanation for human activity [12,18]. A further limitation is that the data is not representative of a population. Existing data shows age-related variation in the profile of Twitter users [19] and regional variation [4], and it is plausible to expect variation in other socio-demographic characteristics, such as class or socio-economic status. In addition, research on Twitter users has shown that there is variation in who uses the geo-location function on Twitter [20] and the volume of tweets per user [4]. Finally, the research on consistent definitions of technical terms is still scarce [18] and there is a need for more research into the validation of the methods [4,18].

Twitter launched to a wide audience in 2007, following a year of exploratory developments. In the 11 years since the launch of Twitter, the estimated number of users rose to 14.1 million in the United Kingdom [19]. In 2014 (the last time Twitter published statistics), 7.2% were aged 55+, 10.6% aged 45–54, 18.8% aged 35–44, 49.9% aged 18–34 and 13.6% aged 17 years and under [19].

11.3 Twitter as a source of transport system data

11.3.1 Accessing Twitter data

In order to analyse tweets for data mining, there are three ways in which Twitter data may be accessed [21]. The first is Twitter's Search API (application programming interface), which allows users to search posted tweets. Search criteria can be keywords, usernames, locations and so on; however, just the most recent 5,000 posted tweets can be queried. Of these, the user can collect only the latest 3,200 messages that match the search criteria. The user is also limited to making 180 searches of the Twitter feed over any given 15-min period [22].

The second is Twitter's Streaming API, a push of data as tweets happen in near real-time. As with Twitter's Search API, users may query these data with a set of criteria, and, as tweets match the criteria, they are pushed directly to the user. However, the Streaming API provides only a sample of tweets; the percentage of total tweets received depends upon the search criteria and the number of tweets currently being sent. Studies have estimated the variation in the percentage of tweets received to be between 1% and 40% due to the lack of Twitter infrastructure to support a 100% feed.

Finally, a guaranteed 100% of tweets matching search criteria may be collected using the Twitter Firehose, a commercial service, pushing data to end users in near real time, but it can be expensive to access all tweets, costing from $0.1 per 1,000 tweets. This service [23,24] does offer less restricted access to historical Twitter data.

By comparing a dataset obtained using the Streaming API with 100 'synthetic' Streaming API datasets created through random sampling of the Firehose [25], it was discovered that the coverage of subjects mentioned in the Streaming API datasets is different from the coverage of subjects mentioned in the tweets collected by random sampling by Firehose; however, the distribution of subjects becomes more representative as the number of tweets collected by the Streaming API increases.

It should be noted that the accuracy of a geotag will vary according to the technology used to obtain the location of the device. The presence of tall buildings and trees can impact upon the quality of the signal between a mobile device and GPS satellites and radio masts, affecting the accuracy of a geotag. It should be noted also that the mobile device used to create a tweet may have no means of establishing its location, even if the 'Tweeting with Location' feature is turned on.

Estimates suggest that approximately 1%–2% of all tweets are geotagged, with just 1% of users generating approximately two-thirds of all geotagged tweets [26,27]. Infrequent users, sending less than 50 tweets per year, and frequent users, sending more than 1,000 tweets per year, geotag a smaller percentage of tweets in comparison to moderate users [28]. The quantity of relevant and usable data in a particular geographic area depends on the volume of users geotagging in that area.

11.3.2 A case study of Twitter for TTI

Some of the qualities and potential of Twitter are illustrated by the following case study based in the Leeds City Region, UK. Three sets of data were used in the study. The first, Dataset A, comprises tweets collected between 9.20 a.m. on the 22nd June 2011 and 9.35 p.m. on the 19th April 2013 using Twitter's Streaming API. The search was restricted to geotagged messages, within a rectangular area bounded by the coordinates 413,100 422,500 to the south west and 447,000 450,200 to the north east, corresponding to the area shown in Figure 11.1. The rationale for focusing on this area is the presence of several strategic roads in the locality, specifically the M1, M62, M606, M621, A1(M), A58(M) and A64(M). In total, more than 2.8 million tweets sent by over 60,000 users were collected.

The second set of data (Dataset B) is made up of geotagged tweets collected using Twitter's Streaming API from the same study area for the period between 5.20 p.m. on 6th June and 6.50 p.m. on 16th August 2014. There are just over 650,000 tweets in this batch, sent by nearly 27,000 Twitter users.

Finally, Dataset C consists of global tweets collected using Twitter's Streaming API; these comprise 8 million tweets sent by 5 million users between 6.44 p.m. on the 9th and 1.26 p.m. on the 11th of November 2014. This dataset includes both geotagged and non-tagged tweets.

Each tweet has a timestamp which was used to order the Twitter feed chronologically. The maximum time gap between any two tweets in the dataset is 1.957 s, with many sent at exactly the same time (to the nearest millisecond). Nearly 1.6% of the tweets in Dataset C obtained from the Streaming API are geotagged; again, consistent with prior research. Of the geotagged tweets in Dataset C, 125 are sent by just over 100 users from the Leeds study area, meaning that, over

Figure 11.1 Leeds region study area

the 42 h monitored, the majority of users in Leeds sent one tweet. Dataset C has been used to verify the outcomes.

11.3.3 Twitter data set characteristics: time and location distribution

The number of tweets sent per minute from within the study area varies between 0 and 90 (Figure 11.2). It can be seen that there is a general upward trend in the number of tweets sent over the period being studied, with several spikes corresponding to one-off events such as the Olympic closing ceremony. However, there are occasions when the number of tweets sent per minute drops to zero for an extended period, and it can be assumed that the Streaming API has stopped collecting tweets, possibly due to a fault.

For the period under study, there were a maximum of 11,700 Twitter users active per month and a maximum of 297,500 tweets sent per month.

The tweets that comprise Dataset B (Figure 11.3) show that there were periods when the Streaming API stopped collecting tweets. It is also interesting to note the 'spikes', some of which correspond to the Tour de France (5th/6th July) or the World Cup. In total, 224,800 tweets were sent by 14,500 Twitter users during June 2014 after 5.20 p.m. on the 6th, 273,500 were sent during July by 16,800 Twitter users, and 154,300 tweets were sent by 11,700 users during August before 6.50 p.m. on the 16th. These figures are comparable with those obtained for Dataset A, suggesting relatively constant Twitter usage over these two periods.

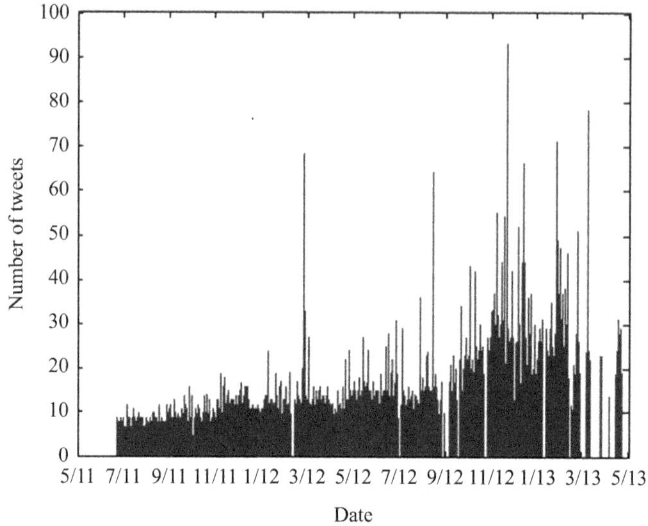

Figure 11.2 Tweets per minute within the study area: Dataset A

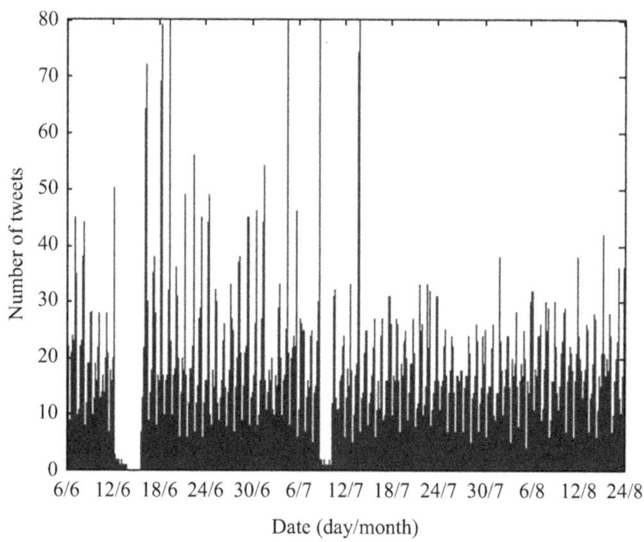

Figure 11.3 Tweets per minute within the study area: Dataset B

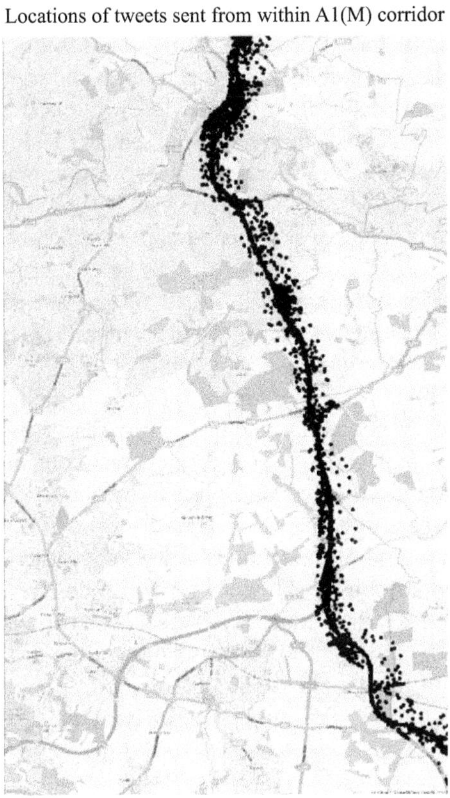

Figure 11.4 Locations of tweets in Dataset A sent from A1(M) corridor

The stretch of the A1(M) road between junctions 42 and 46 was selected for analysis of the tweets in Dataset A (all geotagged). Using ArcGIS processes* to identify the tweets sent from users on the A1(M), 14,000 tweets sent by 2,720 users were selected and are illustrated in Figure 11.4. Note that the selection includes a number of tweets sent from residential areas close to the A1(M), including parts of Wetherby.

Considering the Twitter users active within the A1(M) corridor over the study period, twelve users were responsible for more than 100 tweets each, two of whom post more than 1,000 messages. These 'super users' have been removed from the following frequency analysis in order to highlight the base level of Twitter messaging generated by the vast majority of individuals in the study area. This information is of interest as

*The national road network, as defined by Strategi digital vector data [29], was imported into ArcGIS. A 500 m corridor around the path representing the A1(M) was created using the 'Buffer' command [30]; this distance was chosen to compensate for minor errors in the locations of tweets and with the Strategi data. The 'Intersect' command [31] was used to select those tweets sent from within the corridor defined by the buffer around the A1(M).

the level of messaging may otherwise fluctuate substantially if a 'super user' has a lifestyle change and is no longer contributing messages to the pool. The number of tweets sent per month from within the A1(M) corridor is shown in Figure 11.5, and the corresponding number of people per month using Twitter is shown in Figure 11.6. It can be seen that there is a fall in both tweets being sent and Twitter users from January to March 2013; however, this corresponds to the period during which no tweets were collected by the Streaming API (as shown in Figure 11.2). Aside from that event, it is clear that there is an increasing overall trend in both the numbers of tweets sent and people using Twitter.

Considering only Twitter usage during 2012, the average numbers of tweets sent over the course of a day and week are shown in Figures 11.7 and 11.8, respectively.

Figure 11.7 shows clearly that a large number of tweets were sent between 12.00 noon and 3.00 p.m. There is a further increase in the number of tweets sent during the evening after 9.00 p.m., many of which can be assumed to be sent by Twitter users are at home in one of the residential areas contained within the A1(M) buffer: several tweets refer to television programmes, including the Olympics, or simply say that the user is at home.

The corresponding figure illustrating the weekday distribution of tweets sent from the A1(M) corridor (Figure 11.8) shows that there are a similar number of tweets sent from Monday to Wednesday, with an increase on Thursday, Friday and over the weekend. During the very early hours of the day (03.00–06.00 a.m.), the average number in this period is low, but this corresponds with the pre-peak period for traffic and there may well be sufficient tweets to provide skeletal information.

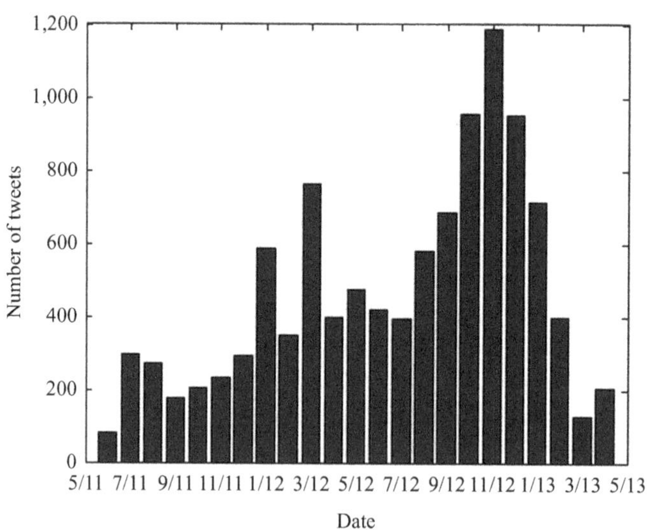

Figure 11.5 Number of tweets in Dataset A sent from within A1(M) corridor

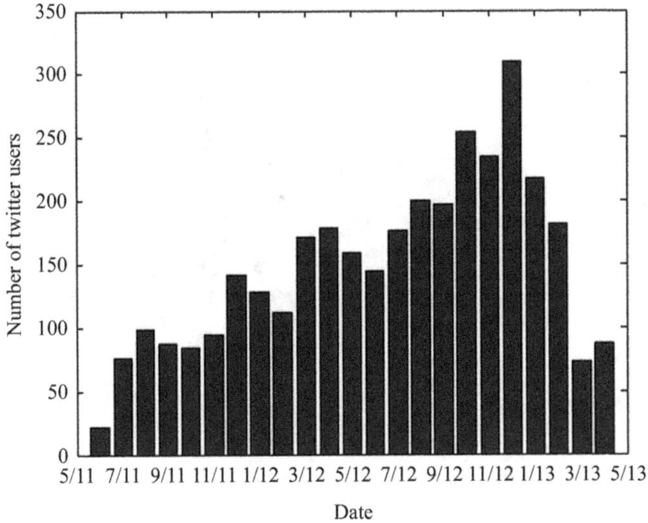

Figure 11.6 Number of Twitter users in Dataset A within A1(M) corridor

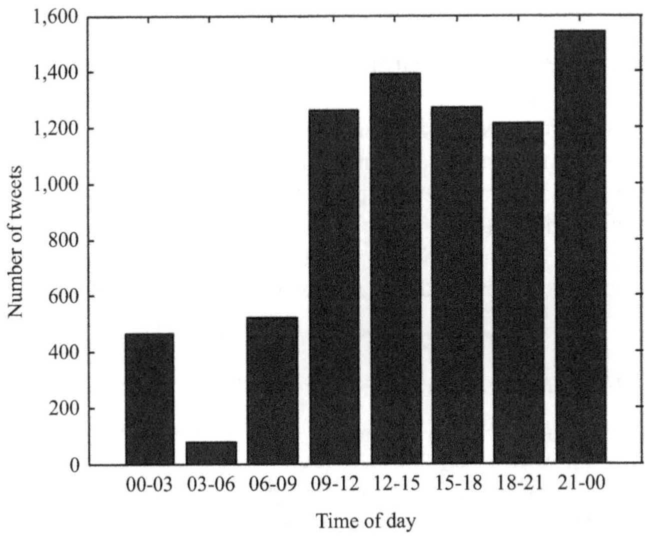

Figure 11.7 Average number of tweets in Dataset A sent from within A1(M) corridor per 3 h over a day from midnight to midnight

Considering the average number of tweets by day provides an early indication that the volume of the data could be sufficient to provide a useful information source, depending on the analyses on travel time (Section 11.4) and textual content (Section 11.5).

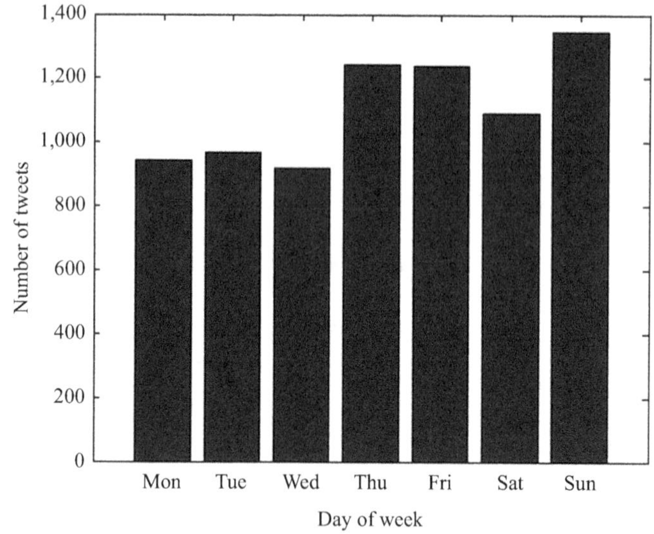

Figure 11.8 Average number of tweets in Dataset A sent from within A1(M) corridor by days of a week

11.4 Travel time estimation from Twitter data

This section describes a study in which vehicle travel time and speed is derived from Twitter data. In the study, data from inductive loop traffic detectors were used as a reference source of traffic speed for comparison. Several inductive loop traffic detectors monitor the A1(M) within the Leeds study area; the sections monitored are listed in Table 11.1. Inductive loop data provide information regarding average travel time, average travel speed, and the total number of vehicles passing over the loops [32].

11.4.1 Calculation of Twitter-based speed

Given a set of geotagged tweets, it is possible to calculate the speed of a Twitter user and the direction in which they are travelling.

The best way to derive speed from geotagged tweets would be to employ a map-matching algorithm to confirm the road locations of the Twitter user, and then use the detailed road geometry to calculate the distance travelled. However, for the purposes of this case study it was judged not to be efficient to spend time implementing map-matching specifically for this Twitter analysis, when the performance of map-matching is not the main focus here. Instead this study takes advantage of the A1(M) in the study area being close to straight, and simply uses a straight-line distance between the two points. Although this assumption will result in underestimated speeds, it is sufficient to allow the analysis to explore the potential of the Twitter data. For an operational implementation, map-matching would be required.

Table 11.1 Sections of A1(M) within the study area monitored by inductive loop traffic detectors

Reference	Section of A1(M) covered	Length of the section (km)	Speed limit (km/h)
LM1056	J42 to J43	6.08	107
LM65B	J42 to J43	8.53	112
LM66B	J43 to J42	7.44	112
LM1057	J43 to J42	6.07	112
LM22A	J43 to J44	3.62	112
LM65A	J43 to J44	6.92	112
LM66A	J44 to J43	7.43	112
LM23A	J44 to J43	4.78	112
LM22	J44 to J45	5.72	112
LM1062	J44 to J45	6.12	112
LM1063	J45 to J44	6.11	112
LM23	J45 to J44	5.19	112
LM1064	J45 to J46	4.75	112
LM1065	J46 to J45	4.69	112

Speed calculation from geotagged tweets

- For each user, find pairs of tweets sent less than 5 min apart
- Calculate the distance between the locations of each pair of tweets
- Calculate the period of time between the timestamps of each pair of tweets
- Speed = distance/time

A 5-min gap between tweets has been selected based upon the average time vehicles take to travel each length of road monitored by the inductive loop sensors.

The direction in which the Twitter user is travelling was used to determine if they are on either the north- or south-bound carriageway.

- Calculate change in easting and northing between the pair of tweets
- Direction of travel = arctan(Δeasting/Δnorthing)

11.4.2 Correlation with loop speed data

Using the algorithms described above, it was found that some calculated speeds were impossibly high. There are two apparent reasons for this: the first is that there may be errors in the location of the phone; the second is that there may be problems with the timestamps of some tweets.

If the location of a mobile phone is obtained from GPS, it should be reasonably accurate; however, if a device does not have GPS enabled or it is not available, its location is determined through triangulation by using information from the mobile network radio masts to which it is connected. However, if there is poor radio mast coverage, such as in rural areas, there is less likelihood for there to be enough masts for triangulation. Further, the distance between a mast and mobile device is likely to be large, resulting in the accuracy with which the device can approximate its location reducing significantly.

A number of speeds were calculated to be infinite because the user appeared to move between sending tweets (with different content, ruling out the user sending the same tweet twice), but the timestamps of the tweets are the same to the nearest second. This appears to be due to the device posting these messages to the Twitter feed at the same time, even if the user has generated the tweets some time apart. This is possibly because the user has emerged from an area of no mobile coverage after the tweets have been generated.

Twitter users, including 'super users', with calculated speeds >10 km/h (jogging speed) and <200 km/h have been included in the following analysis. Approximately 400 pairs of tweets have been found with corresponding speeds of between 10 km/h and 200 km/h. The speeds of the users sending these tweets have been compared to the average speeds of vehicles acquired from the inductive loop data corresponding to that user's direction, and the location and timestamp of the first tweet sent by the user.

The correlation coefficients for the speeds for several sections of the A1(M) are summarised in Table 11.2, together with the mean speeds for these data over the study period. An additional point, corresponding to 0 km/h for both the loop speed and the speed calculated from the Twitter data, has been included in the analysis when calculating the correlation coefficient. These data are also illustrated as graphs in Figure 11.9; each showing the average speed obtained from loop data over a period of 15 min plotted against the corresponding speed calculated from geotagged tweets. In Figure 11.9, the line has been placed through the origin under the working assumption for the study that the loop data represents the true speed of the user, therefore when the loop data indicates zero speed the user is stationary.

It can be seen that there is a reasonable match between the mean overall speeds obtained from the loop data and the Twitter feed, especially given that Twitter speeds are expected to be an underestimate due to the simple assumption used for the study.

However, the correlation coefficients show that the fit between individual speeds calculated from the Twitter data and the corresponding loop speed is relatively poor. It should be noted that the speeds obtained from the inductive loops are averages for vehicles passing the detectors over a 15-min period; in contrast, the speeds obtained from the Twitter feed are for individual users/vehicles.

During the study period, 2,300 tweets were sent from the A1(M) corridor (Dataset B); however, <50 of these meet the selection criteria (tweets sent up to 5 min apart and a calculated speed of between 10 and 200 km/h). This highlights one of the drawbacks to using Twitter data for estimating speed: there may be a relatively small number of data points, with geotagged tweets being sent infrequently.

The A1(M) link between junctions 44 and 45 was selected for more detailed analysis of speeds, this time using MIDAS data (i.e. highway sensor data, based on induction loops spaced at 500-m intervals and used to indicate vehicle movements) averaged over a period of 1 min. This section, including slip roads, is monitored by 14 sets of inductive loops. Figure 11.10 shows the speeds obtained from the MIDAS data over the period of one day. Also shown in this figure are speeds for one user calculated from tweets. A detail of this graph is shown in Figure 11.11 in which each column of dots represents 1-min average speeds collected by MIDAS.

Table 11.2 Mean speeds and correlation coefficient (Corr. coeff.) from inductive loop data and calculated using Twitter data

Section of A1(M)	Length(km)	Mean loop speed(km/h)	Mean Twitter speed(km/h)	No. of data	Corr. coeff.
J42 to J43 north	14.6	103.4	91.4	35	0.618
J43 to J44 north	10.5	101.0	103.6	75	0.439
J44 to J45 north	11.8	109.5	86.5	138	0.230
J45 to J46 north	4.7	104.7	75.1	24	0.200

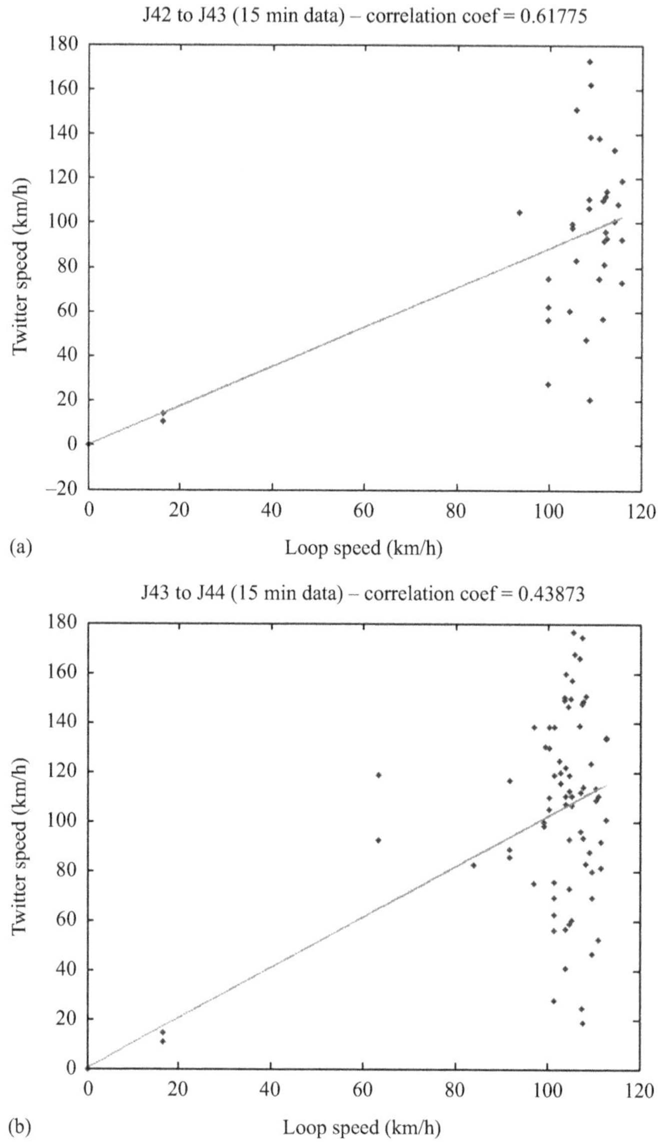

Figure 11.9 Speeds acquired from loops plotted against corresponding speeds calculated from tweets

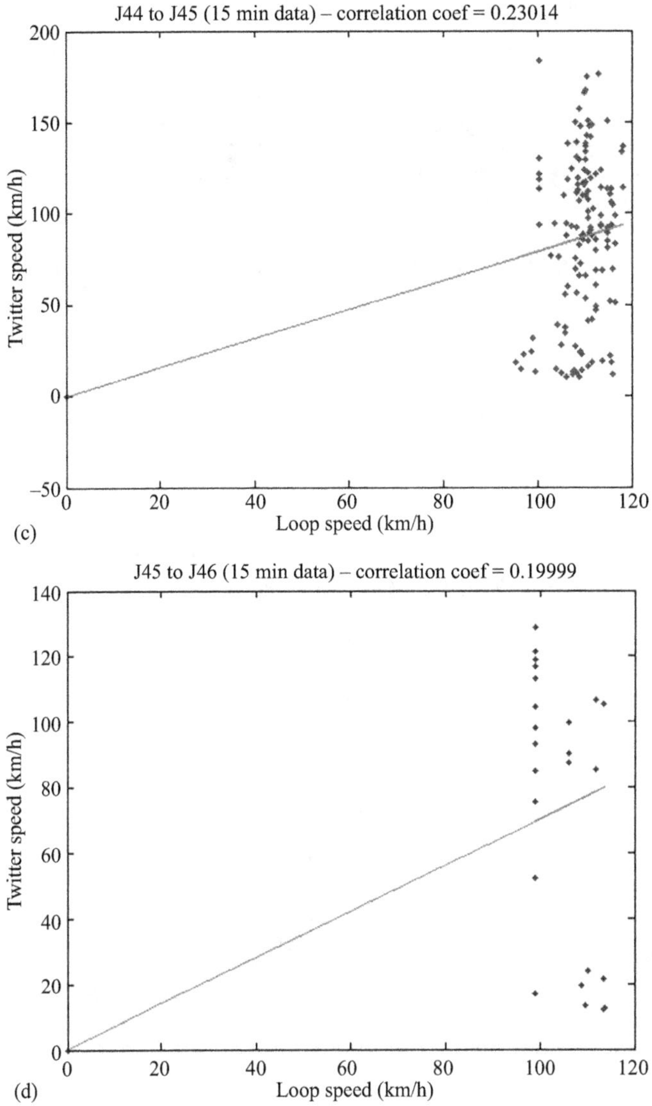

Figure 11.9 *(Continued)*

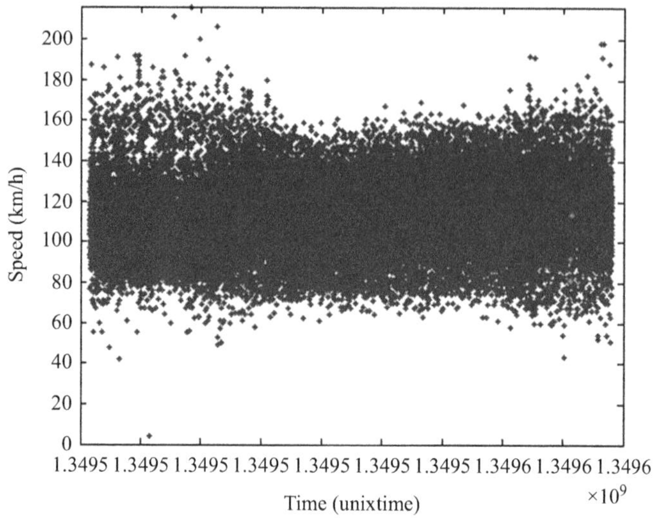

Figure 11.10 *Speeds for one user, calculated from tweets (red dots), compared to speeds obtained from MIDAS loops (blue dots) for 1 day*

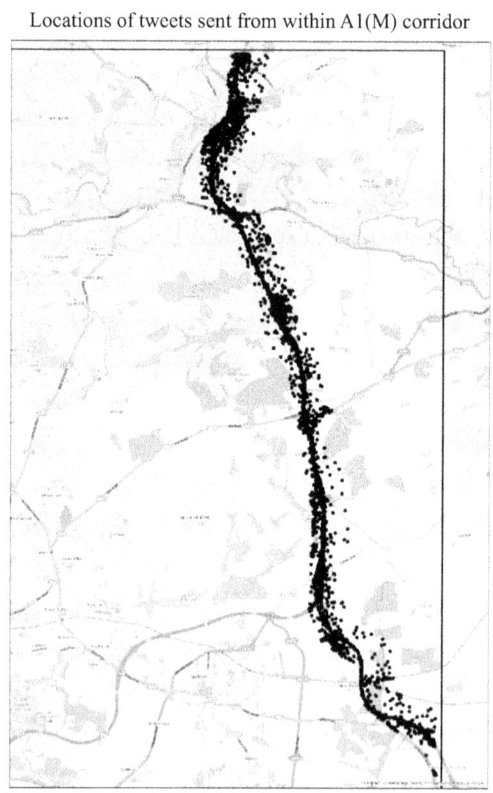

Figure 11.11 *Speeds for one user, calculated from tweets (red dots), compared to speeds obtained from MIDAS loops (detail of 10.10)*

It can be seen that the speeds calculated from Twitter for this particular user correspond well to the speeds obtained from the MIDAS loops. Overall the Twitter-based speeds tend to underestimate the average speed obtained by the MIDAS loops, but this is at least partly due to the simple straight-line distance assumption made in this study.

11.5 Analysis of Twitter content for traffic management relevance

This section describes a study of the textual content of tweets. The case study investigates:

- whether these data contain information relevant to road traffic management;
- if so, whether it is possible to build classifiers that allow relevant messages to be detected in the Twitter feed.

11.5.1 Definition of relevant content

An inspection of tweets in both Datasets A and B found that a large number of road transport-related messages had been sent by the West and North Yorkshire Travel services. These tweets were used to obtain the following list of roads in the study area:

> M1, M606, M62, M621, A1(M), A162, A168, A58, A58(M), A6025, A6029, A6032, A6036, A6037, A6038, A6069, A61, A6110, A6120, A614, A6176, A6177, A6181, A62, A63, A630, A636, A638, A639, A64, A64(M), A641, A642, A643, A644, A645, A647, A649, A65, A650, A651, A652, A653, A6539, A655, A656, A657, A658, A659, A660, B1222, B6117, B6121, B6122, B6123, B6126, B6135, B6136, B6137, B6144, B6145, B6152, B6154, B6155, B6156, B6157, B6159, B6164, B6380, B6382 and B6421.

These road names were checked manually against the road structure in the study area to confirm that they were relevant to the study area. Tweets that had been clearly sent by the travel services were then removed prior to the text mining analysis.

The list of road names relevant to the study area is a clear example of relevant textual content. In analysing the Twitter stream further, it was necessary to refine the criteria for relevance in order to:

- separate the general Twitter feed into relevant tweets and irrelevant tweets (as part of the classification process)
- identify the nature of the content itself.

  The meaning of 'relevant content' was defined by a combination of:
- the stated operational and other objectives and priorities for the national roads authorities,
- an ontology derived to represent the entities and structural relationships for the transport system and highways travel context.

11.5.2 Understanding relevance through stated traffic management objectives

A synthesis of the operational and other objectives identified the relevant data required at the Meta level and the message content for operational value (Tables 11.3 and 11.4) as well as KPI for the long-term policy goals (Table 11.5). In constructing these tables the fact that the data is in textual format rather than other formats of data (digital sensor, images, etc.), together with the requisite that the content has been contributed by users of the system (rather than authorities or operators) has led to the positive inclusion or exclusion of particular items.

11.5.3 Understanding relevance through ontology

The second element used to define relevant content is ontology: 'Ontology' is a formal naming and definition of the types, properties and interrelationships of the entities that fundamentally exist for a particular domain or discourse. Examples of ontology from the transport domain [33] were considered with a view to modifying an existing ontology for use in the study here. However, as is the case with most ontology, these were all purpose-specific and would have required considerable modification. Despite much related research, a comprehensive transport ontology covering all parts of the transport system is not currently available (see also [34]) and it would be resource-intensive to establish this. Fusing existing ontological resources may alleviate this effort and some attempts in this direction have already been made, for example [35].

In terms of general methodology, there is no one correct way to model a domain and a number of viable alternatives will always be possible. The modelling approach adopted here is based on the widely accepted general framework [36], defined by:

- concepts (or classes),
- properties of each concept describing various properties, features, attributes or roles
- role restrictions or facets (possible values).

The framework also proposes the following structure for describing the context of the ontology:

- the domain that the ontology will cover
- the purpose of the ontology
- a set of competency questions that the ontology should be able to answer or address
- a definition of the user of the ontology.

Domain

The fundamental rationale used here is that of the multi-layer infrastructure network.

Table 11.3 Definition of relevance from the meta-attributes of tweets

Meta-attribute of tweet	Relevant content at the meta level	Not relevant
Origin	Individuals – passengers, stationary drivers Travelling public Residents	Agency/services messages Highways England messages
Temporal characteristics of the data	Historical experience Current experience Future intent	
Geographic characteristics	Content originating from or referring to the strategic network and the roads adjacent to it.	Content with unknown origin or where no particular road/highway can be identified

Table 11.4 Data needs for traffic information

Data needs	Examples that might be found in tweets
Analysis of network performance	Speed, intensity of traffic, delay
Road traffic prediction – which can use origin/destination information	Point of access to the highway or egress. Original point of departure or final destination
Event detection and handling	Events on highway, roadside or hard shoulder, wildlife, weather related incidents. Infrastructure related content – equipment and pavement.
Lane management, information on speed and headway	Complaints of inappropriate messaging

Table 11.5 Content relevant for long-term objectives

Relevant to broader objectives	Examples that might be found in tweets
Traveller sentiments	Perceptions of level of service, ease of journey, satisfaction/dissatisfaction, clarity or confusion concerning expected behaviour on highway
Traveller well-being	Stress, unwell, health-related issues

There are many references and adaptations for this, but most characterise the system as comprising

- infrastructure
- traffic characteristics
- operational conditions.

For this study, we have added a fourth class of the 'user' to bring a more modern socio-technical perspective of the system. The domain here is a representation of the physical and other entity types that lie within the scope of the highways management task.

While the ontology design process was conducted separately to the tweet content analysis, there is a loose interface between content and the multi-layer perspective underlying the design: physical or other aspects of the system that most users would be unlikely to see or experience (and therefore would be unlikely to tweet about) were not included in the design.

The level of detail in the ontology (i.e. the number of tiers and subclasses) has been driven by the following factors:

- the ability for a given level detail to address the competency questions (at what level of detail is it 'fit for purpose'),
- the review of the level of other transport ontology,
- the context in this study which focuses on user-generated textual content.

As an example, for the 'Environment and met' subclass 'Air Quality', a user may be unable to distinguish between different forms of air quality, so the next level of detail is not provided.

Purpose

The ontology is intended for the identification and classification of relevant textual information arising as user generated content within social media. This user generated content forms an information input to the task of managing the strategic network. The ontology draws on an understanding of the scope of the highways management task and from the definition of relevant information to support delivery of those objectives.

Competency

The requirements (in the form of questions) that the ontology is designed to address are as follows:

- What are the infrastructural, environmental, traffic system and psychological factors that are relevant to traffic management performance, which users may experience and report during their journey?
- What elements of the strategic network infrastructure (or neighbouring roads) are used by (or exposed to) the travelling public, for which the functionality impacts on their journey?
- Which parts of the highway traffic system are most susceptible to environmental disruption and which types of environmental disruption are of most concern to users?

- Which category of operational conditions are users most likely to contribute information about?
- What aspects of the user experience are most likely to generate positive or negative sentiment?

Some of the questions are answered by the defined hierarchical class structure, while others are answered by analysis of the data arising in practice.

Ontology users

- the intended users (direct and indirect) of the ontology
- analysts of user-generated textual content related to the strategic highways
- those concerned with the management of the strategic highways.

The ontology constructed in this study is presented in Figures 11.12 and 11.13, which shows an example of the class hierarchy, the attributes of the leaf-level classes and their possible values for traffic conditions. While wider review of the ontology would be likely to refine some details of the selected decomposition, the ontology is illustrative of the potential use of Twitter data for TTI.

11.5.4 Classifying tweets as relevant or not-relevant

This section explores whether it is possible to build classifiers that allow relevant messages to be detected in the Twitter feed. In this section, the classification is a simple decision between either relevant or not relevant. From a manual search of the tweets in Dataset A it was clear that the textual information included relevant information to assist traffic management.

The way that the social media data could be used for traffic management depends on the level of automatic information extraction that is practical. For example if it is possible to automatically distinguish between relevant and non-relevant tweets then a service could be built to produce a small filtered stream of relevant tweets to present to a traffic manager for further interpretation. If it is possible to automatically classify further details then extra automation could be built into that service; the most advanced case would be a fully automated traffic event detection service. This case study therefore considers simple classification into 'relevant' or 'not relevant' classes in this section, then considers finer-grained classification subsequently.

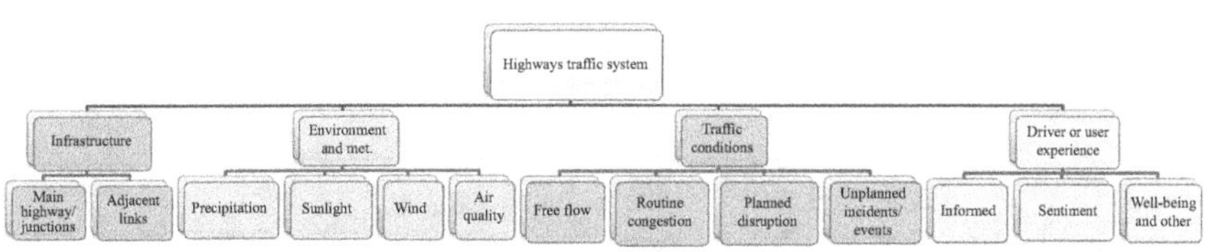

Figure 11.12 Main class hierarchy

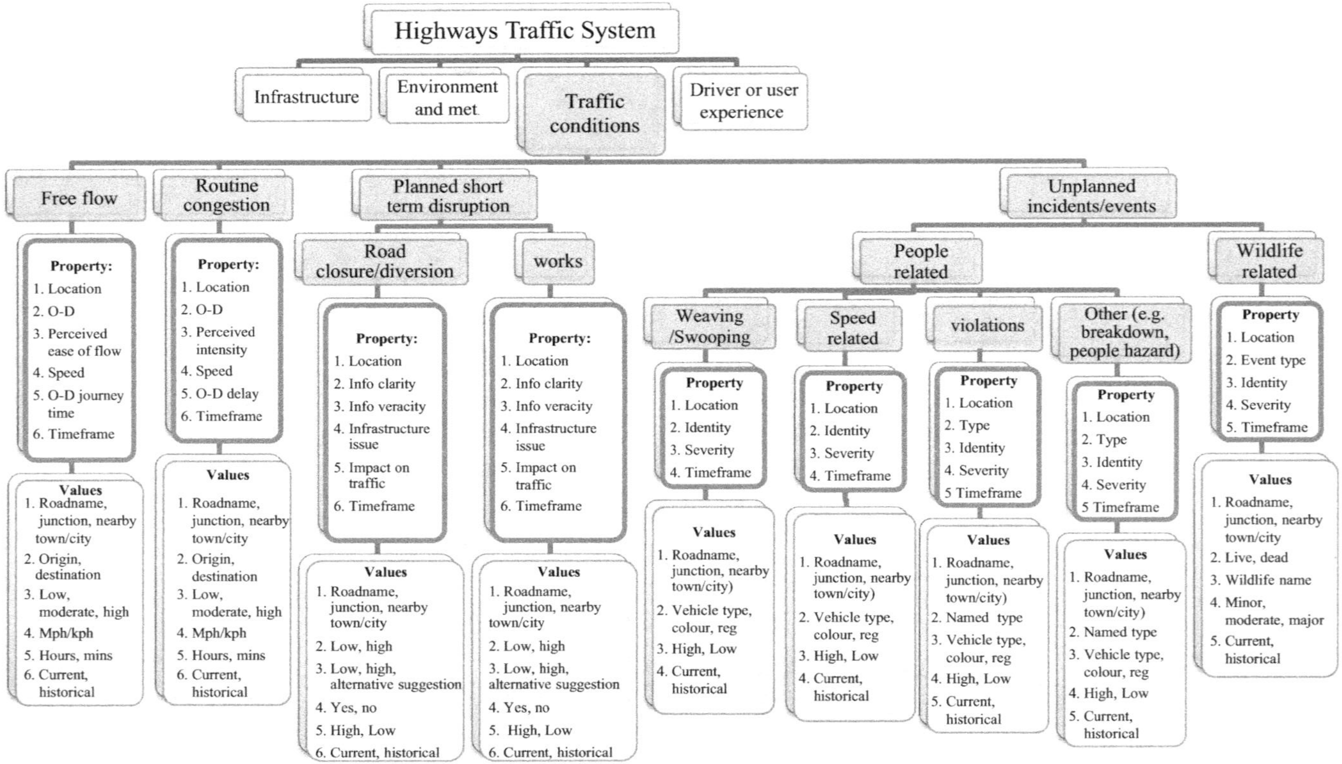

Figure 11.13 Example leaf-level classes for Traffic Management

Data preparation

A manual search of the tweets in Dataset A referring to relevant road numbers was undertaken to identify messages containing relevant travel information. Those tweets considered relevant were highlighted; for example, a tweet such as, 'Love the M62 me #stationary' would be considered relevant, if not particularly useful, and a tweet such as, 'Either I've taken a wrong turning on the way home or there's a New York Taxi on the M62 ?!!?', considered irrelevant. The tweets were then split into three groups of similar size, each containing the roughly the same proportion of useful and irrelevant tweets.

A further set of tweets containing no useful road travel information (for example, 'Don't you hate it when you can't find your car keys?!') were also selected and split into three groups of approximately the same size. These tweets were then appended to the previously established three groups containing road numbers, resulting in the following:

- group 1: 149 useful and 261 irrelevant tweets,
- group 2: 145 useful and 230 irrelevant tweets,
- group 3: 148 useful and 241 irrelevant tweets.

Pre-processing the Twitter data involved removing all non-alphanumeric characters and converting the text to lower case, avoiding the situation where 'west' and 'West' would be considered different words. The order of each set of tweets was then randomised.

Results from automatic classifiers

The open-source program Weka [37] was used for text mining, and a range of classification algorithms provided by the program were chosen for this analysis: naïve Bayes, Bayes network, a stochastic gradient descent (SGD) algorithm for training a support vector classifier (SVC), a sequential minimal optimisation (SMO) algorithm for training an SVC, a pruned J48 decision tree, a random tree, a forest of random trees and a nearest neighbour classifier. These algorithms have been used successfully by other researchers investigating machine learning techniques for text mining; for example, [38].

These algorithms work on matrices of numeric encoded data, so to use them on tweets it is first necessary to create a list of all words that occur in the messages (a 'bag of words'), and represent each tweet with a sparse matrix. The elements in these matrices correspond to the occurrence of words in each message. For example, the tweet, '6 h now stuck on the M62 dayover' (considered to be relevant) is represented by the matrix, {54 1,515 1,892 1,1176 1,1333 1,1358 1,1740 1,1796 1}. In this matrix, the number 54 corresponds to the number '6' in the text message, 515 corresponds to the word 'dayover', 892 corresponds to 'hours', the number 1,176 corresponds to 'M62' and so on. The irrelevant (or 'negative') tweet, 'spotted M62 westbound stobart', is represented by the matrix, {0 negative,1,176 1,1710 1,1729 1,1968 1}. Here, the number 1,176 again corresponds to 'M62', 1,710 corresponds to 'spotted' and so on.

Two further classifying algorithms, both operating directly on text string data, were also selected for this analysis: multinomial naïve Bayes and an SGD algorithm for training an SVC.

In order to evaluate the relative performance of the models obtained using the classifying algorithms, the first group of tweets was employed as a training set, the second formed a test set, and the last group was used for verification of the classifiers prepared using the training and test sets.

Initially, the training set of tweets was split into two subsets, with one subset containing two thirds of the tweets in the training set and the other containing one third of tweets in the training set. The larger subset was used to train the classifiers and the smaller set used to test the classifiers. Next, classifiers were obtained using all of the tweets in the training set. These classifiers were then tested using the test set of tweets. Finally, these classifiers were tested again with the verification set of tweets. The percentage of correct classification for each of the three experiments was recorded, together with the kappa statistic [39]. This statistic is a measure of agreement for qualitative items, taking into account the probability of the agreement occurring purely by chance. If there is agreement, then kappa = 1. If there is no agreement, kappa = 0.

The relative performance of each classifying algorithm is summarised in Table 11.6. The table lists the average and maximum percentages of the correct class being selected by the algorithm for each set of tweets used for training, testing and verification. The corresponding percentages for the incorrect class being chosen are also listed. Finally, the table lists the average and minimum kappa statistic.

The best performing machine learning algorithms identified the correct class (relevant or not relevant) for around 80% of the tweets in the case study datasets. This supports the idea that a service could be built to produce a small filtered stream of relevant tweets to present to a traffic manager for further interpretation.

It can be seen from Table 11.6 that the greatest correct class percentages are achieved using the Naïve Bayes algorithm, SGD algorithm for training an SVC and SMO algorithm for training an SVC. However, taking into account the kappa statistic, the SGD algorithm outperforms the other classifiers investigated. It is interesting to note that the classifying algorithms designed specifically for text strings are outperformed by the Naïve Bayes, SGD and SMO algorithms.

A similar exercise was undertaken with weightings manually assigned to all individual words according to their relevance to road travel on a scale of 0 to 5; for example, 'sliproad' is given a weighting of 5, 'rushhour' scores 3 and 'foggy' is given a score of 1. The annotator was a transport domain specialist with experience of the annotation task. However, it was found that only the naïve Bayes classifier showed any significant change in performance. The results are summarised in Table 11.7.

Comparison of the results in Tables 11.6 and 11.7 shows that there was actually a slight detrimental effect on performance when using weightings for individual words.

The trained classifiers obtained using all the tweets contained in the training set obtained from Dataset A were then tested using a group of tweets selected from Dataset B. This comprised 60 useful tweets and 95 irrelevant tweets. The results are summarised in Table 11.8 in which it can be seen that the SGD and SMO algorithms again outperform the other classifiers considered.

Table 11.6 Performance of text classifying algorithms trained and tested using Dataset A

	Percentage correct avg. (max. in brackets)	**Percentage incorrect avg. (max. in brackets)**	**Kappa statistic avg. (min. in brackets)**
Naïve Bayes	79.28 (81.49)	20.72 (24.46)	0.552 (0.465)
Bayes network	75.08 (76.53)	24.92 (26.62)	0.451 (0.415)
SGD algorithm for SVC	80.35 (80.58)	19.65 (19.79)	0.569 (0.554)
SMO algorithm for SVC	80.25 (81.60)	19.75 (21.58)	0.565 (0.507)
Pruned J48 decision tree	74.41 (76.27)	25.59 (27.34)	0.427 (0.354)
Random tree	67.54 (68.35)	32.46 (33.60)	0.275 (0.262)
Forest of random trees	77.38 (79.18)	22.62 (25.18)	0.456 (0.365)
Nearest neighbour classifier	71.08 (74.10)	28.92 (33.60)	0.344 (0.239)
Multinomial naïve Bayes for text strings	76.92 (78.41)	23.08 (24.46)	0.508 (0.454)
SGD algorithm for SVC for text strings	77.07 (79.47)	22.93 (25.90)	0.482 (0.381)

Table 11.7 Performance of naïve Bayes classifier trained and tested using Dataset A incorporating weightings for individual words

	Percentage correct avg. (max. in brackets)	**Percentage incorrect avg. (max. in brackets)**	**Kappa statistic avg. (min. in brackets)**
Naïve Bayes	79.28 (73.93)	20.72 (26.07)	0.449 (0.308)

Table 11.8 Performance of classifying algorithms trained using Dataset A and tested with Dataset B

	Percentage correct	**Percentage incorrect**	**Kappa statistic**
Naïve Bayes	81.94	18.06	0.583
Bayes network	81.29	18.71	0.543
SGD algorithm for SVM	85.16	14.84	0.650
SMO algorithm for SVM	84.52	15.48	0.633
Pruned J48 decision tree	80.65	19.35	0.530
Random tree	74.84	25.16	0.426
Random forest of trees	78.71	21.29	0.471
Nearest neighbour classifier	73.55	26.45	0.343
Multinomial naïve Bayes for text strings	80.00	20.00	0.565
SGD algorithm for SVM for text strings	78.71	21.29	0.476

The best performing machine learning algorithms identified the correct class (relevant or not relevant) for 80%–85% of the tweets in our datasets. This provides further validation of the potential suggested by Table 11.6.

It is interesting to note that just 40 of the 8 million tweets in Dataset C mention one of the road numbers listed above, six of which are geotagged. Several of the tweets refer to a road in another country (specifically, Australia and Ireland), and many others refer to a completely unrelated subject. Of these 40 tweets, only ten contain relevant information regarding roads in the study area; however, six were generated by road travel services, four of which are geotagged. Recall however that the Twitter data employed here accounts for less than 1% of the total number of tweets sent globally.

In an attempt to improve the performance of the classifiers, phrases containing two words or more (referred to here as n-grams) associated with road travel (for example, 'traffic jam', 'tail back' and 'dual carriage way') were concatenated such that the classifying algorithms would treat each phrase as one word ('trafficjam', 'tailback' and 'dualcarriageway'). The results of this exercise using Dataset A are summarised in Table 11.9, from which it can been seen that there is little or no improvement over classifying tweets using individual words.

11.5.5 Classifying tweets using ontology

This section considers whether it is possible to build classifiers that can distinguish between different categories of relevant content, based on the ontology described in Section 11.5.

A decision was made to use only the first tier of the ontology for this analysis due to the relatively low quantity of data available. The remaining tiers provided in the ontology can of course be used for larger case studies and where more data is available for analysis.

The categories included were the first hierarchical level of the ontology, as follows: infrastructure, environment and meteorology, traffic conditions and user experience. Each of the road travel-related tweets in Dataset A was assigned one of these categories.

As in the analysis described in the previous section, machine learning algorithms were used, this time to automatically categorise tweets into one of the ontological categories. Again, a range of classifying algorithms were chosen for this analysis: naïve Bayes, Bayes network, SMO algorithm for training an SVC, pruned J48 decision tree, random tree, forest of random trees, nearest neighbour classifier, a multi-class classifier and a decision table, each of which operate on a 'bag of words'. A multinomial naïve Bayes classifier for text strings was also selected for this analysis. The number associated with the usefulness of each tweet was used as a weighting for the classifying algorithms.

As with the previous analysis, the training set of tweets was split into two subsets: one containing two-thirds of the tweets and the other containing one third of tweets. The larger subset was used to train the model and the smaller set used to test the model. The model was also tested using 10-fold cross-validation, in which the

Table 11.9: Performance of classifying algorithms incorporating n-grams using Dataset A

	Percentage correct avg. (max. in brackets)	**Percentage incorrect avg. (max. in brackets)**	**Kappa statistic avg. (min. in brackets)**
Naïve Bayes	79.37 (81.49)	20.63 (24.46)	0.555 (0.465)
Bayes network	75.08 (76.53)	24.92 (26.62)	0.450 (0.415)
SGD algorithm for SVM	79.56 (81.33)	20.44 (21.58)	0.547 (0.496)
SMO algorithm for SVM	79.16 (80.80)	20.84 (23.02)	0.542 (0.480)
Pruned J48 tree	74.23 (75.73)	25.77 (27.34)	0.423 (0.354)
Random tree	69.59 (67.87)	30.41 (32.13)	0.320 (0.283)
Random forest of trees	77.45 (78.66)	22.55 (24.46)	0.462 (0.387)
Nearest neighbour classifier	70.83 (74.10)	21.17 (33.33)	0.334 (0.230)
Multinomial naïve Bayes for text strings	77.12 (78.66)	22.88 (25.18)	0.512 (0.441)
SGD algorithm for SVM for text strings	77.08 (79.20)	22.92 (26.62)	0.487 (0.375)

training set is split randomly into ten subsets of equal size. Of these subsets, one is used for testing the model, and the remaining nine subsets are used as training data. The cross-validation process is then repeated ten times, with each of the subsets used once as the validation data. The results are summarised in Table 11.9.

It can be seen from Table 11.10 that the best performing classifiers are the decision table and Bayes network. However, inspection of the kappa statistic reveals that the naïve Bayes actually outperforms both.

In order to improve the performance of the classifiers, all tweets in Dataset A were manually allocated a value between 1 and 5 based on the perceived relevance of the tweet content to the highways traffic management context. Two annotators undertook the task, one being a transport domain expert and the other having no specialist expertise. The inter-annotator agreement rate was 41%, with the expert assigning the tweets a score 0.3 higher than the non-expert on average. This is a relatively low degree of correspondence; hence only the expert weights were used for further analysis. A value of 1 corresponds to messages such as 'I hate the M62!!!!'. A value of 5 corresponds to tweets mentioning the precise location and cause of a traffic jam, specific user experience, or weather conditions; for example, 'Avoid M621 to Wakefield j42-40 traffic is murder in this searing heat'. These values formed weights for each of the tweets in the Dataset. The results of the classification process following the weighting process are summarised in Table 11.11.

Inspection of the table reveals that, in most cases, the addition of weightings improved the performance of the classifiers, although three classifiers show a decrease in performance.

The outcomes in terms of proportion of tweets within each class of the ontology hierarchy are shown in Figure 11.14. The proportions are approximate as some tweets could have been allocated to more than one class and where this happens, the dominant part of the message was used to form an allocation.

The best performing algorithm correctly categorised 72% of the tweets. This does not support the idea of a fully automated traffic event detection service, but does suggest that categorisation could be useful, for example to improve a service offering filtered information for further manual inspection by a traffic manager.

11.6 Conclusions

The first part of this chapter has used a specific case study to demonstrate that geotagged Twitter messages may be used to obtain the speed and direction of users as they travel on the transport network, providing traffic data as an alternative to inductive loops and number plate recognition systems.

The method to calculate speeds used in the case study provided here requires a relatively short time period (<5 min) to have elapsed between consecutive tweets being sent. It was found that there are issues with accuracy of the geotags and the timestamps of tweets, resulting in unrealistic speeds being calculated for a number of users. Realistic speeds were obtained from around 450 pairs of tweets in the

Table 11.10 Performance of ontology classifiers, evaluated using 66% split and 10-fold cross-validation

	Percentage correct	**Percentage incorrect**	**Kappa statistic**
Naïve Bayes	57.43, 59.17	42.57, 40.83	0.277, 0,309
Bayes network	62.16, 60.09	37.84, 39.91	0.112, 0.158
SMO algorithm for SVM	58.78, 56.65	41.22, 43.35	0.254, 0.249
Pruned J48 decision tree	58.78, 53.44	41.22, 46.56	0.252, 0.204
Random tree	48.65, 49.77	51.35, 50.23	0.112, 0.123
Random forest of trees	59.46, 59.40	40.54, 40.60	0.210, 0.183
Nearest neighbour classifier	40.54, 44.95	59.46, 55.05	0.091, 0.060
Decision table	67.57, 63.30	32.43, 36.70	0.285, 0.255
Multi-class classifier	52.03, 53.67	47.97, 46.33	0.214, 0.265
Multinomial naïve Bayes for text strings	15.54, 15.60	84.46, 84.40	0.049, 0.450

Table 11.11 Performance of ontology based classifiers incorporating weightings, evaluated using 66% split and 10-fold cross-validation

	Percentage correct	Percentage incorrect	Kappa statistic
Naïve Bayes	56.46, 65.37	43.54, 34.63	0.145, 0.346
Bayes network	64.05, 64.56	35.95, 35.44	0.181, 0.252
SMO algorithm for SVM	61.52, 60.78	38.48, 39.22	0.237, 0.268
Pruned J48 decision tree	56.46, 54.73	43.54, 45.27	0.224, 0.208
Random tree	49.62, 50.05	50.38, 49.95	0.147, 0.132
Random forest of trees	63.54, 63.57	36.46, 36.43	0.086, 0.097
Nearest neighbour classifier	36.96, 43.01	63.04, 56.99	0.076, 0.104
Decision table	72.41, 66.55	27.59, 33.45	0.330, 0.244
Multi-class classifier	53.16, 58.16	46.84, 41.84	0.223, 0.304
Multinomial naïve Bayes for text strings	11.14, 11.99	88.86, 88.01	0.025, 0.015

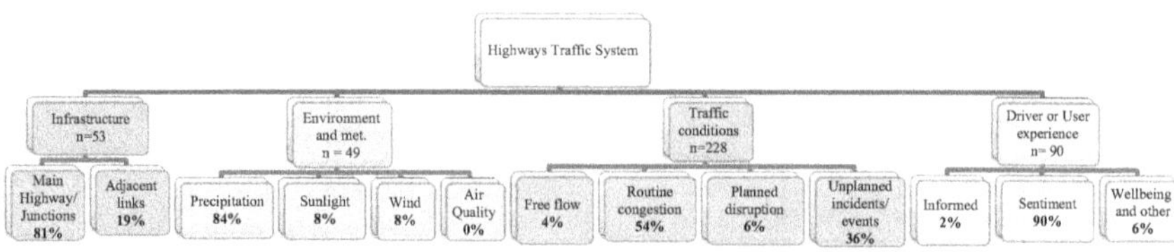

Figure 11.14 Classification outcomes by ontology class hierarchy

study area defined by the A1(M) corridor. This study used a simple method to calculate speed from straight-line distance and hence underestimates speeds, but clearly an operational service would implement a map-matching algorithm to improve the accuracy of speed estimation.

The main drawback of using social media in this application is the limited amount of relevant pairs of tweets. Only a small number of data points are available through free access to the Twitter pool, and for the majority of social media users, tweets are sent relatively infrequently and not usually when driving. Best estimates to date suggest that geotagged tweets currently only account for 1%–2% of the total number of tweets sent globally. The analysis reported here was also based on Twitter data that is freely and publicly available, which is only a portion of the total pool. The partial nature of the sample used to generate these results implies these are only indicative of the results that might be obtained with a greater sample, for example with purchased data. The free and open nature of the data offers an advantage given the potential cost of acquiring speed and travel time in geographic areas that are currently not well instrumented.

The second part of the chapter considered the textual content of the tweets, specifically, the use of text mining in order to automatically select tweets that contain relevant travel information. The best performing machine learning algorithms identified the correct class (relevant or not relevant) for 80%–85% of the tweets in our datasets. This supports the idea that a service could be built to produce a small filtered stream of relevant tweets to present to a traffic manager for further interpretation.

The algorithms were then used to classify tweets according to an ontology specific to road transport. The best performing algorithms could correctly categorise 60%–70% of these messages. It is anticipated that the performance of the algorithms in identifying and correctly categorising tweets could be improved by increasing the number of tweets used in the training and testing process. In addition, careful tuning of the parameters available for each of the classification algorithms may improve performance; however, it is important to avoid 'over-fitting' to training data. These results do not support the idea of a fully automated traffic event detection service, but do suggest that categorisation could be useful, for example to improve a service offering filtered information for further manual inspection by a traffic manager.

The issue regarding the number of relevant data available is again an issue when considering the textual content of tweets; just 40 of the 8 million tweets in Dataset C mention one of the road numbers in the study area. Of these, several of the tweets refer to a road in another country with the same number and many others are unrelated to road travel. It could be assumed that if there were a particularly extreme traffic event, such as a severe delay on a major road or motorway, there would be an increase in the number of related tweets being sent. Note that the Twitter data used in this case study accounts for less than 1% of the total number of tweets sent globally. However, when one considers the global Twitter feed, there appears to be sufficient tweets for a mobile app, Twitraffic [40], to derive information regarding road traffic. This supports the notion that access to Twitter data on a commercial basis may result in an enhanced data source for use in traffic management. The app API queries the global Twitter Firehose in order to collect tweets mentioning the number of a major road in the United Kingdom. An algorithm then scores each tweet using a system of 'scores' for words considered to be relevant to road transport (as with the weighting technique used in this paper), and only those tweets meeting a specific total score are analysed. A 'junk' filter is then used to remove irrelevant messages and those not in English or sent from within the United Kingdom. This is followed by sentiment analysis to remove positive messages regarding traffic, considered to be 'suspicious' by the API. The apparent viability of this app over a period of time demonstrates the potential for using the textual content of Twitter messages (possibly supplemented by speeds calculated using geotagged tweets) to obtain information regarding road traffic. However, the cost associated in accessing the global Twitter Firehose would need to be taken into account.

So while both aspects indicate that useful information can be automatically extracted from Twitter data, the practicality for traffic management applications ultimately depends on the business models that take account of how to access the data.

Acknowledgements

The authors would like to gratefully thank the Conference of European Directors of Roads (CEDR), the 'UNIETD' project and colleagues Ian Cornwell, (Mott MacDonald), Thomas Heinrich, (Transver), Daniel Elias (NAST consulting) and Martin Clarke (University of Leeds) for their support. All views expressed are the authors' own.

References

[1] Voigt C and Bright J M (2016) The lightweight smart city and biases in repurposed big data In: The Second International conference on human and social analytics. (HUSO 16).

[2] McNeill G, Bright J and Hale SA (2017) Estimating local commuting patterns from geolocated Twitter data', EPJ Data Science, 6:24. DOI 10.1140/epjds/s13688-017-0120-x.

[3] Harrison G, Grant-Muller SM and Hodgson FC (2019) New and emerging data forms in transportation planning and policy: Opportunities and challenges for 'Track and Trace' data. Transportation Research Part C: Emerging Technologies, 117, 102672.

[4] Swier N, Komarniczky B and Clapperton B (2015) Using geolocated Twitter traces to infer residence and mobility. Office of National Statistics GSS Methodology Series 41.

[5] Cottrill C, Gault P, Yeboah G, Nelson JD, Anable J and Budd T (2017) Tweeting transit: An examination of social media strategies for transport information management during a large event, Transportation Research Part C, 77,421–432.

[6] Fu K, Nune R and Tao JX (2015) Social media data analysis for traffic incident detection and management. Transportation Research Board 94th Annual Meeting, 2015.

[7] Mai E and Hranac R (2013) Twitter interactions as a data Source for transportation incidents. Proceedings Transportation Research Board 92nd Annual meeting, 2013.

[8] Abbasi A, Rashidi TH, Maghrebi M and Waller S.T (2015) Utilising location based social media in travel survey methods: Bringing Twitter data into the play, ACM, LBSN'15 November 3, 2015.

[9] Manca M, Boratto L, Roman VM, Martori i Gallissà O and Kaltenbrunner A (2017) Using social media to characterize urban mobility patterns: State-of-the-art survey and case-study. Online Social Networks and Media 1 (2017) 56–69.

[10] Yin J, Karimi S, Lampert A, Cameron M, Robinson B and Power R (2012) Using social media to enhance emergency situation awareness. IEEE Intelligent Systems, 2012. 27(6) p. 52–59.

[11] Cameron MA, Power R, Robinson, B and Yin J (2012) Emergency situation awareness from twitter for crisis management, in Proceedings of the 21st international conference companion no World Wide Web. 2012.

[12] Saleem HM, Al Zamal F and Ruths D (2015) Tackling the challenges of situational awareness extraction in Twitter with an adaptive approach. Humanitarian Technology: Science, Systems and Global Impact 2015, HumTech2015, Procedia Engineering 107 (2015) 301–311.

[13] Hasan S and Ukkusuri SV (2014) Urban activity pattern classification using topic models from online geo-location data, Transportation Research Part C: Emerging Technologies, Volume 44, 2014, pp. 363–381, ISSN 0968-090X, https://doi.org/10.1016/j.trc.2014.04.003.

[14] García-Palomares JC, Salas-Olmedo MH, Moya-Gómez B, Condeço-Melhorado A and Gutiérrez J (2018) City dynamics through Twitter: Relationships between land use and spatiotemporal demographics, Cities 72 (2018) 310–319.

[15] Huang A, Gallegos L and Kristina Lerman K (2017) Travel analytics: Understanding how destination choice and business clusters are connected based on social media data. Transportation Research Part C 77 (2017) 245–256.

[16] Kokkinogenis Z, Filguieras J, Carvalho S, Sarmento L and Rossetti RJF (2015) Chapter 12 – Mobility network evaluation in the user perspective: Real-time sensing of traffic information in Twitter messages. In: Advances in Artificial Transportation Systems and Simulation, 219–234. Academic Press Cambridge, MA, USA, 2015, Pages 219–234, ISBN 9780123970411, https://doi.org/10.1016/B978-0-12-397041-1.00012-1.

[17] Maghrebi M, Abbasi A, Rashidi TH. and Waller ST. (2015) Complementing travel diary surveys with Twitter data: Application of text mining techniques on activity location, type and time, in 17th International Conference on Intelligent Transportation Systems (ITSC) 2015, IEEE: Las Palmas, Spain.

[18] Chen C, Mab J, Susilo Y, Liu Y and Menglin Wange M (2016) The promises of big data and small data for travel behaviour (aka human mobility) analysis, Transportation Research Part C 68 (2016) 285–299.

[19] Statista (2019) https://www.statista.com/statistics last accessed 16/9/19.

[20] Sloan L and Morgan J (2015) Who Tweets with their location? Understanding the relationship between demographic characteristics and the use of geoservices and geotagging on Twitter. PLoS ONE 10(11):e0142209.

[21] Twitter Inc. (2014) https://dev.Twitter.com/streaming/overview. Accessed 20/11/14.

[22] Twitter Inc. (2015) https://dev.Twitter.com/rest/public/rate-limits. Accessed 11/3/15.

[23] Gnip Inc. (2013) http://gnip.com/products. Accessed 20/11/14.

[24] DataShift (2014) http://datasift.com/platform/datasources. Accessed 20/11/14.

[25] Morstatter F, Pfeffer J, Liu H and Carley K.M (2013) Is the sample good enough? Comparing data from Twitter's Streaming API with Twitter's Firehose. Proceedings of the International Conference on Weblogs and Social Media, Boston, USA.

[26] Gelernter J and Mushegian N (2011) Geo-parsing messages from microtext. Transactions in GIS, 15(6), 753–773.

[27] Leetaru K, Wang S, Cao G and Padmanabhan A (2013) Mapping the global Twitter heartbeat: The geography of Twitter. First Monday, 18(5).

[28] Georgia Tech Research Corporation (2012) Years in review report https://gtri.gatech.edu/public/prod/2017-07/GTRI_2012_AnnualReport_r8. pdf accessed 20/12/2019

[29] Ordnance Survey (2014) Strategi. http://www.ordnancesurvey.co.uk/business-and-government/products/strategi.html. Accessed 12/12/14.

[30] ArcGIS (2013). Buffer (Analysis). http://resources.arcgis.com/en/help/main/10.1/index.html#//000800000019000000. Accessed 12/12/14.

[31] ArcGIS (2012). Intersect (Analysis). http://help.arcgis.com/en/arcgisdesktop/10.0/help/index.html#//00080000000p000000. Accessed 12/12/14.

[32] Taylor NB (2013) HATRIS JTDB Data GuideV7. Transport Research Laboratory.

[33] Grant-Muller SM, Galtzur A, Minkov E, Kuflik T, Nocera S and Shoor I (2015) Integrating transport data collection through social media sources: methods, challenges and opportunities. IET Intelligent Transport Systems. doi: 10.1049/iet-its.2013.0214

[34] Grosenick S (2012) Real-time traffic prediction improvement through semantic mining of social networks. Thesis (Master's) University of Washington, 2012 URI: http://hdl.handle.net/1773/20911

[35] Yang WD and Wang T (2012) The fusion model of intelligent transportation systems based on the urban traffic ontology. Physics Procedia, 2012, 25, pp. 917–923

[36] Noy NF and McGuinness DL (2001) Ontology development 101: A guide to creating your first ontology.

[37] Hall M, Frank E, Holmes G, Pfahringer B, Reutemann P and Witten IH (2009) The WEKA data mining software: An update. SIGKDD Explorations, 11(1).

[38] Sokolova M and Lapalme G (2009) A systematic analysis of performance measures for classification tasks. Information Processing and Management, 45(4), 427–437.

[39] Cohen J (1960). A coefficient of agreement for nominal scales. Educational and Psychological Measurement, 20(1), 37–46.

[40] Telent Technology Services Ltd. (2013) http://api.twitraffic.co.uk/how-twitraffic-works/. Accessed 5/3/15.

Chapter 12
Social media services and business models

Frances Hodgson[1]

12.1 Introduction

This chapter considers the business models and availability of services to support the uptake of social media data in traffic management. The focus is on Twitter as an example of a commonly available type of social media data. As we have seen in previous chapters, social media data, such as Twitter when accompanied by geographical information may be useful for estimating traffic characteristics. Information on traffic flow, delays, infrastructure, and environment-related traffic issues may be obtained from studying the textual content of the messages.

The chapter provides an assessment of the potential of these social media data with relevance to the needs of road administrations, particularly in the context of traffic management. Applying a widely available business model canvas we are able to explore two scenarios: 'outsourced' and 'in-house' social media data processing and analysis. This analysis identified areas for potential growth within each scenario: including expertise and skillsets, key partnerships with social media data analysts and customer segmentation within the roads authorities. In addition the analysis characterised areas of weakness in the value proposition and identified areas of investment to direct and exploit the potential of social media data. Finally the analysis identified issues of public acceptability, governance and monetisation as significant issues associated with new vulnerabilities of using future data.

12.2 Background and context: ecosystems and added value

The use of social media data within traffic management is a relatively new area of application [1–5]. However, it is clear that to exploit these new forms of data one of the key collaborations is that of road authorities and transport operators and managers and social media analysts. Our consideration of the business models for using Twitter data in traffic information shows the development of new skills, partnerships and the identification of new value. The aim of this section is to explore issues in the business

[1]Institute for Transport Studies, University of Leeds, Leeds, UK

models that may accompany the use of Twitter data in the generation of travel time information.

It is clear that there is a growing value ecosystem [6,7], around social media harvesting and analytics, with a growing set of available services available, both internationally and nationally. We have identified that many of the social media analytic enterprises have 'off-the-shelf' business analysis tools available that will provide specific analyses that focus on the message, e.g., the tweet; the user, e.g., the poster of the tweet; the location of the messenger; and the content of the message; the linkages across social media platforms; and identifying where tweets feature in a value chain of purchasing actions. However, there is less evidence on the expertise of applying social media analytics in traffic management or travel time information generation.

Two business model scenarios are considered here, namely (i) extensive outsourcing, in which it is assumed that the social media search, identification and analysis will be done by social media analysts who will provide input to road authorities; and (ii) in-house development of resource and expertise in which it is assumed that the road authorities will perform the search, identification and analysis themselves, and use that as input to traffic management solutions. In the discussion we also consider a variation of these models, i.e. a co-creation business model whereby the roads authority and others (such as the travelling public) work collaboratively in generating and analysing the data.

The business models are highly dependent on assumptions about the availability of social media data, and the extent to which it is 'free'. Social media companies own the social media data and may choose in the future to charge app developers or users for access to that data. In addition, users themselves may in the future become more aware of the commercial value and the potential for monetisation of their own data.

In addition as the specification of the tweet changes, for example, changes to the collection or specification of the meta data, or a shift in the balance of use of graphical content (gifs, photographs) as opposed to textual content, the value of the data can change. This has significant implications for future use and indicates some of the vulnerabilities inherent in making use of non-specific datasets in contrast with those that have been deliberately constructed for a specified transport need.

It is also clear that the immediacy of traffic management and the necessity to apply Twitter data as it is being generated is not an area where many social media data analyst enterprises have extensive experience.

12.3 Alternative models for access to social media services

A business model describes the underlying principle of how an organisation constructs, generates, conveys and captures value [8]. Business models can vary but commonly nine fundamental elements are used to build a picture of the business rationale. These elements cover four main areas of business including: the offer, customers, infrastructure and financial viability. A UK practitioner review [9] concludes that one of the strategic longer term activities should be the development of new data-driven business models. This would assist with the exploitation of transport–related data which is generally considered to be a game-changer for transport management. Key considerations are: data, value proposition and adaption.

Here, we are interested in the value proposition that is offered by social media data analytics for the purposes of traffic information. In this regard we are interested in the value proposition from the position of the *customer* (i.e. the road authorities, who act as consumers of the service provided) and its usefulness. Most business models are primarily focussed on the *selling* of a product and as such they look at the value proposition from the perspective of the *provider*, i.e. the enterprise that is selling a product, in this case a social media analysis service.

However, it should be noted that the position of *customer* versus *provider* is not straightforward because there is the possibility of value being added by the road agencies. This comes from the potential to change and add to the product and service in the data analytic process, for example by combining datasets such as automatic number plate recognition with the social media analysis. The implications are that there is a complex value proposition, cost structure and revenue streams. The significance of combinatorial approaches is being recognised by Governmental agencies by the Office of National Statistics and the Catapult initiative, 'Web-connected fixed and mobile sensors, plus crowd-sourcing are the emerging transport data collection mechanisms. In particular, personal/vehicle location data are the likely 'game-changers' for intelligent mobility' [9].

Further added value can arise from any processes of data validation which road agencies develop to work with social media analysis. Data validation has been identified as a key problem of social media data utility and usefulness [9–11], and may be necessary to identify and weed out malicious content. Developing protocols to validate social media data within the field of traffic management would add value and advance the field of social media analytics.

Looking to the future, there may be new sources of data which can also be used in tandem with social media data which impact on the business model and the value proposition. A specific example is that of drone technology and the possibilities for use in both routine provision of data on traffic and in situations of immediate incident management. Future data sources and combinational approaches both have an impact on the business model.

Additional value may also be present in the way that data can affect pricing policy elsewhere within the transport network in order to better manage the network. As an example, Singapore uses data on traffic volume and demand to determine road charging costs. Future implementations may allow demand driven parking pricing policies or car club hire process that may fluctuate in accordance with demand. Alternatively a revenue stream may be to sell information derived from social media analysis to other users, such as public transport operators to develop demand driven pricing.

A further example is the use of a tailor-made mobile phone app specifically for crowd-sourcing information on movement. This can be used to inform decisions for traffic management as well as *post hoc* evaluation of the traffic management strategy. 'Personal/vehicle location data from crowd-sourced and mobile connected sensor networks are not yet being fully exploited. The greater level of insight they provide into the behaviour of people and movement of things (e.g. vehicles and goods) means they are expected to be a cornerstone on which intelligent mobility services will be developed and delivered in the future.' [9]

The multiplicities of data sources and analyses together with the potential avenues of future development, coupled with a rapidly developing, rapidly expanding international business domain add to the complexity of the business model development. This complexity is exacerbated by the level of future uncertainty and creative innovation in the social media analytics sector. Whether the business model is based on outsourcing or in-house capabilities, the data is the most significant element.

12.4 Business model canvas

The business model used here (Figure 12.1) has all of the common elements disaggregated further into 9 different elements as follows:

- *Product*: value proposition
- *Customer interface*: target customer; distribution channel; relationship
- *Infrastructure management*: key activities; key resources; partnership
- *Financial aspects*: cost structure; revenue model.

This model will be applied to each of two business model scenarios: 1. extensive outsourcing and 2. in-house. In this investigation, each of the main four business model elements will be examined to identify the significant factors in each scenario. For each of the business model scenarios a set of key issues is raised. The issues are derived directly from the business model canvas allowing us to assess each potential business model scenario, and are given in Table 12.1.

The criteria developed in the context of generating travel time information will provide some input into determining the value proposition in each of the business models. This will cover the standard and optional functionalities, the ability to provide 'tailored' solutions, the scope of social media sources used, the ability to look in more detail at message strings, data ownership, data storage, retrieval, cost packages etc. The exercise will contribute to an understanding of the value of the raw and processed data available in the market.

12.5 Outsourcing/Proprietary services – accommodation of transport as a specialism by typical proprietary services

In this scenario (Table 12.2), the assumption is that the utilisation of social media data in traffic management by roads authorities has been outsourced to social media analysts. This will involve the use of social media analyst enterprises that are specialist in sectors such as marketing. As a result, we assume that outsourcing is to social media analysts that do not have experience of traffic management. It is also

<table>
<tr>
<td rowspan="2">Key partners
Some activities and resources are outsourced and acquired from outside the enterprise.
Alliances should contribute to the optimisation of the business model.
Partnerships fall into four types:
• Non-competitor strategic alliance;
• Competitor strategic alliance;
• New business development through joint ventures;
• Buyer-supplier alliances to assure reliable supplies/sales
Activities in this section of the business model canvas require identification of:
Key partners, key suppliers; key resources and activities required from them.</td>
<td>Key activities
Include production, problem-solving or platform.
Required to make the business model work.</td>
<td rowspan="2">Value propositions
Customer satisfaction and problem-solving.
Refers to the combination of products and services that creates value for each customer segment.
What is being satisfied or solved with the product or service?
Recent advances are customisation and customer co-creation.</td>
<td>Customer relationships
The type of relationships a company establishes with specific customer segments.
There are a number of different types of relationship including:
Automated, self-service, co-creation, communities, personal assistance, dedicated personal assistance.</td>
<td rowspan="2">Customer segments
The different groups of people or organisations an enterprise aims to reach and serve.
For whom are we creating value?
Segmentation of the customer base is determined by a number of axes including:
Different customer needs, different distribution channels, relationship variation, differing profitability, willingness to pay for parts of the offer.</td>
</tr>
<tr>
<td>Key resources
These are the most important assets required to make the business model work and provide value propositions and ultimately the revenue streams. Key resources can be physical, financial, intellectual or human.
Key resources can be owned, leased by the company or acquired through key partners.</td>
<td>Channels
Communication, distribution and sales channels for delivering value propositions to each customer segment.</td>
</tr>
<tr>
<td colspan="2">Cost structures
The cost structure is the result of the business model elements.
This structure can be cost driven or value driven.</td>
<td colspan="3">Revenue streams
Revenue streams are the result of a successful offer of a value proposition to a customer segment.
It refers to the income generated from each Customer segment. A business model distinguishes multiple revenue streams. Each revenue stream has a pricing mechanism. There are a number of different pricing mechanisms, e.g., fixed list prices, bargaining, auctioning, market dependent, volume dependent, or yield management.
There are two main types of revenue stream: transaction revenues associated with one-off customer payments, and recurring revenues, which are usually the result of on going payments for the delivery of a Value Proposition or the provision of post-purchase customer support.
There are several types of revenue payment including: asset sale, usage fee, subscription fee(s), lending/renting/leasing, licensing, brokerage fee, a dvertising. Each may have a different pricing mechanism which may be fixed or 'dynamic'.</td>
</tr>
</table>

Figure 12.1 Business model canvas (Source: Osterwalder and Pigneur [8])

Table 12.1 The different business model elements and key issues for each element

	Business model issues for each element
Product	Value proposition • What combination of products and services creates value for each customer segment? • What is being satisfied or solved with the product or service? • What role, if any, for customisation and customer co-creation?
User/customer interface	Target customer • Who are the different groups of people or organisations an enterprise aims to reach and serve? • For whom are we creating value? • How can the customer base be segmented? the customer base is determined by a number of axes including: Different customer needs, Different distribution channels; Relationship variation; Differing profitabilities; Willingness to pay for parts of the offer Distribution channel • What are the communication, distribution and sales channels for delivering value proposition to each customer segment? Customer relationship • What types of relationships is required for each customer segment: automated; self-service; co-creation; communities; personal assistance; dedicated personal assistance?
Infrastructure management	Key activities • What are the production, problem solving or platform activities? Key resources • What are the physical, financial, and intellectual or human resources required? • Are they owned, leased by the company or acquired through key partners? Partnership What partnerships are necessary: suppliers, partners, resources, activities? What type of partnerships are necessary: • Non-competitor strategic alliance; • Competitors strategic alliance; • New business development through joint ventures • Buyer–supplier alliances to assure reliable supplies/sales?
Financial aspects	Cost structure • Is the cost structure cost driven or value driven? Revenue model • What are the income streams and are they transaction or recurring? • Is it possible to distinguish pricing mechanism e.g., fixed list prices, bargaining, auctioning, market dependent, volume dependent, or yield management? • Is it possible to distinguish which types of revenue payment including: • Asset sale, • Usage fee, • Subscription fee(s), • Lending/renting/leasing, • Licensing, • brokerage fee, • advertising? each may have a different pricing mechanism which may be fixed or 'dynamic'.

Table 12.2 *BMC: the outsourcing business model scenario*

Key partners	Key activities	Value propositions	Customer relationships	Customer segments
Key partnership between social media analysts and road authorities, particularly if the latter are providing the transport expertise. Partnership could be in the form of a non-competitor buyer–supplier alliance. This would ensure exchange of expertise and increase the value proposition for each customer segment, including the efficiency and ergonomics of the software for the road authority users.	The principle activities associated with infrastructure management include: • Development of a knowledge management platform to use input data from social media analysis; • Development of a knowledge management platform for lexicon input to sentiment analysis; • Development of a knowledge management platform for location-based analysis; • Evaluation process.	Reliant on the quality and volume of social media content generated that is relevant to the road network. Impacted by any systemic bias in social media engagement and content, for example by age group. Reliant on the lexion skillsets required to identify the relevant social media messages and to perform sentiment analyses of the content. Reliant on analysis of content relevant to specific locations. Reliant on skillsets to determine the significance of the timing in interaction with the location on the network, for example, is it upstream or downstream of a 'hot spot' area or incident, and impact on demand. Reliant on enabling roads authorities to improve their understanding of network performance.	Influenced by appropriate customer segmentation within road authorities and the distribution channel. For example, an automated warning service of traffic at hot spots on the network using content and geotag analysis of social media. Reliability and accuracy in social media data and the frequency patterns for utilisation. High risks and costs associated with error and need for trained personnel to mitigate errors. In the short term, a 'personal assistance' relationship such as a 'bespoke' help desk role for technical expertise. The drawback is high investment.	Principal customer segment: roads authorities engaged in the traffic management of inter-urban road network. Subsequent customer segmentation dependent on specification of traffic management problems. Segmentation granularity dependent on the criteria/axes of customer need. Further influenced by the mix of relevant social media content generators such as car traveller, road haulier, bus and other p.t. operator, bus and other p.t. traveller. Requires significant inside knowledge from the roads authorities.
	Key resources Road authorities have to provide the transport expertise (specifically road management expertise) to be applied with social media analytical techniques.		Channels Social media analyst enterprises expertise in using social media for communication appropriate to each customer segment within the roads authorities. Immediacy of	

(Continues)

Table 12.2 (Continued)

<table>
<tr><td></td><td>Skills needed in:
• lexicon development for sentiment analysis;
• spatial network analysis;
• network performance;
• understanding of social media data, platforms and analysis to be able to establish meaningful dialogue with the social media analysts.</td><td></td><td>communication is significant to customer segments that need immediate real-time information for traveller and network congestion management.
Immediacy and appropriate time-lags in analysis of social media data to the roads authorities is significant.
Lack of reliability in the speed of communication channel is a significant consideration.</td></tr>
<tr><td colspan="2">Cost structures
The financial aspects include the cost structure and the revenue model.
Decisions would have to be taken as to whether the commissioning of the social media analysts and the subsequent service was cost driven or value driven.
Value driven would mean that the extent to which the aims associated with the aims of the value proposition was the driver behind the expenditure on the service.</td><td colspan="2">Revenue streams
Cost share or revenue share from the added value of taking what is currently ‘free’ data and creating value from that free data.
Additional opportunities from the creation of datasets of ‘clean’ or processed data and the sale of that to other concerns.
Revenue streams designed around subscription service for guaranteed early information.
Other revenue streams include renting out advertising space.
Revenue streams could be derived from both recurring and transaction payments.
Key partnerships with other forms of transport provision (for example parking) could provide revenue streams through the implementation of ‘preferred customer’ information on parking availability or demand driven parking pricing.
Revenue payment from the road authorities to the social media analysts for the services including: subscription fees, usage fee, and lending/renting/leasing.
Pricing mechanisms including volume dependent, yield management, fixed or dynamic.</td></tr>
</table>

reasonable to assume that they do not have extensive experience within the transport sector. This will mean a much more intensive partnership in order to supply, aggregate and interrogate the social media data for the aim of traffic management. The business model analysis is done from the viewpoint of the road authorities.

12.5.1 Outsourcing scenario: user/customer interface

The definition of the customer segment for the business model is the first step. In this study the principal customer segment is the roads authorities engaged in the traffic management of inter-urban road networks.

Further study is required to identify and define subsidiary segmentation of the principal customer and to define the benefits of such segmentation. This will involve extensive knowledge of the specificities of the traffic management problems on the road network in order to develop customer segmentation on the basis of customer need. The level of granularity for this segmentation will require further study, in order for example, to establish the criteria/axes for determining customer need. For example, segmentation may depend on what the roads authorities are ultimately trying to achieve in the management of specific sections of the road highway and in tackling problems such as persistent morning peak hour congestion for specific off ramps. This could be further influenced by the mix of relevant Twitter content generators such as car traveller, road haulier, bus and other public transport operator, bus and other public transport traveller.

An extensive outsourcing scenario can only identify the customer segments and deliver for each of the segments needs with significant insider knowledge from the roads authorities. In the outsourcing business model scenario, roads authorities would also have to provide extensive transport expertise into the development of the segmentation of the relevant social media generators.

There is uncertainty around whether the transport expertise would be something that would be acquired by the social media analysts' enterprise through hiring relevant personnel or if it could be provided by close working partnerships with the roads authorities. These are significant areas of risks and significant areas of costs and require further investigation.

12.5.1.1 Distribution channel

One of the strengths of the extensive outsourcing business model scenario is that the social media analyst enterprises bring considerable levels of social media expertise. These could be used to develop a multiple set of communication channels across a range of platforms using multiple devices for the distribution of the value proposition appropriate to each customer segment within the roads authorities. A principal consideration with the distribution channel between the social media analyst enterprise and the roads authorities will be the immediacy of the communication. This will require consideration of the customer segment needs particularly if segmentation has been on the basis of different areas and problems of the road network. To make outsourcing succeed, establishing how immediate the analysis has to be and the time lag in the output of the analysis to the roads authorities will be an important issue. Social media analysts will have to show that output from analysis of Twitter data can be communicated to the roads authorities to provide positive input to the real-time management of the road highway. Further study is required to establish the reliability and the risks associated with any lack of reliability in the speed of that communication channel. This is separate to the consideration of whether the right analysis has been done at the correct critical stage and time in the network management to impact positively and significantly on the road management solutions.

12.5.1.2 Customer relationship

The relationship between each customer segment, that is, the roads authorities and segments within the roads authorities, and the social media analysis enterprises can take a variety of forms including: self-service, automated, personal assistance, co-creation and community working. It is also clear that the form of the customer relationship is influenced by decisions about the most appropriate customer segmentation within road authorities and the distribution channel. For example, a fully automated relationship providing a warning service of traffic build-up at specific hot spots on the road network using analysis of social media data (covering both content and location geotag), would require a level of clear cut specification and understanding of the reliability of the message content and location that can only come from observation and experimentation.

The risks of getting it wrong and subsequent impacts on roads authorities, plus the need to have trained personnel available to find a solution in that event may increase costs.

In the short term, it might be most appropriate to establish a relationship of personal assistance to provide technical expertise. The drawback is obviously the investment required to develop this relationship. It would open the possibility that the social media analysts work with the roads authorities to design and establish automated services in the mid-term. This would provide the flexibility required to ensure that the organisational form, customer segments within the roads authorities and the distribution channels were appropriate to the ultimate aims of the roads authorities in managing the roads, the unreliability of the data sources and the need to test in the operational environment.

12.5.2 Outsourcing scenario: value proposition

In order to establish the value proposition it is necessary to identify what is being satisfied or solved with the product or service for each customer segment. Typically roads authorities overall aim can be described as improve, maintain and manage the road network in an efficient way.

The value proposition for outsourcing is reliant on the quality and volume of social media content generated about the road network. It is also influenced by any systemic biases that may be present in the distribution of social media engagement and content (see previous chapter). In addition it is reliant on the quality of the skillsets available to create the lexicon required to identify relevant social media messages and to perform sentiment analyses of the content. The outsourcing business model scenario is also reliant on the quality of the skillsets to determine the significance of the timing of a social media message, and the significance of the timing in interaction with the location on the network, for example, whether it is upstream or downstream of a 'hot spot' area or incident, and whether it impacts on surges in levels of demand.

12.5.3 Outsourcing scenario: infrastructure management

12.5.3.1 Key activities

Infrastructure management in a business model scenario refers to production, problem solving and platform activities. In the outsourcing business model, the principle activities associated with infrastructure management include:

- Development of knowledge management platform to use input data from Twitter analysis;
- Development of knowledge management platform for lexicon input to sentiment analysis;
- Development of knowledge management platform for location-based analysis;
- Evaluation process.

12.5.3.2 Key resources

The outsourcing scenario requires considerable intellectual and human resources to be used in collaboration with the social media analysts. The road authorities have to provide the transport expertise (specifically road management expertise) to be applied with social media analytical techniques. Early sections demonstrated the skills underlying lexicon development for sentiment analysis and the analysis of location-based geotagged tweets. In addition there will be a need for specialist understanding of social media data, platforms and analysis to be able to establish meaningful dialogue with the social media analysts.

12.5.3.3 Partnership

Clearly the extensive outsourcing business model scenario requires a key partnership between social media analysts and road authorities particularly if the latter are providing the transport expertise. One form for this partnership would be as part of a non-competitor buyer–supplier alliance to ensure exchange of expertise and to increase the value proposition for each customer segment, including the efficiency and ergonomics of the program for the road authority users.

12.5.4 Outsourcing scenario: financial aspects

12.5.4.1 Cost structure

The financial aspects include the cost structure and the revenue model. Decisions would have to be taken as to whether the commissioning of the social media analysts and the subsequent service was cost driven or value driven. Value driven would mean that the extent to which the aims associated with the aims of the value proposition was the driver behind the expenditure on the service.

12.5.4.2 Revenue model

Within the outsourcing business model scenario it is possible to perceive the outsourcing of the work as merely a cost. Increasingly however commercial enterprises are exploring opportunities to establish a cost share or revenue share within the partnership. The opportunity for this arises from the added value of taking what is currently 'free' data and creating value from that free data. Thus there may be opportunities from the creation of datasets of 'clean' or processed data and the sale of that to other concerns. These revenue streams could be designed around the willingness of each customer segment to pay for the value proposition, for example, subscription service for guaranteed earlier information. Other revenue streams could include renting out advertising space. Revenue streams could be derived from both recurring and transaction payments.

In addition it is possible to see that key partnerships with other forms of transport provision (for example parking) could also provide revenue streams through the implementation of preferred customer information on parking availability or demand driven parking pricing. Obviously potential income streams may be more or less publically acceptable.

In addition it is possible to see a number of different types of revenue payment from the road authorities to the social media analysts for the services including: subscription fees, usage fee, and lending/renting/leasing. In addition there may be different pricing mechanisms including volume dependent, yield management, fixed or dynamic.

12.5.5 In-house provision

In this scenario (Table 12.3), the assumption is that the utilisation of social media data in traffic management by road authorities has been carried out in-house. This will involve the use of social media analysis software within road authorities. This will necessarily involve decisions about the amount and range of product and customer support. For the purposes of this scenario we will assume that use will be made of a mixture of 'off the shelf' and bespoke software by road authorities' staff to provide information to enhance the management of road traffic. For consistency the business model analysis is carried out from the viewpoint of the roads authorities.

12.6 In-house scenario: user/customer interface

In this scenario, the main segment is the same as in the outsourcing scenario, that is the roads authorities engaged in the traffic management of inter-urban road network. However further segmentation may occur within the roads authorities. This further segmentation may be based on skillsets, or on other criteria defined by the organisational form and management processes, for example there may be data processes, data analysts and road management and solution providers or by regions. The change in the scenario of who analyses the social media dataset and develops solutions does not necessarily impact on any end-user segmentation that road authorities may decide is necessary. In this scenario however, a further segment is the road authorities using the social media data within the task of managing the road network.

In the in-house business model scenario it is expected that road authorities would make use of 'off the shelf' software and services. The in-house scenario means that the transport expertise (necessary for customer segmentation

Table 12.3 BMC: the in-house business model scenario

Key partners	Key activities	Value propositions	Customer relationships	Customer segments
Social media analysts and transport experts. One form of partnership would be as non-competitor buyer–supplier alliance to ensure exchange of expertise and to increase the value proposition for the customer segment, including the efficiency and ergonomics of the software for the road authority users. Would require social media analysts to provide training in the use of social media software, rather than providing expertise and output in the form of data analysis. Less intense contact from social media analysts.	Key activities can become part of the everyday tasks in road network management. Training schedule: software comprehension and use. Series of activities related to the production, problem solving and platform for the use of social media data, including development of knowledge management platforms: – for lexicon input to sentiment analysis; – to use input data from social media analysis; – for location-based analysis; – search pattern templates; – knowledge management processes to search for, identify and harvest social media data; – Evaluation process. Combining social media analysis with other traffic management data.	The principal elements are the improving, maintaining and managing the road network in an efficient way. Time-efficiencies: quick availability of transport expertise makes it more probable that the traveller end-user will be satisfied with the service or outputs. This becomes increasingly critical if there is segmentation *within* the roads authorities. Value proposition reliant on road authorities: (i) developing access to appropriate social media data; and (ii) determining the appropriate parameters for social media data selection based on temporal and spatial criteria. Can ensure that solutions are consistent with policy plans and principles which are of particular value at local and smaller spatial scales. An obstacle is the software responsiveness to future information needs.	Road agencies can maintain existing types of relationship and the benefits of that, including smooth communication channels between any segments within the roads authorities. Traffic management may require segmentation by spatial scale, so being able to communicate immediately within the roads authorities at the critical time in road network management is a significant feature for effective implementations. This is one significant advantage of the in-house scenario over the outsourcing scenario.	Roads authorities engaged in the traffic management of inter-urban road network. Further segmentation within the roads authorities based on skillsets, or management processes, for example by regions. Transport expertise necessary for customer segmentation appropriate for the spatial scale and road network is easier to guarantee. Risks mostly arise because the 'off the shelf' programmes for data analysis have largely been developed for the marketing sector. As a result pre-programmed customer segmentation is often not as sensitive to location-based analyses, which form the basis for traffic management.
	Key resources Requires considerable financial, intellectual and human resources. Increasing human capital in this area could deliver a clear business benefit and advantage within the market place both		**Channels** Communication channels between the social media analysts and the roads authorities. The distribution channel would be limited to customer support. It is expected that many of the social media analysts would	

(Continues)

Table 12.3 (Continued)

	for the road authorities and the individual staff members. Human capital investment in the right area could result in market leader opportunities.		offer tutorials and that these could be delivered within the roads authorities to provide skillset development. Communication channels would include social media and also orthodox means. Appropriate team to team or person to person support. Profile of support outside normal working hours, over the contract or software lifetime.
Cost structures Road authorities have control over the cost structure and the decision to be cost driven or value driven. More control over costs and the avoidance of creeping costs. More opportunity to ensure a close relationship between the cost structure and the value proposition.		Revenue streams Potential revenue streams: (i) the provision of information; or (ii) with the provision of solutions for implementation, for example, subscription service for guaranteed earlier information by customer segment. Other revenue streams could include renting out advertising space. Revenue streams could be derived from both recurring and transaction payment. Opportunity to develop innovative partnerships with other service providers within and outwit the transport sector e.g., crowd-sourcing data through loyalty card systems-based incentives.	

appropriate for the spatial scale and road network) is much easier to guarantee. In addition, it is equally feasible that the expertise necessary to be able to determine the value proposition for each customer segment is easier to identify and to utilise in the social media data analysis and subsequent road and traffic management implementations and solutions. However, there are significant risks associated with this scenario. The risks mostly arise because the 'off the shelf' programmes for data analysis have largely been developed for the marketing sector. As a result pre-programmed customer segmentation is often not as sensitive to location-based analyses, which are very significant for traffic management.

There are additional needs dependent on the extent to which the software-based approaches need to be compatible with current platforms, models and other software used by the roads authorities. Further work is required to determine the necessary types and extent of technological compatibility needed by the roads authorities. In addition, there is further work around the ergonomics of the process of using the social media analysis software.

12.6.1 Distribution channel

The distribution channel refers to the communication channels between the social media analysts and the roads authorities. In the in-house business model, the distribution channel would be limited to customer support. There would be attendant issues of costs, communication vs. training, team-based response or dedicated individual specialists, and non-social working hours availability, the volume of support and calls required and how that would be profiled over the lifetime of the contract and or software. Together these resources would be utilised to develop the appropriate range of channels across a range of social media platforms and devices for message distribution and engagement.

12.6.2 Customer relationship

The in-house scenario provides the opportunity for road agencies to be able to maintain existing types of relationship and the benefits of that, including smooth communication channels between any segments within the roads authorities. Traffic management may require segmentation by specific parts of a road network, so being able to communicate immediately within the roads authorities at the critical time in road network management is a significant feature for effective implementations. This is one significant advantage of in-house scenario over the outsourcing scenario.

12.6.3 In-house scenario: value proposition

The principal elements of the value proposition are the same as in the outsourcing scenario: namely the overall aim of improving, maintaining and managing the road network in an efficient way. The difference between the outsourcing and in-house scenarios is in the time delay in bringing transport expertise to analyse social media content and 'geotag' data and to build solutions. It is expected that there are time-efficiencies associated with applying transport and road management expertise to analysis and solution development. The in-house scenario has a greater probability of being able to leverage transport expertise immediately and to be able to match more quickly and efficiently the appropriate expertise for the problem. This immediacy and timeliness is a fundamental element of the value proposition for roads authorities. It becomes increasingly critical if there is segmentation within the roads authorities. The potential for transport expertise to be quickly available makes it more probable that the relevant traveller end-user will be satisfied with the product.

It is expected that road authorities would be able to develop access to appropriate Twitter data. It is also expected that the roads authorities would be able to determine the appropriate parameters to be able to select Twitter data based on the temporal and spatial criteria, for example identification of time appropriate tweets for the appropriate spatial area.

The potential for immediate transport expertise in the social media analytics and solution development is an obvious advantage for the in-house scenario. It is this expertise that has the most significant impact on ensuring the value proposition for the customer segments. However, the biggest obstacle will be the extent to which the software is bespoke and flexible enough to grow and respond to developing future information needs.

12.6.4 In-house scenario: infrastructure management

12.6.4.1 Key activities

There are a series of key activities required to establish the full cycle of generating social media datasets and using them to establish the state of a road network, then to develop solutions for implementation. The in-house scenario provides the opportunity for these key activities to become part of the regular everyday task of road network management.

These key activities will also have to include a meaningful partnership between transport expertise and social media data analysts in order to ensure that training in software use and comprehension is up to date.

As in the outsourcing scenario there are a series of activities related to the production, problem solving and platform for the use of social media data. The in-house business model scenario identified the following activities:

- Development of search pattern templates;
- Development of knowledge management platform for lexicon input to sentiment analysis;
- Development of knowledge management processes to search for, identify and harvest social media data;
- Development of knowledge management platform to use input data from social media analysis;
- Development of knowledge management platform for location-based analysis;
- Evaluation process.

In the in-house scenario, there are additional activities associated with the actual search and identification and interrogation of the social media data. In addition there is the opportunity for more flexibility to work in real-time and faster to match immediate real-time knowledge of the state of the road network with the patterns of social media postings.

12.6.4.2 Key resources

The development of the in-house scenario requires considerable financial, intellectual and human resources. However, there is a clear opportunity to increase human capital in this area to deliver a business benefit and advantage within the market place both for the road authorities and the individual staff members. Human capital investment in the right area could result in market leader opportunities.

12.6.4.3 Partnership

The in-house scenario requires key partnerships between social media analysts and transport expertise. However, the main difference between the scenarios is the length of the interaction and support. The in-house scenario would require social media analysts to provide training in the use of social media software rather than provide expertise and output in the form of data analysis. In addition, there may be differences in the duration of the contract. The profile of the duration and intensity of the contact between the road authorities and the social media analysts may vary by business model. It is expected that the intensity of contact would be less throughout the duration of the contract with the in-house scenario.

12.6.5 In-house scenario: financial aspects

12.6.5.1 Cost structure

The financial aspects include the cost structure and the revenue model. With the in-house scenario road authorities would have control over the cost structure and the decision to be cost driven or value driven. Arguably the in-house scenario allows more control over costs and the avoidance of additional costs creeping in, although neither business model scenario can mitigate additional unforeseeable costs. In addition the in-house scenario provides more opportunity to ensure a close relationship between the cost structure and the value proposition.

12.6.5.2 Revenue model

For both the in-house and the outsourcing business model scenario it is possible to see a series of potential revenue streams, all of which require investment. The in-house scenario provides road agencies with the opportunity to develop those revenue streams for their own profit or to improve the value proposition for end-user travellers. Potential revenue streams can be identified with (i) the provision of information; or (ii) with the provision of solutions for implementation. In the in-house business model scenario, revenue streams could be designed around the willingness of each customer segment to pay for different aspects of the value proposition, for example, subscription service for guaranteed earlier information. Other revenue streams could include renting out advertising space. Revenue streams could be derived from both recurring and transaction payment.

In addition the in-house scenario provides the opportunity to develop innovative and creative partnerships with other service providers both within and outside the transport sector. One example is the potential for collaboration with loyalty card systems providing incentives for crowd sourcing data.

12.7 Conclusions

This chapter has focused on the business models associated with introducing Twitter data into traffic informatics. This involved the application of a business model analytical approach for the application of Twitter data in traffic informatics in two scenarios. The first scenario assumed the social media data processing and analysis was 'outsourced' and the second assumed that the analysis was performed 'in-house' by transport system managers. The business model addressed the use of both message content and meta data in application of traffic informatics.

The business model analysis is in an environment where social media analytics is changing rapidly and that future specialism in the transport sector is an area of potential growth. The lack of expertise in the area of social media within the transport sector means that key partner analysis is a significant step in future collaborations. There are opportunities for growth in applied analytic services, training and customer support services.

It is clear that one of the cornerstones of future development of social media data in the management of traffic is the skillsets that need to be developed by the road agencies. The skillsets required are considerable and vary according to the different business model scenarios. The skillsets required for the in-house business model are more extensive and also involve considerable specific skillsets around the use of different software such as Hootsuite or Conversocial. The business model scenarios all involve the development of key partnerships. The nature of the partnerships with the social media data analysts varies with each scenario. The relationship with the greatest intensity is that of the outsourcing business model. In this analysis, the main customer segment is stated as roads authorities but it is clear that there may be further customer segmentation within the roads authorities which may develop over time as needs change and the use of the social media analysis is refined.

Analysis of different business model scenarios shows that the value proposition with regard to utilising the Twitter message content needs a lot of investment if there are many variations in the articulation of traffic problems from those who tweet. There is an obvious opportunity to the value proposition through improved branding specification for incidents or road types and locations. However, there is also a requirement to anticipate changes in message content and the increased use of graphics and the attendant need to have graphic analytics.

Analysis of the business models has also shown that future data availability and data alignment is a significant aspect for the development of all business model scenarios. There are three main aspects of this issue. The first is the cost to the road authorities of obtaining the data from the social media companies. For some forms of the data this is free, however, this is not uniformly the case and may not be the case in the future. In addition it is not clear that the specification of the data will remain the same. Future changes could include alterations of the meta data, restrictions to the allowable content, or restrictions on the collection and processing of the data resulting in non-collection or partial collection of each tweet. The final issue is that of public acceptability. It is not clear whether people will continue to be happy about the provision of free data indefinitely into the future. Public attitudes to marketisation of their data, or the altruist 'gifting' of their data, and privacy are long-term issues to resolve [9,12]. The underlying issue being that repurposing social media generates new vulnerabilities associated with data alignment and availability.

It is possible to see that social media data applications in the management of traffic may generate new value and possibilities for new revenue streams. Possibilities arise from added value associated with dataset combinatorial approaches. In addition added value may also arise from a Business model based on 'co-creation'. That is, a community-based approach to the use of Twitter data, whereby users and other stakeholders, e.g., app developers, actively engage in the generation and application of Twitter data in traffic informatics.

References

[1] Wojtowicz, J., and Wallace, W. A. (2016). "Use of social media by transportation agencies for traffic management". *Transportation Research Record*, 2551(1), 82–89. https://doi.org/10.3141/2551-10

[2] Chaniotakis, E., Antoniou, C., and Pereira, F. (2016). "Mapping social media for transportation studies". *IEEE Intelligent Systems*, 31(6), 64–70. https://doi.org/10.1109/MIS.2016.98

[3] Torre-Bastida, A. I., Del Ser, J., Laña, I., Ilardia, M., Bilbao, M. N., Campos-Cordobés, S. (2018), "Big Data for transportation and mobility: recent advances, trends and challenges". *IET Intelligent Transport Systems* (2018), 12(8):72. http://dx.doi.org/10.1049/iet-its.2018.5188

[4] Ketonen-Oksi, S., Jussila, J. J., and Kärkkäinen, H. (2016). "Social "media-based" value creation and business models". *Industrial Management & Data Systems*, 116(8), 1820–1838. doi:http://dx.doi.org/10.1108/IMDS-05-2015-0199

[5] Sloan L., Morgan, J., Savage, M., *et al.* (2015). "Who Tweets with their location? Understanding the relationship between demographic characteristics and the use of geoservices and geotagging on Twitter". *PLoS ONE* 10 (11):e0142209.

[6] Kim, D. (2016). "Value ecosystem models for social media services". *Technological Forecasting & Social Change*, 107, 13–27. https://doi.org/10.1016/j.techfore.2016.03.010

[7] Gal-Tzur, A., Grant-Muller, S. M., Kuflik, T., Minkov, E., Nocera, S., and Shoor, I. (2014). "The potential of social media in delivering transport policy goals". *Transport Policy*, 32, 115–123.

[8] Osterwalder, A. and Pigneur, Y. (2010) *Business Model Generation: A Handbook for Visionaries, Game Changers, and Challengers*, Wiley: UK, ISBN: 978-0-470-87641-1

[9] Catapult (2015). *The Transport Data Revolution, Investigation into the data required to support and drive intelligent mobility*, March 2015, https://ts.catapult.org.uk/wp-content/uploads/2016/04/The-Transport-Data-Revolution.pdf.

[10] Swier, N, Komarniczky, B., and Clapperton, B. (2015). *Using Geolocated Twitter traces to infer residence and mobility*. Office of National Statistics GSS Methodology Series 41. HMSO; London. Available online: https://www.ons.gov.uk/methodology/methodologicalpublications/generalmethodology/currentmethodologyarticles

[11] Chen, C., Mab, J., Susilo, Y., Liu, Y., and Wange, M. (2016). "The promises of big data and small data for travel behaviour (aka human mobility) analysis", *Transportation Research Part C* 68 (2016) 285–299.

[12] SUNSET, "Deliverable 3.3: Impact of Incentives". A deliverable of the EU project SUNSET available from www.sunset-project.eu.

Chapter 13
Traffic and travel information into the future

Alan Stevens[1]

13.1 Introduction

This chapter aims to look into the future to see how traffic and travel information (TTI) is likely to develop. This is, of course, uncertain and, as Niels Bohr commented, 'prediction is very difficult, especially if it's about the future'. However, there are some fundamentals and clear trends to start from and we prefer to acknowledge the more positive note of the science fiction writer William Gibson who said, 'The future is already here – it's just not evenly distributed'. But even if we can see glimpses of the future around us there are still important questions about which trends will catch on and then how widely and how quickly they will spread. There are naturally also 'unknown unknowns' concerning unforeseen developments that might arise.

This chapter is structured as follows:

Section 13.2 considers some major strategic issues (also called 'megatrends') which provide a relatively stable or predictably changing context in which to examine transport developments. Section 13.3 then looks at future travel demands with a focus on surface transport modes and Section 13.4 describes some emerging and future transport technology developments. This is followed by Section 13.5 on 'mobility trends' which acknowledges an increasing focus on journeys rather than the mode of transport through which they are accomplished. Recognising the central importance of data upon which TTI is built, Section 13.6 focuses on new data sources and contains a discussion on some of the emerging issues in processing vast new amounts of data including artificial intelligence (AI) and issues of ownership and privacy. Then, to examine the thinking of influential organisations about future needs and developments, Section 13.7 reviews, so-called, 'roadmaps' (aspirational futures) from three organisations which provide some indications, through scenarios and other descriptions, of likely future investment and research directions.

In Section 13.8, we reflect on the key conclusions from all the contributed chapters and finally Section 13.9 concludes with what all the developments and trends identified in this chapter can tell us about TTI into the future.

[1]Transportation Research Group, University of Southampton, UK

13.2 Major trends (Megatrends)

Megatrends are far-reaching sustained and relatively certain forces that will substantially affect the way society and markets function in the future. They present both challenges and opportunities and will have global implications for transport infrastructure, services and information needs. A number of the megatrends most likely to influence transport are summarised below. Further information is available from sources such as the European Environment Agency [1].

13.2.1 Shift in global economic power

Global growth is increasingly shifting from developed western countries east and south to countries in Asia Pacific, China and South America. This is expected to increase market size and competition as well as emphasising the need for global standards in all areas including transport.

13.2.2 Demographic changes

The global population is expected to increase to around 9.5 billion by 2050 although different countries around the world have distinct demographic trajectories. Some societies' populations are shrinking, and their workforces will be constrained as a share of the total population. Other societies are young and growing, which will create ever larger labour forces and consumer markets. The UK's population, for example, is expected to grow to around 73 million by 2040 [2] which will increase pressures on transport systems. Another global trend is an increase in what might be called the 'middle class' which has implications for lifestyle choices, including transport.

In Europe and the United Kingdom, a significant demographic trend is populating aging. From 2016, over 80% of population growth in the United Kingdom to 2041 will be in the over-65 age group, with the number of people over 85 almost doubling from 1.6 million to 3.2 million over the same period [3]. Ageing populations are expected to have impacts on the design of mobility solutions with older generations perhaps having markedly different needs, concerns and preferences from their previous generations. Many of them will also have the wealth to make satisfying their needs attractive.

13.2.3 Increasing urbanisation

In the 1950s, less than 30% of the world's population lived in cities. Currently, that proportion is around 50% and expected rapid urbanisation means that by 2050 around 70-75% of the global population will be urban dwellers. Globally there are more than 20 'Megacities' (cities with populations of over 10 million).

Urbanisation has important implications for transport because different transport modes flourish in different contexts. Urban areas are characterised by a substantial high-density population of travellers and freight deliveries; this engenders efficient use of public transport and shared use of micro-mobility modes such as cycles. Targeted development such as light rail schemes can unlock areas for economic development and the availability of low-cost final-mile transport can increase the number of people that can benefit from it. By contrast, rural areas have a much lower population density and this represents a particular challenge for provision of accessible and affordable public transport. There we are seeing a reduction in services and an increase in ticket prices for bus journeys, undermining the feasibility and attractiveness of buses as a mode of travel. In rural settings, on-demand transport is likely to be more appropriate than fixed schedule transport and small-scale shared transport may be the key to making this affordable to more demographics of the population. A third context is inter-urban which is characterised by relatively long distances between travel origins and destinations and where fast motorised transport such as rail and private vehicles are likely to be the main modes.

13.2.4 Rise of technology

The growing pace of technological change will be one of the main drivers for the transport sector and hence will greatly influence needs for TTI which will change as transport itself changes. On the one hand, automation and other technologies could reduce the need for basic information; on the other hand, the plethora of options and how they are delivered could increase information needs. New transport and mobility applications are described in more detail below; here, the underpinning wider technology trends in hardware and software are noted. Key developments in hardware include:

- Materials and manufacturing
- Consumer electronics (displays including virtual and augmented reality)
- Video and other sensors
- Emerging and future communications
- Cloud computing
- Energy storage
- Robotics.

Software developments most relevant to transport include:

- Machine learning and AI more generally
- Social media
- Internet of Things (IoT)
- Big Data and data analytics
- Blockchain and distributed ledger technology for data security.

13.2.5 Climate change

The effects of climate change are expected to be complex and wide-ranging with more frequent and intense extreme weather events. Increasing temperatures, rising sea levels and more intense storms could increase the risk of disruptions, damage and failure of transport systems. Tunnels, bridges and transport power supplies might be particularly vulnerable although it can also be anticipated that changes to design, maintenance and operation of these assets will also develop to improve resilience. The importance of reliable and timely travel information therefore seems likely to increase.

13.2.6 Sustainability and resource security

Transport is responsible for around a third of all final energy consumption in Europe and for more than a fifth of greenhouse gas emissions. Furthermore, transport has a serious impact on the landscape because it divides up natural areas (e.g. with linear built features) with serious consequences for animals and plants.

Road building and maintenance can be identified as one of the significant contributors to greenhouse gas emissions. There are emissions from electricity used in fabricating steel and producing cement, and there are chemical emissions from the cement binding process itself. There are also petrol and diesel emissions from all the construction and maintenance vehicles and other plant equipment. The construction and maintenance industries are increasingly aware of their impact and are seeking commercially viable solutions to mitigate these effects. Road traffic can also be highlighted as currently responsible for a large share of urban air pollution as well as noise nuisance. However, this is likely to be mitigated by adoption of electric and hydrogen transport over the next two decades.

The resources required to sustain current levels of economic growth may not be available over the coming decades. As well as food and water, more specific concerns have been expressed about scarcities or other difficulties with supplies of minerals for electric vehicle batteries including nickel, copper lithium and cobalt [4]. In some parts of the world there is increased environmental awareness of the need for sustainable development. This and regulation (where present and enforced) are influencing individuals and businesses to adopt more sustainable patterns of behaviour [5].

At an individual level, an increased focus on health and well-being has included concerns about air quality and greater adoption of active travel modes such as cycling and walking. Another trend is shared economy of service provision rather than product ownership (e.g. use of taxis, sharing, pooling and car hire rather than personal vehicles).

13.3 Future demands for transport

Even in relatively developed countries the forecast for travel demand continues to increase over the coming decades. However, in many countries, including the United Kingdom, this is largely driven by population growth; people are travelling less per person now than one or two decades ago. This seems to be as a result of less commuting (an increase in flexible and home-working) and less shopping trips (an increase in on-line shopping). Some notes on demand by mode, which will presumably affect demand for TTI, are provided below.

13.3.1 Passenger car

In the United Kingdom, United States and many other developed countries, road vehicle demand has essentially plateaued over the past two decades and, even though predictions for future car traffic vary widely, some commentators have argued that we have reached 'peak car'. However, government forecasts in the United Kingdom predict that traffic will grow by between 17% and 51% over the period 2015 to 2050 [6], driven by increased population and car ownership, declining fuel cost (due to increased fuel-efficiency) and this is expected to result in an increased proportion of traffic to be in congested conditions.

The size of the passenger car fleet is also forecast to grow, from 30 million cars in 2010 to 44 million by 2040 [7].

13.3.2 Bus

Factors such as frequency, journey times and cost tend to dominate decision-making regarding bus use. Once these are overcome, real-time information, better-trained drivers and CCTV at bus stops can increase their use. Nevertheless, bus use is expected to continue to decline in the United Kingdom (at least outside of London) up to 2040.

The UK Department for Transport (DfT) is promoting the Bus Open Data Service in order to digitally transform bus services – high-quality open data will help passengers plan journeys, find best value tickets and get real-time service updates.

13.3.3 Micro-mobility

Personal micro-transport (including cycles and scooters, whether electric or not) have been taken up enthusiastically in some parts of Europe, Asia and the United States; they are also beginning to become popular in the United Kingdom. Cycle hire schemes are available in many urban areas. The 'e' in e-Bikes and e-Scooters brings wider public appeal, enhancing the potential for these modes to progress from early adopters towards larger populations of travellers. Anecdotally, many of those trying these devices for the first time find the experience enjoyable. They can replace short individual car journeys or provide the first- and last-mile of journeys on public or shared transport which might otherwise only be convenient by individual vehicles. As the micro-mobility environment becomes more attractive, social norms are changing and cycling increases.

13.3.4 Rail

The rail network has a limited geographical coverage, and many journeys do not start or end near a rail node. However, people tend to see trains as safer, less stressful and higher status than buses. They also like trains because timetables, real-time information and stops are widely perceived as easy to obtain. Rail differs from road transport predominantly because it is used for longer distances. The geography of rail use is also different as 64% of rail journeys start or end in London in the United Kingdom, with lower rail use elsewhere. Over half of rail journeys (56%)

are made for commuting, a further 10% for business and 23% for leisure [8]. Over the past 20 years, rail demand per person in the United Kingdom has increased sharply and modelling suggests further growth in passenger rail demand of 45–66% up to 2040, allowing for constraints [9].

13.3.5 Summary of future UK transport demand

Concerning people and the transport system in the United Kingdom, the future of transport report [10] summarises it thus:

- People are making fewer trips and travelling less per person compared to 20 years ago. However, population growth means that the total distance travelled in the United Kingdom each year has increased.
- Younger people are travelling less within the United Kingdom, a trend that appears to continue throughout their lives. This is probably due largely to causes outside the transport sector, including declining home ownership, increasing living costs, digital connectivity replacing some need to travel, and the trend towards starting a family later in life.
- As personal transport expenditure rises relative to income, the transport burden increases. Among the poorest households, this hardship is often exacerbated by the need to own and run a car, which is costly.
- Walking has declined over the past 20 years, while cycling trips per person have remained broadly stable. Both modes of transport suffer from latent demand: people walk and cycle more when better infrastructure is available.
- Simultaneous changes in technology, society and transport have markedly affected transport behaviour over the last 20 years and this is likely to continue. The changing nature and location of work, land use, housing and online retail growth have transformed transport behaviour.
- People and companies make transport decisions as part of wider choices including accessibility, lifestyle and connectivity, and are strongly influenced by external changes in society. Practical factors, such as the built environment, location, cost, journey times and availability of infrastructure, currently dominate decision-making and this will likely continue to 2040.
- Car dependence has grown partly because the necessary infrastructure has improved, partly because buses are infrequent (or absent) but also as it is seen as a 'superior' or 'easier' mode of travel. Car use is habitual for many.
- Freedom of choice leads to travel behaviours that are good for the individual or group, but not necessarily for the system. The best outcome for the environment, society or the economy is likely to differ from the outcomes from unconstrained individual choices.

13.3.6 Managing transport demand

For historical reasons, transport governance is often complex and fragmented which makes it challenging to integrate outcomes for different modes and regions, and to coordinate investments to support a joined-up transport system.

Nevertheless, with the 'predict and provide' model now seen as inappropriate in a world of finite resources, attention has largely turned to making the best use of existing infrastructure – and this requires making value judgements; these could be about optimising passenger flows, making freight movements more reliable, increasing economic impact, or achieving health, environmental or social outcomes. However, there are often conflicting priorities for different modes of travel and these vary between organisations and individuals, the poorest and the richest, and between walkers, cyclists and other vehicle users. There is limited road space, and the 'best' allocation is contested. One major strain on the UK's transport infrastructure is peak travel demand, when the most people are using the system. This varies with time, season and geography, but generally corresponds to morning and evening rush hours.

Managing demand can be achieved through a range of 'soft' and 'hard' approaches. At the softer end is information about the state of the transport networks including likely travel time and cost of travel and advice about timing and mode of travel. Softer options also include gamification, which means using elements of gameplay to encourage particular behaviours in non-game contexts. At the harder end of demand management are the applications of various pricing mechanisms for travel, access and parking which are intended to affect behaviours such as choice to travel or the mode and timing of a trip. Information to fully inform travellers concerning the implications of their choices will, of course, be required.

13.4 Emerging and future transport technology

Historically, new transport technologies have always had both positive and negative impacts, though government has often intervened to regulate and mitigate some of the negative impacts. Examples include the growth of railways, developments in aviation and the rise of privately-owned road vehicles. Emerging and future technologies are exciting, but their impacts are uncertain. In this section, the developments considered most likely to impact on mobility, and hence on the need for TTI, are described. A later section considers mobility developments, more generally, when issues around business models and the interactions between different transport technologies are considered.

13.4.1 Automation in transport

Automation is technology that allows a process or procedure to be performed with minimum human assistance and there is considerable potential for increased automation in transport. The convergence of developments such as sensor technology machine learning (a form of AI) and real-time data will increase automation in multiple areas and are expected to provide new ways to improve infrastructure, vehicles and the whole transport system. TTI will increasingly be used directly by automated systems with the human travellers only requiring summary or exception information.

There is already significant use of automation within transport systems. For example, automated 'driverless' metro trains have been operating for many years and aircraft 'auto-pilots' can operate for much of a flight. Cross-modal experience may be valuable [11] and further automation can be expected in all modes where financial and safety improvements can be shown; however, the most dramatic changes are expected in the road sector where autonomous buses, trucks, pods and cars are currently topics of intense research and development. With wages currently amounting to around 60% of costs for bus companies, future driverless buses could halve the current cost of operations which could transform the commercial dynamics of public transport.

13.4.2 Road vehicle powertrains

Responding to concerns about vehicle emissions and air quality there is a gathering trend away from diesel and petrol internal combustion engines (ICEs) towards alternatively fuelled vehicles. Principally the initial shift is towards electric (and thereafter hydrogen) vehicles [12]. Example developments illustrating this trend include:

- Swedish carmaker Volvo's announcement that all new models will have an electric motor from 2019
- BMW announced that a fully electric version of the Mini will be built at the Cowley plant in Oxford, UK from 2019
- In July 2017, the UK Government announced a plan to ban the sale of all diesel and petrol cars by 2040 which in 2020 was brought forward to 2035 [13].
- President Emmanuel Macron announced similar plans to phase out diesel and petrol cars in France, also from 2040.

The pace of change is variable across the world. Currently, as a percentage of new car registrations, electric cars only represent around 1.5 % cent of the total new car market in the United Kingdom [14]. Other countries, notably Norway, are further advanced. To support this development, public charging infrastructure is being deployed in addition to home/workplace charging. Nevertheless 'range anxiety' is a key concern for drivers and re-charging takes considerably longer than fuelling ICE vehicles.

As well as electric vehicles, other alternatives include vehicles fuelled by liquefied natural gas and hydrogen. These may be particularly suitable in ports and for heavy goods vehicles. Hydrogen powered buses operate in London and other cities.

The emergence of alternatively fuelled vehicles suggests new requirements for travel information. Drivers of electric vehicles want information including about charge point technology, location, availability and costs, and may wish to book charging slots in advance. Another potential development, particularly being developed for trucks, is inductive re-charging technology built into the road infrastructure. This would allow vehicles to recharge whilst driving at normal motorway speeds and if this became available, drivers of suitable electric vehicles might want an option of incorporating such roads into their route. Similarly, hydrogen (or other) powered vehicles will need information on refuelling services.

13.4.3 Connectivity and IoT

Telecommunication technologies provide seemingly continuing improvement in bandwidth, data transfer speed, latency (connection delay) and network availability/reliability. 5G (fifth generation) is the latest major phase of mobile network connectivity, although there are other communication technologies being developed, from short-range links to those involving satellite links.

The IoT (Internet of Things) refers to a network of physical objects capable of detecting and communicating information between each other using embedded sensors, software and connectivity. IoT layers on the existing internet by creating a pervasive and self-organising network of unique, identifiable and addressable physical objects. Data may be transferred 'Machine-to-Machine' without human intervention. The impacts and implication of connectivity and IoT are expected to be wide-ranging with implications for transport including improved communications and interactions between mobile devices, vehicles and the transport infrastructure.

Previous chapters in this book have noted the emergence of connected vehicles and cooperative intelligent transport systems; for example, all new European vehicle models (not all new vehicles) from April 2018 onwards will be connected, due to the mandatory eCall function [15]. Connected vehicles can provide data as a source of TTI and can also receive TTI for display to drivers and passengers. Even if connectivity directly to vehicles takes some time to achieve, connectivity is available through smartphones and a recent survey has demonstrated that nearly 9 in 10 smartphone users (87%) use their phones for travel purposes, with navigation and route planning being the most popular uses [16].

13.4.4 Autonomous road vehicles

'Autonomous', 'Automated' and 'Driverless' represent a range of overlapping descriptions for technologies which automate some or all aspects of the human driving task. They are expected to become a significant feature of future road traffic although there is some confusion over terminology/features and timescales for implementation. One classification progresses from manual driving through assistance aids to partial and highly automated driving. There are also various operational driving domains including, for example, off-road, 'campus' and motorway each with different constraints and opportunities. Roadmaps for automated vehicle deployment have been elaborated and the most consolidated ones for Europe are provided by European Road Transport Research Advisory Council (ERTRAC) [17].

Technologies such as adaptive cruise control, lane keeping and parking assistant all provide partial automation and also information for drivers to supervise and intervene when necessary in the vehicle control tasks. More advanced systems may undertake all driving tasks under certain conditions, again with full driver supervision. The (so-called) 'Level 3' automation provides unsupervised vehicle control but requires drivers to re-take control when requested. Higher levels of automation will increasingly be capable of handling a wider driving domain and will come to a safe stop even without driver intervention.

Autonomous vehicles will still need to interact with their passengers (and potentially other road users and traffic systems) so new forms of TTI will be required. Users of these vehicles will need information about travel times, journey progress and parking, for example, in order to plan and change their travel.

A likely early application for highly automated driving will be 'Highway Autopilot including Highway Convoy' on interurban roads. Convoys, also known as platoons, are currently being developed and evaluated, particularly for trucks where the business case for saving driver hours and reducing fuel consumption (by reducing air resistance drag) is particularly strong; see, for example, the European ENSEMBLE project [18]. Future drivers may then seek information about potential platoons and opportunities to join 'convoys' for sections of their journey.

13.4.5 Parking technology

Automated (car) parking systems [19] aim to store vehicles more efficiently than manual drivers can and hence minimise the area and/or volume required for parking cars. They make use of mechanical robotic systems that move vehicles around either on a surface or using lifts within multi-story towers.

Users of such systems require electronic interactions allowing faster booking, storage, location and retrieval of their vehicles. However, even those wanting to make use of conventional parking arrangements may benefit from connectivity and information. Surveys [20] suggest that the time taken to search for parking and parking itself is taking an increasing proportion of total journey time, particularly in urban areas. This suggests that better information about availability of off-street and on-street parking would be welcomed. Drivers may also be interested in other information about parking such as prices, payment options (pre- and post-parking, cash/card/on-line) and booking or amendments to booking. There are now a number of parking Apps available with a range of these features.

13.4.6 Road access and charging

Many countries implement some form of charging for access to specific areas or stretches of their road networks including tolled roads, bridges and tunnels. A more recent phenomenon has been charging of some vehicle categories for access into city areas in order to implement low emission zones or to curb congestion. The technology for charging and enforcement is constantly developing and includes short range and cellular communication, satellite positioning systems and video number-plate reading.

With decreasing tax revenues from fuel duty and from taxes on more polluting vehicles (due to more efficient vehicles and a trend to electric vehicles) the case for more general pay-as-you-drive taxation is increasing. Such charges, if introduced, could depend on time, distance and place and offer opportunities for drivers to change their behaviour to minimise motoring costs [21].

There therefore seems to be an increasing need for information (in advance for travel planning and in real-time for routing) about when and where vehicle can go and the likely charges they will incur. This information could be presented to drivers in some form or could be automatically incorporated into routing and payment systems.

13.4.7 Micro-mobility

Micro-mobility refers to an increasing trend, particularly in urban areas, for innovative active travel using self-powered or electrically assisted/powered personal transport systems such as bikes, scooters and segways [22]. These micro-vehicles can provide lower energy and emission alternatives to short car journeys and have been taken up enthusiastically in some parts of Europe, Asia and the United States.

Their adoption depends on the local context including legal issues (e.g. scooters are not legal on roads or pavements in some countries), physical infrastructure and environmental conditions. Micro-mobility share schemes can be very efficient in dense urban areas and can reduce transport energy demand by providing the first and last-mile of journeys on public or shared transport which might otherwise only be convenient by individual vehicle. Initially, most share schemes required 'docking' of the micro-vehicle after use, but 'dock-less' schemes (enabled through smart access and vehicle tracking) allow micro-vehicles to be collected and deposited with fewer restrictions.

Electrically assisted micro-mobility is relatively new and, as with all disruptive technology, raises new concerns and possibilities; e-bikes and e-scooters may be used for longer journeys than their manual counterparts and may encourage new ridership. However, their hybrid nature challenges conventional transport corridors which have reserved spaces for pedestrians and other spaces for motorised vehicles.

Information needs could include safer routes for micro-vehicles, buses or other vehicles that will transport personal micro-vehicles, storage facilities (location, cost, occupancy, booking), as well as location and booking systems for shared micro-mobility services.

13.4.8 Future modes of transportation

Human creativity, supported by increasing technological capabilities, has generated many new transport ideas and options. Some examples are briefly presented below.

Unmanned aircraft and drones are widely used for recreational, professional and military purposes and drones are also beginning to be used in specialist freight delivery applications. Their use as flying personal transport is being developed but cost and other factors are expected to limit their widespread use.

Droids are pavement-going automated delivery robots that are typically used for delivering food and other small packages. Droids operated by the company Starship have been trialled in Greenwich and Milton Keynes in the United Kingdom but their wider use and development path in transportation is as yet unclear.

Hyperloop [23] is a mode of passenger and/or freight transportation using a pod travelling at high speed through a sealed tube with much reduced air resistance and friction. There have been limited trials and proposals for pilot systems in the United States, Saudi Arabia and Dubai.

It is unclear what impact, if any, such new transport modes might have on TTI but if they are adopted, their users and other affected transport systems will presumably need information concerning their operation.

13.5 Trends and developments in mobility

This section examines some key developments in mobility that are expected to impact on the way that users interact with transport systems in the future and may shed light on future information needs. These developments are made possible by advances in communications and other technologies discussed above. Their impact depends on the extent to which they are successful and are used by travellers and so the important issue of business models is also discussed. With so many options and possibilities for how transport (and society) might develop, a number of governments and strategic advisory organisations are adopting scenario planning to identify possible mobility futures and to explore how policy and commercial organisations might need to respond.

13.5.1 Shared mobility

There is a general anticipation of a trend from individually owned vehicles towards various forms of shared ownership and use [24]:

Sharing access allows users to make independent journeys with specific vehicles (including cars, bikes and e-scooters). The vehicles may be owned by individuals (or by a group of individuals) and lent out on a peer-to-peer basis via an intermediary platform, or form part of a fleet owned and maintained by companies or transport authorities. Individuals then pay a fee each time to have temporary access to a vehicle without the costs and responsibilities of ownership. Individuals typically access vehicles by joining an organisation that maintains a fleet of vehicles deployed at specific locations. Companies like car2go, DriveNow and Zipcar belong to this category. ZipCar estimates that 800,000 Londoners (15% of those who drive) could be active car club members by 2025 [25].

Ride sharing (car/van pooling) is formal or informal shared rides among drivers and passengers with similar origin-destination pairings. Companies such as BlaBlaCar are part of this category and the pooling arrangements can be organised in near real time.

Ride hailing (also known as Transportation Network Companies or ride-sourcing) provides prearranged and on-demand transport services for a fee, which connect drivers of personal vehicles with passengers. Companies like Lyft or Uber are included under this category.

On-demand public transport potentially combines the cost effectiveness of bus travel with the convenience of a personal vehicle. There are no fixed routes, with journeys determined by where passengers want to go within an area. Passengers can 'order' and track a vehicle from an app, which provides them with a guaranteed fare and allows them to choose their pick-up point and reserve a seat. Computer algorithms match passengers travelling in the same direction, dynamically routing vehicles in real time to find the optimal route for their trip.

The extent to which shared mobility will be adopted is debateable [26] and research has identified a range of social and behavioural barriers, including concerns about safety, loss of comfort and privacy and longer and less reliable journey times [27]. Vehicle manufacturers generally expect to continue as manufacturers and distributors of personally owned vehicles whilst also exploring new business models including becoming providers of mobility services. It therefore seems most likely that there will be a blend of public, shared and private vehicles in the future. Nevertheless, information about these shared modes will be of great use to at least some sectors of transport users.

13.5.2 'End-to-end' mobility services

In their 'Future of Mobility: Urban Strategy' [28], the UK DfT describes an outline vision for how a future mobility ecosystem could operate. This framework includes an 'OPTION SELECTION AND PURCHASING LAYER' which provides the best possible information on journey options (price, speed, modes, routes), ideally including seamless purchasing of end-to-end journeys incorporating local charges or regulations.

A commercial form of this is already available and known as Mobility-as-a-Service (MaaS). It is an example of a user-centric approach to mobility which operates through an online interface, comprising an intermodal journey planner, a single payment portal and a booking system for entire end-to-end journeys. Full MaaS systems integrate multiple modes of transport to provide a single mobility solution. Partial forms include ride-sourcing, route-planning and ticketing apps.

MaaS is still at a relatively early stage of development. Its cost and scalability, and who will develop it, are unknown. It is probably most viable in areas of high population density, where scaling up is more likely and commercial returns higher.

13.5.3 The importance of sustainable business models

New business models are taking advantage of data and digital connectivity and are indirectly driving social changes including use of transport. Business models are often based on, or described by, a 'value chain' which identifies how different entities can collaborate to deliver services and all share and benefit from the revenues derived from the services. Increased business collaboration and sharing of information in real-time can provide greater clarity on transport services such as when services will arrive, travel times, comparative costs and emissions, air quality, step-free access and calories burned. The ability to optimise journey time, mode and routes based on such factors has value and allows users to tailor their trips to meet personal needs.

It is important to appreciate that for a business model to be sustainable all the entities required as part of the value chain must derive some benefit from participation. If there is a disparity between where costs are necessary and where benefits are derived, the value chain becomes unsustainable. It is also vital for sustainability that the revenues (the benefits to participants within the value chain) ultimately exceed the costs. In this context, it is interesting to note that most of the companies which provide new mobility options such as micro-mobility and pay-as-you-use services have never made a profit (in spite of their high valuation), with some of them recently declaring bankruptcy. The MaaS concept is also yet to be proven as sustainable. It may be that the structural, legal and contractual issues will be fully overcome (and/or that governments may intervene to support their resolution) or it may be that the business model is not sustainable in most places and cases.

13.5.4 Scenarios of future mobility

Scenarios thinking is one approach that helps to consider the uncertainty surrounding the impacts of new technology and future mobility developments. The scenario approach does not attempt to predict the future but to explore a spectrum of possibilities. Scenario planning can help decision-makers explore how policy choices will play out in different futures; it also helps to make policy decisions more resilient. Similarly, providers of commercial services can consider how their current offerings can be made more resilient to future changes or where new products and services might be offered. The methodology of scenario planning has similarities with evidence gathering via questionnaires and for roadmaps. It usually involves soliciting opinions from a range of experts and then distilling their aspirations, predictions and concerns into a small set of descriptions or stories illustrating aspects of society, transport options and journeys. These may represent extreme situations with the most likely future being some blend of these extremes. Recent examples of UK future transport scenarios are from the RAND corporation [29] and the Government Office for Science [10].

13.6 Trends and developments in transport data

It is widely believed that future transport activities will generate significant amounts of new data and that they will also benefit from the vast volumes of data generated elsewhere in society. Increasing levels of data are already stimulating advances in machine learning, a branch of AI allowing systems to learn and improve by identifying patterns in data, without being explicitly programmed.

In this section, we examine some of the newer sources of data but also the wider issues of data quality, trust in processed data and data governance, including privacy concerns. All of these issues are expected to impact on the processing and provision of TTI in the future.

13.6.1 Data from connected vehicles

Automatic vehicle location systems, widely used for buses and some freight vehicles, have long been used as a source of data about traffic in real time and have also been used to influence traffic signal priority systems. More recently, information about traffic flows has been derived using automatically generated anonymised signals from a wider range of 'floating' or 'probe' vehicles. Initially the data was used specifically with in-vehicle sat-navs or web-based maps to avoid congestion. It has, more recently, been used to improve accuracy in incident detection and for modelling and predicting traffic flows as a road management tool. Several commercial services are available which provide information to drivers and road managers by combining 'Floating Vehicle Data' from connected vehicles, other data sources and mapping systems.

As vehicle sensors develop, a wider range of data is potentially available (such as about local weather and road condition). This data has the potential to support both existing services and stimulate new services and its commercial value is increasingly being appreciated. Which organisations can access and exploit the data will determine how this market develops. Vehicle manufacturers seek to control access to vehicles through the On-Board Diagnostics (OBD) interface [30] in order to better manage vehicle security but this would restrict vehicle data available to third party service providers.

Three technical solutions for the access to in-vehicle data and resources have been identified [31]:

- Data server platform
- In-vehicle interface
- On-board application platform.

Key features of the data server platform concept are that the data from the vehicle is sent to a back-end server where it can be made available. Therefore, both the vehicle data and the application using the data are outside the vehicle system. Access to data using an in-vehicle interface is enabled via an upgraded OBD interface inside the vehicle; any application using data would run outside the vehicle system, either on an external device or on a layer on the interface itself. Finally, the on-board application platform would allow access to vehicle data and the execution of applications inside the vehicle environment.

There are calls for government regulators to intervene with legislation but the challenge is in balancing the demands of safety and security with fair and undistorted competition.

13.6.2 Public transport data

Large amounts of data have long been potentially available from public transport operations. This includes sales, ticket information, smart-card use, crowd-sourced app data, passenger counts, timetable performance and traffic information. Typically, the information has been used locally and commercially but is increasingly being required to become more open and available to third parties. This availability, coupled with other technological advances, is allowing data to be integrated and exploited in new ways.

Further developments can be anticipated in data collection, real-time information provision, modelling and prediction with many benefits to passengers through better TTI.

13.6.3 Social media

The role of social media within TTI is expected to grow in the future. Transport users typically interact with social media on portable devices and hence use it on (or whilst waiting) public transport and within personal vehicles but it can also be used on desktop devices in the home.

Social media enables and incentivises direct engagement between users and providers of transport services and it is expected that social media will increasingly be used to communicate with and control devices linked through the IoT. Public consultations, traditionally carried out by questionnaire have begun to use sentiment mapping techniques based on social media to provide rich new information sources. The dominant business model is expected to remain 'freemium' where the basic platform is free but additional services or freedom from advertisement is provided for an additional fee.

13.6.4 Other data sources

Potentially a wide range of data from other systems and devices could supplement or contribute to TTI. Some specific examples are provided below.

- Major incidents (e.g. fire, terrorism)
- Weather data
- Connectivity between transport modes
- Location-based advertising (e.g. special offers)
- Estimated start and finish times of sporting or cultural events
- Placing orders for goods, services, food, etc.
- Dynamic information on fuel availability and prices
- IoT.

13.6.5 Big Data and trustworthy AI

Big Data is a broad term referring to the generation, collection, storage, processing and visualisation of vast amounts of data. Big Data offers new opportunities and challenges in many areas including transportation. Current challenges include data security and privacy and there have been several high-profile examples of data breaches which have been a particular focus in Europe since the introduction of the General Data Protection Regulations [32] which came into force on 25th May 2018.

Big Data has also stimulated advances in machine learning, a branch of AI allowing systems to learn and improve by identifying patterns in data, without being explicitly programmed. This has also generated controversy where limited training data sets have essentially caused AI systems to implement biased or prejudiced decisions. This has reinforced the need for trustworthy systems which provide information or are part of automated systems.

An EC Group has published guidelines [33] which put forward a set of 7 key requirements that AI systems should meet in order to be deemed trustworthy. A specific assessment list aims to help verify the application of each of the key requirements:

Human agency and oversight: AI systems should empower human beings, allowing them to make informed decisions and fostering their fundamental rights. At the same time, proper oversight mechanisms need to be ensured, which can be achieved through human-in-the-loop, human-on-the-loop and human-in-command approaches.

Technical robustness and safety: AI systems need to be resilient and secure. They need to be safe, ensuring a fall-back plan in case something goes wrong, as well as being accurate, reliable and reproducible. That is the only way to ensure that unintentional harm can be minimised and prevented.

Privacy and data governance: Besides ensuring full respect for privacy and data protection, adequate data governance mechanisms must also be ensured, taking into account the quality and integrity of the data, and ensuring legitimised access to data.

Transparency: The data, system and AI business models should be transparent. Traceability mechanisms can help achieving this. Moreover, AI systems and their decisions should be explained in a manner adapted to the stakeholder concerned. Humans need to be aware that they are interacting with an AI system, and must be informed of the system's capabilities and limitations.

Diversity, non-discrimination and fairness: Unfair bias must be avoided, as it could have multiple negative implications, from the marginalisation of vulnerable groups, to the exacerbation of prejudice and discrimination. Fostering diversity, AI systems should be accessible to all, regardless of any disability, and involve relevant stakeholders throughout their entire life cycle.

Societal and environmental well-being: AI systems should benefit all human beings, including future generations. Hence it must be ensured that they are sustainable and environmentally friendly. Moreover, they should take into account the environment, including other living beings, and their social and societal impact should be carefully considered.

Accountability: Mechanisms should be put in place to ensure responsibility and accountability for AI systems and their outcomes. Auditability, which enables the assessment of algorithms, data and design processes, plays a key role therein, especially in critical applications. Moreover, adequate and accessible redress should be ensured.

13.6.6 Data standards development

There are many established and emerging standards which support and enable data and information exchange in transport systems. For example, as a standardised e-language for traffic and travel data exchange, DATEX II, plays an important role in the world of established and mature transport services and has been operating successfully for many years. Acknowledging this, the European Commission refers to DATEX II as a preferred standard in several Delegated Acts supplementing the ITS Directive 2010/40/EU. 18 Member States participate actively in the DATEX II organisation, which is hosted by CEDR [34].

As one example of future standards development and integration, the DATEX model and standards will be developed to support new connected vehicle services being developed and promoted through the C-Roads Platform [35]. The aim is to harmonise and align messages in the two domains to promote more efficient exchange of information.

As well as updating and enhancing existing standards (e.g. as new data sources and interfaces become available) there will be a greater need to incorporate data from domains beyond transport and so involve other actors including end users. Also, mechanisms for accessing standards, including those concerning TTI, may evolve from static digital files to more dynamic accessible ones.

13.6.7 Cybersecurity

Cybersecurity (information technology security) is the protection of computer systems from theft or damage to hardware, software, or electronic data in order to provide uninterrupted services. Cybersecurity has become a key issue for major infrastructure including transport networks.

There have been some high-profile ‘hacks’ of vehicles and data breaches which have raised concerns about safety and security. Transport vehicles, such as cars, are examples of ‘cyber-physical’ systems i.e. systems that can have a physical interaction with the world (as opposed to computer databases, for example). If these systems are compromised by erroneous or malicious data the results can have immediate physical impacts on transport users. So, traceability of data is increasingly important: where did it come from and to what extent can it be trusted?

The European Commission has recently released the EU Cybersecurity Act [36] in order to improve cyber resilience, increase trust in digital systems and scale up the EU’s response to cyber-attacks. Physical security and cyber security often focus on high-impact events such as terrorism. Users expect (and need to be assured) that such issues are fully considered. However, on a day-to-day basis the related issue of resilience is of concern by users who typically ask:

- Will the transport system work and get me to where I want to go?
- Will failures make the journey too onerous?
- Will I and my personal details be safe and secure?

Resilience includes provision of suitable information in a timely manner to travellers about transport network performance, future performance and alternatives in case of disruption.

13.6.8 Data governance for transport platforms

Over the last decade, transport platforms have started to emerge as tools to store, combine and add value to transport data which can then be supplied to users, both public and private. They may combine flexible individual means of transport with less flexible time- and route-bound collective public transport services. Typically, they will charge end users for their services or may enter into advertising or revenue sharing arrangements with different transport providers. They may also develop new markets (unrelated to transport) for the data that they hold.

A central or geographically dominant transport platform can accommodate many competing transport service providers, for instance ride-hailing and taxi services, but a future concern might be that consumers and transport service providers become dependent on them as the only effective exchange channel. On the other hand, multiple platforms may undermine efficient management by creating 'walled gardens' which limit choices available to users.

Transport platforms may be provided by a private for-profit firm or a public authority and can regulate transport by means of a central prescriptive approach or by setting the rules to achieve better coordination. Some form of governance over transport platforms and their data may be required in order to avoid market failures and to ensure efficient access to traffic and transport data. For example, a city authority may decide to make the participation of consumers and transport providers in its transport platforms mandatory and impose a number of rules to improve the efficiency of transport management.

Open data and open standards are most likely to drive innovation and improve users' experience of transport systems.

13.7 Strategic future plans

Some insight into future development in transport generally, and hence TTI, may be gleaned from published strategic plans and 'roadmaps' from various influential organisations although these have to be critically analysed. At their best, roadmaps are a useful tool to develop broad consensus on challenges and objectives and to clarify the role of different stakeholders in future transport scenarios. They may also provide scientific information around topics or industries, contribute substantial input into funding decisions by national or regional governments, and support investment business cases. At their worst, roadmaps become a R & D wish-list or are cynically deployed for headline-grabbing publicity purposes with heroically optimistic sector growth prospects and grandiose, but vague, statement of intent.

Here, plans and roadmaps from just three organisations are examined; one from the UK government, one from Highways England (HE) and one from a respected European transport research collaboration organisation.

13.7.1 UK Government Industrial Strategy and future of mobility study

At the end of 2017, the UK government published its Industrial Strategy [37] which set four Grand Challenges to put the United Kingdom at the forefront of the industries of the future. There are:

- Ageing society
- Clean growth
- AI and data economy
- Future of mobility.

All of these have implications for the development of transport services and hence for TTI requirements. The needs of an ageing society and for clean growth have already been noted as global megatrends, but as UK Grand Challenges, additional focussed support and R & D activity can be expected.

In the 'Future of Mobility' Grand Challenge, one key focus is on automated vehicles where the UK government wants to see fully self-driving cars on UK roads by 2021. Public transport is another major focus with new business models including Mobility as a Service being highlighted and where a flexible regulatory framework to encourage new modes of transport and new business models will be established.

Another Grand Challenge, which will certainly support future developments in transport data and information, is 'Growing the Artificial Intelligence (AI) and Data-Driven Economy'. A section earlier in this chapter discussed data and implications for TTI.

As part of the UK government approach, missions to tackle these Grand Challenges are being developed to focus on specific problems which bring together stakeholders from government, businesses and other organisations.

13.7.2 Strategy papers from HE

Highways England (HE) is responsible for operation of 4,400 miles (over 7,000 km) of the Strategic Road Network in England. To support future investment over the period 2020 to 2025 (known as Road Investment Strategy Period 2) HE has published a number of planning and consultation documents. Their 'Connecting the country' document [38] begins by analysing demand, infrastructure and vehicle trends and then develops four future scenarios.

The 'Strategic Road Network Initial Report' [39] sets out proposals and recommendations for Road Period 2 (2020–2025). One development area is better information, where HE states:

> We want to continue to improve the quality of information and digital messaging we provide to our customers, as well as continuing to provide real-time traffic updates, keeping pace with advancements in communications technology, and investigating providing integrated travel information. Our roadside Variable Messaging Signs (VMS) are an important source of

> information for our customers and we propose to continue working with the Department for Transport to investigate whether greater delegation of freedom of action would maximise the benefits VMS could bring.

Over this period, and further into the future, HE is working to create a network of digitally connected, smart roads where the hard shoulder can be in operation, speed limits can be altered depending on conditions and more information can be provided to drivers on issues such as journey time planning and road works. As the connected network is developed to communicate with drivers, conventional road infrastructure could be reduced and ultimately this has been described as the 'naked motorway' concept.

13.7.3 ERTRAC future strategy

ERTRAC is the European Technology Platform for Road Transport which is supported by the European Commission and which brings together road transport stakeholders to develop a common vision for road transport research in Europe. ERTRAC has developed its Strategic Research Agenda through research roadmaps [40], which present topics for research, development and the deployment of innovation. These include Integrated Urban Mobility (2017), Connected Automated Driving (2019), Safety (2019) and Electrification of Road Transport (2017).

All the roadmaps point to areas where future R & D effort is most likely to be targeted. Although provision of information to transport users features in all the roadmaps, the most detailed research needs are identified in the Integrated Urban Mobility roadmap which proposes research in three broad topics: better understanding of current and emerging societal trends and challenges; creating or improving frameworks that remove barriers to enhance more efficient mobility provision; and finally, identifying innovative urban mobility solutions and services to tackle mobility challenges at short, medium and long term.

Specific topics include:

- Defining the role of local authorities in the open data market and how to deliver data in an efficient and cost-effective manner
- Developing standard platforms for collecting data from private operators and public authorities and exploiting new forms of data and systems integration
- Researching privacy issues and data ownership
- Exploiting real-time data (e.g. crowdsourcing, social media, sensors, C-ITS vehicle data, Big Data approaches) to support network management, while addressing legacy systems, new processes and interfaces
- Developing forecasts of future user demand and flow based on Big Data
- Providing enhanced travel information:
    - Before travel (travel query, travel products booking and payment, changing requests)
    - During travel (including along the connection path between two transport modes for transferring passengers)

- After a travel incident (service delay or disruption) has occurred (re-routing, contact points, assistance services)
- Injured passengers assistance

- Developing on-street real-time information about the level of traffic congestion and opportunities to use public transport and nearby off-street parking
- Developing broad reliable network coverage for communication with mobile devices
- Developing new tools to connect individuals and new mobility providers of 'on-demand' services such as car-sharing and on-street bike hire
- Integrating ticketing for the various travel products and improving practices to provide integrated real-time information (e.g. travel route, ticketing, alarm, via smart phone or other devices for all modes)
- Supporting services (like passenger guidance by a trip tracker or assistance for persons with reduced mobility) to connect travellers from a point of arrival of a given transportation mode to the point of departure of the next transportation service.

13.8 Reflections from contributed chapters

In the introduction to this book, we noted that the COVID-19 pandemic has changed our social interactions and the behaviours within society, at least on a temporary basis and will likely further change society in ways we cannot yet imagine. However, the need to travel will remain and so TTI will always have a part to play and will need to develop and adapt to new requirements.

In Chapter 1 on road traffic data collection, we saw how inductive road loops have been supplemented with a broad range of technologies to collect data about vehicles and traffic movements. Video camera technology, and in the future, remote sensing from satellites, are likely to remain important but, increasingly, it is expected that vehicles and road users will become key data generators themselves. This data will be produced by sensors on-board connected vehicles and by road users with smart phones and wearable devices that can inform about both physical and emotional issues.

Chapter 2 provided an historical perspective on airborne observation of road traffic and clearly described how commercial business cases need to be developed and maintained if particular TTI services are to continue. Services such as Capital's 'Flying Eye' were very successful in their time but were overtaken by both technology and user preferences.

Chapter 3 described how location referencing was fundamental to provision of TTI and how many different methods and standards are evolving to suit different applications and needs. One of the future challenges is the need to interoperate between different location referencing systems to support more integrated transport mobility solutions. A key issue, not yet fully resolved, is how the quality of location referencing can be described and how it is affected by transformations within different systems. Noted within the chapter are anticipated future development brought about by increased deployment of connected and automated road vehicles. As vehicle sensing and communications improve, more data about status and changes on the network can be 'crowd sourced' and there is greater opportunity for creating and sharing navigation databases. Automated vehicles are likely to need much greater precision in order to navigate, so it is expected that location referencing will be developed to support fine-grained detailed high-definition maps. There is also a recognition that some traditional aspects of managing the road network require modernisation; in particular, the development of information about traffic regulations and local restrictions in a digital format.

Chapter 4 explained how the need to deliver TTI in the preferred language of the traveller led to information being digitally encoded. In digital form, data from multiple sources can be combined, filtered by type, location and relevance, and be disseminated through diverse media. Further integration of data derived from vehicles and the road infrastructure is envisaged in the future; data on the vehicle's position and status and the conditions surrounding the vehicle will be automatically shared with the infrastructure and other travellers. Another increasingly important source of information will be smartphones applications (such as Waze) to both collect and share traffic information. It is expected that these new data sources will eventually replace conventional infrastructure-based sensors. In terms of collation, exchange and transmission of traffic information, it is expected that the backbones and message sets of DATEX 2, ALERT C and TPEG will still be used but together in more integrated systems (rather than being stand-alone) and with dissemination through social media and C-ITS services.

Chapter 5 on European developments in TTI foresaw a future more integrated transport environment and a continuing role for TISA, a market-driven membership association with worldwide scope, focussed on proactive implementation of TTI services and products based on existing standards. It seems likely that language independent standards such as RDS-TMC will continue to develop whist retaining backward compatibility. The requirements for easy exchange of comprehensive travel information across borders will ensure that message sets and data exchange standards will continue to be developed by transport authorities, infrastructure managers and service providers. By using standardised message sets and location referencing facilities, emerging and new services based on Cooperative ITS will be able to coexist with legacy systems. Gradual integration is foreseen towards 'Connected Cooperative Automated Mobility' (CCAM) although much development of individual components is still required. For example, the management of traffic regulation and signing databases will require standardised message sets of signs and road markings; this is where standards like ISO 14823:2017 – Intelligent transport systems – Graphic data dictionary can make a significant contribution, although the pace and way in which standards are updated and delivered will also need to develop to support future requirements.

Chapter 6 explained the pioneering role of the private commercial sector in developing road traffic information from a public service, mostly informing about road incidents and planned events, to a service able to support more personal and targeted applications including dynamic navigation and route guidance. In particular, the chapter discussed RDS-TMC where the author foresaw that TMC services over RDS would be phased out over 5-10 years. Nevertheless, with the integration of technology and of travel services, a continuing strong commercial market for TTI is still anticipated, with information being consumed via digital communications and internet services.

Chapter 7 described dynamic signage for motorway and other environments. It envisaged that conventional fixed plate traffic signs will continue to be required even as vehicles become more automated and will be read by a vehicle's on-board camera technology. Also, with advances in communications, digital traffic sign posts are foreseen which will transmit TTI wirelessly to vehicles, with the information displayed automatically on in-vehicle displays. This information may change over time, reflecting real-time traffic conditions.

Chapter 8 discussed the development and evolution of 'Smart Motorways' which manage traffic by influencing driver behaviour through the use of a mix of advisory and mandatory signs and signals. As noted in the chapter, the main challenge is maintaining credibility in order to maximise compliance and this means educating drivers about how to respond to the signs and signals. This credibility has recently been challenged in the United Kingdom and a review of safety is underway. Even if they are found to be safer than conventional motorways (as was certainly the case at their introduction), the issue of public perception and hence credibility may restrict or modify their wider implementation. What seems more certain is that infrastructure to vehicle communications will become an additional channel to offer TTI directly to drivers. Initially, this new channel will supplement physical signs and support driver assistance systems. The increased bandwidth not only creates an opportunity for improved information but also creates challenges for system operators in ensuring that information is coherent across multiple channels. Further into the future TTI may bypass the driver altogether and be conveyed directly to automated vehicle systems.

Chapter 9 discussed how Cooperative Intelligent Transport Systems will influence TTI and offers three specific future prospects. Firstly, refinement of the existing TTI provision and its development through commercial services are discussed. Secondly, development of new information services for passenger car user and also public transport and fleet and logistics services; examples being information on parking and loading, hazardous loads and routes, truck routing within areas such as ports, tunnel management and passenger information. Thirdly, TTI will increasingly be used by vehicles themselves rather than by their drivers, for example, to adjust vehicle speeds and optimise engine performance. Many further automated vehicle services have been proposed on the assumption that C-ITS communications can be made sufficiently secure, reliable and rapid and that legal issues such as liability can be solved.

Chapter 10 which focussed on multi-modal TTI underlined the importance of mapping and locating infrastructure assets in order to support travel planning. It also described the fundamental work required to develop comprehensive and real-time public transport information. This work has enabled the on-going development of what is now being called Mobility as a Service (MaaS) and it can be envisaged that as this matures, TTI from multiple modes (car, bus, coach, ferry, flights etc.) will be increasingly integrated. The future seems to be an inexorable increase in bespoke and personalised services, that are location and context specific and are also cognisant of the facilities and services available nearby and which meet the preferences of the individual.

Chapter 11 explored the use of publicly available social media data as a potential additional source of traffic information. A case study demonstrated that geotagged Twitter messages may be used to obtain the speed and direction of users as they travel on the transport network, providing traffic data as an alternative to inductive loops and number plate recognition systems. The textual content of tweets can also be examined for relevant travel information. Here, AI (specifically, machine learning) looks to be a promising approach for data mining to produce a filtered stream of relevant tweets to present to a traffic manager for further interpretation.

Chapter 12 considered the business models and availability of services to support the uptake of social media data in traffic management. Future business model scenarios all involve the development of partnerships, particularly between data users, principally road authorities, and social media data analysts. The chapter identified opportunities for growth in applied analytic services, training and customer support services but cautioned that there may be public concern about new vulnerabilities when social media is intercepted, integrated and repurposed.

13.9 Conclusions and outlook

This section attempts to summarise some key conclusions to reflect on what the developments and trends identified above can tell us about TTI in the future.

13.9.1 Transportation and TTI

People and organisations make travel decisions as part of wider choices and are strongly influenced by external changes in society. These decisions are largely determined by practical factors such as the built environment, location, cost, journey times and availability of transport infrastructure. Despite the rise in ubiquitous communications and a small decrease in travel demand, it does not seem likely that peoples' need and desire to travel both locally and internationally will be extinguished in the foreseeable future. Travel to and in urban areas may actually intensify as more people live and congregate there. Although mobility is becoming increasingly individual, it does not mean that individual ownership of the means of mobility (e.g. personal car) will increase. Indeed, there is evidence to suggest that personal ownership will decrease with more use made of hired, shared and public transport.

A clear trend is towards transportation becoming more connected, integrated and automated but even with this, individual travellers will still want both pre-trip and on trip information. Pre-trip information will need to cover all travel options and explain how they interconnect, including the monetary and environmental costs of different routes. TTI may also develop to offer a broader range of advice including risk of disruption, contingency, comfort, crowding, air quality, on-board facilities and how special needs can be met. On-trip information will need to include real-time versions of the pre-trip information to allow for dynamic developments and changes in travellers' schedules or requirements.

A range of policy, finance and business considerations will determine how comprehensive and reliable future transportation becomes. If the reliability of individual transport modes or services decrease it could be argued that individual travellers will demand even more of TTI in order to implement their own contingency plans.

13.9.2 Gathering, processing and disseminating TTI

Traditional means of gathering traffic data, such as road loops, are being phased out and likely to be maintained only in very specific locations. It is clear that sources of TTI will continue to evolve as both technology and service offerings develop. There is a clear trend for data to come from vehicles and mobile devices rather than fixed sensors within the transport infrastructure. Although some sources may dominate it seems likely that multiple diverse data sources will be used to help build a more comprehensive and real-time picture of transport operations.

Processing of traffic and transport data will continue to develop and as data volumes increase, more processing will become automated. Historical data and trends will provide a starting point but real-time dynamic data is likely to become increasingly important. Huge advances in computing power and software techniques including AI will allow data from diverse sources to be rapidly assimilated, collated and processed to satisfy a variety of requirements. Individual travellers will be one type of user and they will expect personally tailored information relevant to their own needs and location. Other users will include infrastructure operators, logistics providers and travel service providers.

Dissemination of TTI already occurs through multiple technologies and media and this diversity is expected to increase. As well as publicly available signage, digital personal assistants, cognisant of the needs and the preferences of individuals, will tailor TTI in a way that is also location and context specific. This will be presented through a range of mobile, wearable and (possibly) biologically implanted devices. Human-machine interaction will continue to evolve in order to deliver TTI in ways that are appreciated by users. There will also be a growth in 'machine to machine' dissemination of TTI as the information is used by automated processes such as traffic control systems and driverless cars.

A recent initiative, which may have broad implications for the future of TTI dissemination throughout Europe, is the drive for each Member State to provide a digital layer interlinking their transport data. Practically this involves each country setting up their own national access point (NAP), accessible on a non-discriminatory basis, to facilitate the reuse of data. The NAPs may take various forms, such as a database, data warehouse, data marketplace, register or web portal. In the United Kingdom, for example, the DfT plans to create a catalogue of metadata concentrating initially on road and traffic information.

13.9.3 Standards, business models and policy

Standards will continue to be important to facilitate collection, processing and dissemination of TTI so it is expected that message sets, location referencing and data exchange standards will continue to be developed both in depth and breadth in response to changing needs. It is also expected that the pace and way in which standards are updated, delivered and accessed will evolve to support future real-time and machine to machine uses.

TTI has a cost to produce even if it appears free at the point of use, so 'Business models' are vital to the existence of TTI. The extent to which TTI is supported through public funds is a policy decision and the extent of government intervention varies considerably by country. Nevertheless, the private sector has become increasingly important and its participation looks to increase further. Data has become valuable (if it can be suitably processed into useful information) which is increasingly driving innovation and new TTI services.

Governments tend to support competition so it seems likely that they will seek to intervene if particular commercial entities are heading towards a monopoly position. It therefore seems more likely that there will be a diversity of commercial TTI services available with no single business model dominating. Policy intervention may also occur to ensure that basic safety information remains available to all users and that TTI supports social inclusion and steps are taken to ensure individual privacy and choice.

With increasingly diverse travel options, the role of TTI, particularly real-time information, is likely to become increasingly important in helping individuals and organisations plan travel and then monitor progress during journeys. The development of technology has profound effects on society and, judging by recent years, the pace of change in transportation is likely to accelerate further. Hence it seems an appropriate time to have reviewed the development of TTI thus far and it will be very interesting to see how it develops, and how transportation develops, in the coming years.

References

[1] European Environment Agency. Global megatrends. *European Environment Agency.* [Online] [Cited: 9 August 2019] https://www.eea.europa.eu/soer-2015/global.

[2] Office for National Statistics. National Population Projections: 2016-based statistical bulletin. [Online] 2017. www.ons. gov.uk/peoplepopulationand community/populationandmigration/populationprojections/bulletins/nationalpopulationprojections/ 2.

[3] Office for National Statistics. Living longer – how our population is changing and why it matters. *UK Office of National Statistics.* [Online] www.ons.gov.uk/people-populationandcommunity/birthsdeathsandmarriages/ageing/articles/livinglongerhowourpopulationischangingandwhyitmatters/2018-08-13.

[4] Reuters. [Online] 2019. https://uk.reuters.com/article/uk-usa-lithium-electric-tesla-exclusive/exclusive-tesla-expects-global-shortage-of-electric-vehicle-battery-minerals-sources-idUKKCN1S81QY.

[5] United Nations. [Online] 2019. https://sustainabledevelopment.un.org.

[6] DfT UK Department for Transport. Road traffic forecasts. [Online] 2018. https://www.gov.uk/government/publications/road-traffic-forecasts-2018.

[7] Brand, C, Cluzel, C and Anable, J. *Modelling the uptake of plug-in vehicles in a heterogeneous car market using a consumer segmentation approach.* s.l.: Transport Research Part A: Policy and Practice, 2017, Vol. 97, pp. 121–136.

[8] Department for Transport. Rail Factsheet. [Online] 2017. from https://assets.publishing.service.gov.uk/government/uploads/ system/uploads/attachment_data/file/663116/rail-factsheet-2017.pdf.

[9] Blainey, S and Preston, J. Transport systems assessment. [ed.] J W Hall, R J Nicholls, M Tran, and A J Hickford. *The Future of National Infrastructure: A System-of-Systems Approach.* Cambridge: Cambridge University Press, 2016, pp. 88–113.

[10] Government Office for Science (UK). Government Office for Science – Future of Mobility. [Online] 2019. https://www.gov.uk/government/publications/future-of-mobility.

[11] IET. Advancing Autonomy in Transport – Identifying the opportunities for cross-modal learning across the aerospace, automotive, rail and maritime sectors. [Online] August 2019. https://www.theiet.org/media/4281/advancing-autonomy-in-transport.pdf.

[12] OLEV. Uptake of ultra low emission vehicles in the UK. [Online] August 2015. http://webarchive.nationalarchives.gov.uk/20161126194333/https:/www.gov.uk/government/uploads/system/uploads/attachment_data/file/464763/uptake-of-ulev-uk.pdf.

[13] BBC. Petrol and diesel car sales ban brought forward to 2035. [Online] 26 July 2017. https://www.bbc.co.uk/news/science-environment-51366123.

[14] Next greencar. [Online] 2017. http://www.nextgreencar.com/electric-cars/statistics/.

[15] EC. eCall in all new cars from April 2018. [Online] 2015. https://ec.europa.eu/digital-single-market/news/ecall-all-new-cars-april-2018.

[16] Kantar. *Transport and Transport Technology: Public attitude tracker. Wave 3 report.* s.l.: Kantar, 2019.

[17] ERTRAC. Automated driving roadmap V. 7.0. [Online] 27 May 2017. http://www.ertrac.org/uploads/documentsearch/id48/ERTRAC_Automated_Driving_2017.pdf.

[18] ENSEMBLE project. ENSEMBLE project. [Online] 2019. https://platooningensemble.eu/.

[19] Wikipedia. Wikipedia. *Automated parking system.* [Online] https://en.wikipedia.org/wiki/Automated_parking_system.

[20] British Parking Association. [Online] https://www.britishparking.co.uk/News/motorists-spend-nearly-four-days-a-year-looking-for-a-parking-space.

[21] Walker, J. *Road Pricing – Technologies, economics and acceptability.* s.l.: IET, 2018.

[22] Ognissanto, F, Hopkin, J, Stevens, A, Millard, K, and Jones, M. *Innovative active travel solutions and their evaluation*. s.l.: TRL, 2018. PPR 877.

[23] Wikipedia. Hyperloop. *Wikipedia.* [Online] 2019. https://en.wikipedia.org/wiki/Hyperloop.

[24] Shaheen, S, Chan, N, Bansal, A, and Cohen, A. *Shared Mobility: Definitions, Industry Developments, and Early Understanding Bikesharing, Carsharing, On-Demand Ride Services, Ridesharing, Shared Mobility*. 2015.

[25] Zipcar. [Online] https://www.zipcar.com/en-gb.

[26] BCG.perspectives. What's ahead for car sharing. [Online] 2016. https://www.bcgperspectives.com/content/articles/automotive-whats-ahead-car-sharing-new-mobility-its-impact-vehicle-sales/.

[27] Kantar. Future Roads public dialogue – exploring the public's reaction to future road technologies. [Online] 2018. https://assets.publishing.service.gov.uk/government/uploads/system/uploads/attachment_data/file/721722/future-roads-public-dialogue-exploring-the-publics-reactions-to-future-road-technologies.pdf.

[28] DfT. future of mobility urban strategy. [Online] 2019. https://www.gov.uk/government/publications/future-of-mobility-urban-strategy.

[29] RAND corporation. Travel in Britain 2035 – Future scenarios and their implications for technology innovation RAND corporation). [Online] 2015. https://www.rand.org/content/dam/rand/pubs/research_reports/RR1300/RR1377/RAND_RR1377.pdf.

[30] Wikipedia. *On-Board Diagnostcs.* [Online] https://en.wikipedia.org/wiki/On-board_diagnostics.

[31] TRL. Access to in-vehicle data and resources. [Online] 2017. https://ec.europa.eu/transport/sites/transport/files/2017-05-access-to-in-vehicle-data-and-resources.pdf.

[32] European Commission. GDPR. [Online] https://eugdpr.org/.

[33] European Commission. Trustworthy AI. [Online] 2019. https://ec.europa.eu/digital-single-market/en/news/ethics-guidelines-trustworthy-ai.

[34] CEDR. DAREX II. [Online] https://datex2.eu/.

[35] European Commission. C-Roads platform. [Online] https://www.c-roads.eu/platform.html.

[36] European Commission. [Online] 2019. https://ec.europa.eu/digital-single-market/en/eu-cybersecurity-act.

[37] UK Government. Industrial strategy. [Online] 2017. https://www.gov.uk/government/publications/industrial-strategy-building-a-britain-fit-for-the-future.

[38] Highways England. Connecting the country – Planning for the long term. [Online] https://assets.publishing.service.gov.uk/government/uploads/system/uploads/attachment_data/file/666876/Connecting_the_country_Planning_for_the_long_term.pdf.

[39] Highways England. Strategic Road Network – Initial Report. [Online] December 2017. https://www.gov.uk/government/publications/highways-englands-strategic-road-network-initial-report.

[40] ERTRAC. Roadmaps. [Online] https://www.ertrac.org/index.php?page=ertrac-roadmap.

Index